STATISTICAL and COMPUTATIONAL PHARMACOGENOMICS

CHAPMAN & HALL/CRC
Interdisciplinary Statistics Series

Series editors: N. Keiding, B.J.T. Morgan, C.K. Wikle, P. van der Heijden

Published titles

AN INVARIANT APPROACH TO STATISTICAL ANALYSIS OF SHAPES	S. Lele and J. Richtsmeier
ASTROSTATISTICS	G. Babu and E. Feigelson
BIOEQUIVALENCE AND STATISTICS IN CLINICAL PHARMACOLOGY	S. Patterson and B. Jones
CLINICAL TRIALS IN ONCOLOGY SECOND EDITION	J. Crowley, S. Green, and J. Benedetti
CORRESPONDENCE ANALYSIS IN PRACTICE, SECOND EDITION	M. Greenacre
DESIGN AND ANALYSIS OF QUALITY OF LIFE STUDIES IN CLINICAL TRIALS	D.L. Fairclough
DYNAMICAL SEARCH	L. Pronzato, H. Wynn, and A. Zhigljavsky
GENERALIZED LATENT VARIABLE MODELING: MULTILEVEL, LONGITUDINAL, AND STRUCTURAL EQUATION MODELS	A. Skrondal and S. Rabe-Hesketh
GRAPHICAL ANALYSIS OF MULTI-RESPONSE DATA	K. Basford and J. Tukey
INTRODUCTION TO COMPUTATIONAL BIOLOGY: MAPS, SEQUENCES, AND GENOMES	M. Waterman
MARKOV CHAIN MONTE CARLO IN PRACTICE	W. Gilks, S. Richardson, and D. Spiegelhalter
MEASUREMENT ERROR AND MISCLASSIFICATION IN STATISTICS AND EPIDEMIOLOGY: IMPACTS AND BAYESIAN ADJUSTMENTS	P. Gustafson
META-ANALYSIS OF BINARY DATA USING PROFILE LIKELIHOOD	D. Böhning, R. Kuhnert, and S. Rattanasiri

Published titles

Interdisciplinary Statistics

STATISTICAL and COMPUTATIONAL PHARMACOGENOMICS

Rongling Wu
University of Florida
Gainesville, Florida, U.S.A.

Min Lin
Duke University
Durham, North Carolina, U.S.A.

CRC Press
Taylor & Francis Group
Boca Raton London New York

CRC Press is an imprint of the
Taylor & Francis Group, an **informa** business

A CHAPMAN & HALL BOOK

Chapman & Hall/CRC
Taylor & Francis Group
6000 Broken Sound Parkway NW, Suite 300
Boca Raton, FL 33487-2742

International Standard Book Number-13: 978-1-58488-828-4 (Hardcover)

Library of Congress Cataloging-in-Publication Data

Wu, Rongling.
   Statistical and computational pharmacogenomics / Rongling Wu, Min Lin.
     p. ; cm. -- (Chapman & Hall/CRC interdisciplinary statistics series)
   Includes bibliographical references and indexes.
   ISBN 978-1-58488-828-4 (hardcover : alk. paper)
   1. Pharmacogenetics--Statistical methods. 2. Pharmacogenetics--Mathmatical models. I. Lin, Min, 1971- II. Title. III. Series: Interdisciplinary statistics.
   [DNLM: 1. Pharmacogenetics--methods. 2. Chromosome Mapping--methods. 3. Genetic Markers. 4. Models, Statistical. QV 38 W959s 2008]

RM301.3.G45W8 2008
615'.7--dc22
                                                  2008013037

Visit the Taylor & Francis Web site at
http://www.taylorandfrancis.com

and the CRC Press Web site at
http://www.crcpress.com

Dedication

For RW

To Helen and Louie, and Mei, too

For ML

To my parents, Xiang and Kenny

Preface

Although pharmacogenetics or pharmacogenomics, the study of inherited variation in patients' responses to drugs, is still in its infancy, tremendous accumulation of data for genetic markers and pharmacodynamic tests have made it one of the hottest and most promising areas in biomedical research. The central goal of pharmacogenetics is to understand the association of interpatient variability in drug response with specific genomic sites through the use of powerful statistical tools. Traditional approaches for such association studies are based on the statistical inference of putative quantitative trait loci (QTLs) that underlie a phenotypic trait using known linkage or linkage disequilibrium maps. With the advent of high-throughput genotyping assays for single nucleotide polymorphisms (SNPs) and the completion of a haplotype map (HapMap) constructed from SNP data, it has now become possible to characterize concrete quantitative trait nucleotide (QTN) combinations that encode a complex phenotype, and ultimately document, map and understand the structure and patterns of the human genome linked to drug response.

Most existing mapping strategies analyze genetic variation in a static trait, but these are often not effective relative to the complex nature of drug response. A drug acts in patients' bodies through a complex pharmacokinetic (PK) and pharmacodynamic (PD) process. For this reason, clinical trials for pharmacological studies have been designed to investigate a time course of drug deposition, drug metabolism, and drug function in the body and to test the time-dependent relationships between drug effects and drug concentration. In this sense, genetic mapping of a drug response should consider the longitudinal or dynamic nature of pharmacological interactions that determine the manner in which the drug affects the body over time. During the past five years, there has been an increase in interest to develop and apply statistical methodologies for genetic mapping of dynamic traits that incorporate the biological or biochemical principles underlying trait formation and progression. These statistical methods, simply called functional mapping, hold a great promise for the detection of genetic loci, or DNA sequence variants, associated with drug response on the scale of dosage and time, and, ultimately, for the design of an optimal drug, optimal administration time, and optimal dosage schedule for individual patients based on their specific genetic makeups. However, these methods generally have various objectives and utilities and are sporadically distributed across a massive amount of literature. A single volume synthesizing statistical developments for genome mapping may be helpful for many researchers, especially those with a kcen interest in building a bridge between pharmacogenetics and statistics, to acquaint themselves with this expanding area as quickly as possible, and this is why we wrote this book.

This book intends to provide geneticists with the tools needed to understand and model the genetic variation for drug response based on genomic data collected in a mapping population and to equip statisticians with the motivation and ideas used to explore genomic data. This book also intends to attract researchers, especially young scientists, towards multidisciplinary research and to introduce them to new paradigms in genomic science. The statistical and computational theories necessary for the genetic mapping of dynamic traits are developed hand-in-hand and numerous examples displaying the implications of statistical genomics are introduced. It should be pointed out that many examples from plant and animal studies were used to demonstrate some statistical methods because of our temporary lack of similar data in pharmacogenetics, but these methods described will find their immediate applications in drug response with the availability of pharmacogenetic data.

There are 15 chapters in this book. Chapter 1 provides an overview of the study design for mapping QTLs and haplotyping QTNs. In Chapters 2 and 3, the strategy for sequence mapping of QTNs for complex traits is described for a natural population (Chapter 2) and an experimental population (Chapter 3). Chapter 4 introduces a general population and quantitative genetic model for analyzing haplotype diversity, distribution, and effects. In Chapter 5, basic principles of functional mapping, including parameter estimation and test procedures, are reviewed and discussed. The methods and principles of genetic mapping of pharmacokinetics and pharmacodynamics and their links are the topics of Chapters 6 and 7. In Chapter 8, we present a computational framework for the detection of clock genes and its potential application to "chronotherapy," the practice of drug therapy at the body's right time. Chapter 9 discusses statistical models for characterizing DNA sequence variants by integrating allometric scaling theory, which are extended to consider developmental aspects of pharmacological response in Chapter 10. The description of models that can discern differences of genes that control drug efficacy and drug toxicity is done in Chapter 11. Chapters 12 and 13 cover two important types of interactions that occur between different genes (Chapter 12) and between genes and environments (Chapter 13). The last two chapters (14 and 15) discuss the generalization of functional mapping by nonparametric and semiparametric modeling of functional data on drug response. The computer code or software for the statistical methods described in this book have been made freely available at the web address http:\\statgen.ufl.edu for researchers worldwide.

The writing of this book was inspired by our rapidly growing research program in statistical genetics at the University of Florida. A number of faculty, post-doctoral researchers, and graduate students have contributed substantially to the program in different ways, which has made it possible for us to collect material and publish this book. The work by the following people is given special recognition: Arthur Berg, George Casella, Yuehua Cui, Wei Hou, Bong-Rae Kim, Hongying Li, Qin Li, Tian Liu, Fei Long, Chang-Xing Ma, Chenguang Wang, Zuoheng Wang, Song Wu, Jie Yang, Mark Yang, John Yap, Wei Zhao, and Yun Zhu. Their contributions have greatly inspired and stimulated our methodological research in this fantastic area. We are also grateful to our collaborators at the University of Florida and elsewhere—James Cheverud, Roger Fillingim, Minren Huang, Julie Johnson, Matias

Kirst, Brian Larkins, Juan Medrano, Daniel Prows, Roland Staud, Rory Todhunter, Eduardo Vallejos, Peggy Wallace, Lang Wu, and Jun Zhu, who have provided their invaluable data, problems, and comments for our statistical genetics research. We are especially thankful to Professor Hartmut Derendorf and the following students in his department, Qi Liu, Jiang Liu, Mei Tang, Yaning Wang, Kai Wu, and Hao Zhu, for providing us with their insightful views about pharmacodynamics and pharmacokinetics. We thank anonymous reviewers for their constructive comments on the earlier version of the book.

A general frame of this work was designed at the University of Florida and Duke University, and many details were added and refined at Princeton University where R. Wu spent his sabbatical leave. R. Wu thanks Professor Jianqing Fan for his generous hospitality at Princeton and admires his groundbreaking nonparametric work which can potentially foster and polish many aspects of current pharmacogenetic studies. Particular acknowledgements are due to the Department of Statistics within the Institute of Food and Agricultural Sciences at the University of Florida and the Department of Biostatistics and Bioinformatics at Duke University for continued support on this and other projects. Part of this work has been supported by NSF grant (0540745) and Duke Clinical and Translational Sciences grant (UL1 RR024128). Finally, we thank Editor Rob Calver for inviting us to publish a statistical pharmacogenetics book with Chapman & Hall/CRC, which stimulates our deeper understanding of this promising emerging science. Excellent editorial and technical assistances by Michele Dimont and Shashi Kumar have favorably improved the presentation of this book.

Rongling Wu
Gainesville, FL, and Princeton, NJ
Min Lin
Durham, NC

Contents

1

Designs and Strategies for Genomic Mapping and Haplotyping

Almost every biological phenomenon or process including patient's response to medications involves a genetic component. Thus, by deciphering the detailed genetic architecture of a biological trait, we can well understand its developmental mechanisms and further predict its performance before it is formed or obtain a desired performance of that trait. In the past, the genetic analysis of a trait has been a daunting task because no appropriate resources and analytical tools are available. With the advent of novel molecular marker techniques and computational tools, it has now been possible to map and haplotype specific genes that control a biological trait and estimate their main and interaction effects on phenotypic variation.

In this chapter, we will first provide basic genetic knowledge used for genomic mapping and haplotyping, describe the types and designs of genetic mapping populations, review fundamental strategies for detecting, estimating and locating individual QTL in the experimental and natural populations, and point out the application of these models to study the genetic architecture of a quantitative trait. In particular, we will highlight the application of genetic mapping to study the pharmacogenetic and pharmacogenomic basis of drug response as a complex dynamic or longitudinal trait. For a complete coverage of genetic mapping theory and methods, the readers are referred to the statistical genetics textbook by Wu et al. (2007a).

1.1 Fundamental Genetics

1.1.1 Chromosomes and Map

Genetics is the study of heredity or inheritance. Genetics helps to explain how traits are passed from parents to their offspring. Parents pass traits to their young through gene transmission. The fundamental physical and functional unit of heredity is a *gene*, which was first revealed by Gregor Mendel's pea experiments and mathematical model in 1865. Genes are composed of deoxyribonucleic acid (*DNA*), a double-strand helix of nucleotides. Each nucleotide contains a deoxyribose ring, a phosphate group, and one of four nitrogenous bases: adenine (A), guanine (G), cytosine (C), and thymine (T). In nature, base pairs form only between A and T and between G

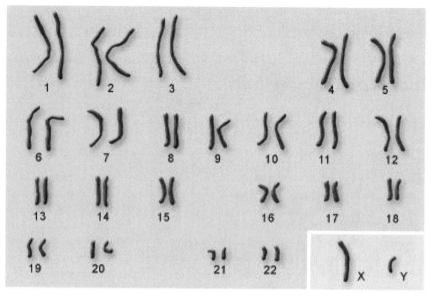

autosomes   sex chromosomes

FIGURE 1.1
Twenty-three pairs of chromosomes in the human genome. Among these, 22 pairs
are autosomes and 1 pair is sex chromosomes. A male carries sex chromosome pair
XY, whereas a female carries *XX*.

and C due to their chemical configurations. It is the order of the bases along DNA
that contains the hereditary information that will be passed from one generation to
the next.

A single DNA molecule condensed into a compact structure in a cell nucleus is
called a *chromosome*. The chromosomes occur in similar, or in homologous, pairs,
where the number of pairs is constant for each species. In humans, there are 23 pairs
of chromosomes, carrying the entire genetic code, in the nucleus of every cell in the
body. For each pair, one chromosome is inherited from the mother and the other from
the father. The entire collection of these chromosomes are referred to as the human
genome. One of the chromosome pairs in the genome is the sex chromosomes (typically denoted by **X** and **Y**) that determine genetic sex. The other pairs are *autosomes*
that guide the expression of most other traits (Fig. 1.1).

1.1.2   Genotype and Phenotype

A gene is simply a specific coding sequence of DNA and may occur in alternative
forms called *alleles*. A single allele for each gene is inherited from each parent,

termed maternal and paternal allele respectively. The pair of alleles constructs the *genotype*, which is the actual genetic makeup. If a given pair consists of similar alleles, the individual is said to be *homozygous* for the gene in question; while if the alleles are dissimilar, the individual is said to be *heterozygous*. For example, if we have two alleles at a given gene of an individual, say A and a, there are two kinds of homozygotes, namely AA and aa, and one kind of heterozygote, namely Aa. Therefore, three different genotypes, AA, Aa, and aa, are formed with a single pair of alleles.

In comparison, *phenotype* represents all the observable characteristics of an individual, such as physical appearance (eye color, height, size, etc.) and internal physiology (disease, drug response, etc.).

1.1.3 Molecular Genetic Markers

Molecular genetic *markers* are readily assayed phenotypes that have a direct 1:1 correspondence with DNA sequence variation at a specific location in the genome. In principle, the assay for a genetic marker is not affected by environmental factors. Genetic markers that represent DNA sequence polymorphisms have many different types. Restriction fragment length polymorphisms (RFLPs) are the first genetic markers that were widely used for genomic mapping and population studies. RFLPs are codominant markers for which homozygotes and heterozygotes can be distinguished phenotypically. The polymerase chain reaction (PCR) provides a useful way to obtain genetic markers. PCR-based anonymous markers include random amplified polymorphic DNA (RAPDs) and amplified fragment length polymorphisms (AFLPs). Although they are dominant markers for which homozygotes and heterozygotes cannot be distinguished, RAPDs and AFLPs have been commonly used in practice because of their cheapness and great availability. Polymorphisms in the lengths of tandemly repeated short sequences, called microsatellite markers, have also become very popular in terms of their capacity to detect multiple alleles at a single marker.

One of the fruits of the Human Genome Project is the discovery of millions of DNA sequence variants in the human genome. The majority of these variants are single nucleotide polymorphisms (SNPs; Fig. 1.2**a**), which comprise approximately 80% of all know polymorphisms, and their density in the human genome is estimated to be on average 1 per 1200 base pairs. SNPs, as the newest markers, have been the focus of much attention in human genetics because they are extremely abundant and well-suited for automated large-scale genotyping. A dense set of SNP markers opens up the possibility of studying the genetic basis of complex diseases by population approaches, although SNPs are less informative than other types of genetic markers because of their biallelic nature. SNPs are more frequent and mutationally stable, making them suitable for association studies to map disease-causing mutations, especially useful in personalized medicine for their association with disease susceptibility, drug treatment response, and nutritional needs.

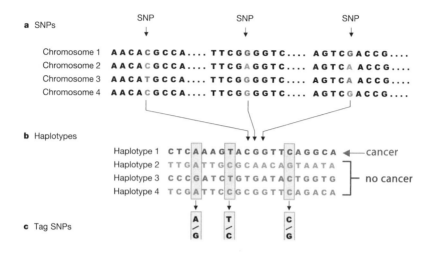

FIGURE 1.2

A short stretch of DNA from four versions of the same chromosome region in different humans. (**a**) Single nucleotide polymorphisms (SNPs) are identified in DNA samples from the four people. (**b**) Alleles of adjacent SNPs on the same chromosome that are inherited together are compiled into "haplotypes," and (**c**) Tag SNPs within a haplotype block are identified that uniquely identify haplotype diversity. Adapted from Gibbs et al. (2003).

1.1.4   The HapMap

The International HapMap Project has recently released a haplotype map (HapMap) of the human genome which describes the patterns of human DNA sequence variation based on SNPs (Gibbs et al., 2003, 2005; Frazer et al., 2007). It is estimated that there are approximately 10 million SNPs in the human genome which make one human different from the other in DNA sequences. Figure 1.2**a** illustrates such sequential differences among four hypothesized humans in SNPs, three of which are highlighted in a particular chromosomal region. The sequence of SNPs in a chromosome or a chromosomal region is described by the *haplotype*, that is, a combination of alleles at closely linked SNPs on one chromosome which tend to be inherited together. A haplotype may refer to as few as two loci or to an entire chromosome. In Fig. 1.2**b**, the haplotypes of the four humans are shown over an extended region of chromosome. Genetic haplotyping aims to detect the association between haplotypes and phenotypes of a complex trait. For example, people who carry haplotype 1 [CTCAAAGTACGGTTCAGGCA] may be predisposed to cancer, whereas people with the other haplotypes 2, 3, or 4 are not susceptible to cancer.

While SNPs in the human genome were previously assumed to be randomly distributed, several recent empirical studies suggest that the distribution of SNPs follow a particular pattern in terms of the extent of linkage disequilibrium (Daly et al., 2001; Patil et al., 2001; Dawson et al., 2002; Gabriel et al., 2002; Phillips et al., 2003). This is shown by the fact that the structure of haplotype on a chromosome can be broken into a series of discrete haplotype blocks. In each haplotype block, consecutive SNPs are in complete (or nearly complete) linkage disequilibrium with each other and there is limited haplotype diversity due to little inter-SNP recombination (coldspot). Adjacent blocks are separated by sites that show evidence of historical recombination (hotspot). Based on a study of the whole chromosome 21 (Patil et al., 2001), 35,989 observed SNPs can be classified into different blocks within each of which there is very low haplotype diversity. Also, because of low haplotype diversity within a block there is a possibility that very few haplotype-tagging SNPs (htSNPs or tag SNPs) can be identified to represent an entire length of haplotypes uniquely for this block by accounting for an adequately large portion of haplotype diversity. For example, Patil et al. (2001) found that 80% of the variation in chromosome 21 can be described by only three htSNPs per block. The number of htSNPs that contain most of the information about the patterns of genetic variation in the human genome is about 300,000 to 600,000, which is far fewer than the 10 million common SNPs.

The discovery of htSNPs provides an efficient way of detecting genetic variants involved in human diseases and drug response with no need to genotype all expensive SNPs (Wall and Pritchard, 2003). In Fig. 1.2c, it is shown that haplotype diversity among the four individuals can be adequately explained by three tag SNPs. This is because if a particular chromosome has the pattern [ATC], [ACG], [GTC], or [ACC] at these three tag SNPs, this pattern matches that determined for haplotype 1, 2, 3, or 4, respectively. Thus, association studies between haplotype and cancer risk can be undertaken simply on the basis of these three htSNPs. Association studies with a much less number of htSNPs will ultimately enhance our ability to choose targets for therapeutic intervention.

The HapMap provides a resource of sequence variation in human beings, including what the sequence variants causing haplotype diversity are, where they occur in the human genome, and how they are distributed among people within populations and among populations in different parts of the world. But it does not purport to establish connections between particular genetic variants and diseases or drug response. To make such connections a reality, there is a pressing need on the development of powerful statistical tools that can jointly model the HapMap and phenotypic data by considering their characteristics.

1.2 Pharmacogenetics and Pharmacogenomics

Pharmacological response to medications displays tremendous interindividual variation. For example, some patients favorably respond to a drug, whereas other patients may have no, or even adverse, response to the same drug. Although such variability in drug effects may be attributed to the pathogenesis of the disease being treated, drug interactions, the individual's age, nutritional status, renal or liver function, there is increasing evidence for influences of genetic differences in the metabolism and disposition of drugs and the targets of drug therapy (such as receptors) on the efficacy and toxicity of medications Evans and Johnson (2001). For this reason, increasing attempts have been made to identify candidate genes that influence pharmacological responses, leading to the emergence of pharmacogenetics and pharmacogenomics as an across-disciplinary science to study the relationship between genetic variation and drug response (Evans and Relling, 1999; Evans and McLeod, 2003; Johnson, 2003; Weinshilboum, 2003; Marsh and McLeod, 2006).

In principle, the terms pharmacogenetics and pharmacogenomics can be interchangeably used, although they were coined at different times. Pharmacogenetics has been used for more than 40 years to depict the study of single genes and their effects on interindividual differences in drug metabolizing enzymes, and pharmacogenomics, coined in connection with the human genome project, is to depict the study of not just single genes but the functions and interactions of all genes in the genome in the overall variability of drug response. Current pharmacogenetic approaches used have been based on candidate gene approaches (Ring and Kroetz, 2002). Genes that have known functions in regulating one or more particular steps of pharmacological pathways are associated with phenotypic values of drug response. The examples include genes involved in drug transport (e.g. polymorphisms in the gene encoding P-glycoprotein 1 and the plasma concentration of digoxin), genes involved in drug metabolism (e.g. polymorphisms in the gene encoding thiopurine S-methyltransferase and thiopurine toxicity), and genes encoding drug targets (e.g. polymorphisms in the gene encoding the β_2-adrenoceptor and response to β_2-adrenoceptor agonists) (Johnson, 2003). A number of polymorphic sites, such as SNPs, within or near these candidate genes can be genotyped. The genotyped SNPs, especially those that occur in gene regulatory or coding regions (cSNPs), can be associated with phenotypic traits to detect genetic variants causing pharmacological response variability.

However, drug response is typically a complex trait, with multiple genes and various biochemical, developmental, and environmental factors contributing differently to the overall phenotype (Watters and McLeod, 2003). This thus needs an approach for conducting genome-wide association studies between SNPs and drug response. With the availability of the HapMap, a systematic scan for significant associations can be made throughout the genome. Unlike candidate gene approaches, genome-wide association studies do not need prior knowledge about gene function, but the number and distribution of significant associations can be estimated and identified.

Another consideration for pharmacogenetic research is about the way individuals respond to varying drug dosages or concentrations that change with time since drug administration. In statistics, such pharmacodynamic and pharmacokinetic responses present a standard longitudinal or functional problem. Figure 1.3 illustrates the pattern of genetic control of a hypothesized gene over pharmacokinetic and pharmacodynamic processes of medications. This gene is assumed to determine inher-

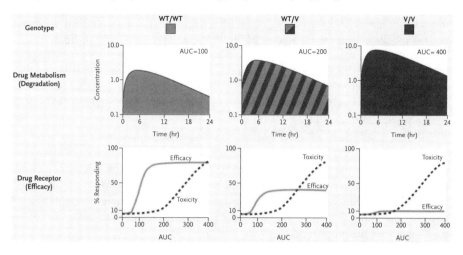

FIGURE 1.3

Genetic control of drug response as a longitudinal characteristic. AUC is the area under the plasma concentration-time curve. Adapted from Evans and McLeod (2003).

ited differences in drug disposition (e.g., metabolizing enzymes and transporters), expressed as pharmacokinetic reactions (top, Fig. 1.3), and in regulating drug targets (e.g., receptors), expressed as pharmacodynamic reactions (middle, Fig. 1.3). Patients are different in drug clearance (or the area under the plasma concentration-time curve) and receptor sensitivity, depending on their genotypes. The patients who are homozygous for the wild-type allele (WT/WT) tend to depose a less amount of drug during a time course than those who are heterozygous for one wild-type and one variant (V) allele (WT/V), or have two variant alleles (V/V) for the two polymorphisms. Also, the clinical efficacy (favorable effect) of the drug is more pronounced for the WT/WT patients than for the WT/V and V/V patients, although the toxicity (adverse effect) of the drug displays no difference among different genotypes. At the bottom of Figure 1.3 are shown the 9 potential combinations of drug-metabolism and drug-receptor genotypes and the corresponding drug-response phenotypes calculated from data at the top, yielding therapeutic indexes (efficacy:toxicity ratios) ranging from 13 (65 percent:5 percent) to 0.125 (10 percent:80 percent).

Genetic mapping based on the nonrandom co-segregation and association between different genes in a population has proven to be a powerful means for a systematic genome-wide search for genes that control a pharmacological reaction process, their number, genomic distribution, genetic effects and interactions with other genes and environments (Wu et al., 2007a). However, to understand more fully the genetic basis of drug response, new approaches are needed with which specific genes responsible for a whole process of pharmacological reactions or different stages of reactions can be identified. Also, a simple application of genetic mapping to pharmacogenetic research is not sufficiently enough when the pharmacokinetic and pharmacodynamic mechanisms of drug response are not integrated. Functional mapping, which embeds mathematical aspects of biological and biochemical principles within a mapping framework, holds a great promise to map and haplotype genes that determine the discrepancies in the extent and pattern of different persons to respond to drugs. Genetic information gained from functional mapping can improve our ability to prospectively identify patients at risk for severe toxicity, or those likely to benefit from a particular treatment.

The inherited nature of such variation in drug disposition and effects must be clearly elucidated to provide theoretical principles for optimizing drug therapy on the basis of individual patients' genetic constitutions. A greater understanding of the genetic determinants of drug response has the potential to revolutionize the use of many medications, to deliver the right drug to the right patient at the right dose in terms of the patient's genetic makeup, and provide individualized molecular diagnostic methods that improve drug therapy.

1.3   Genetic Designs

The pharmacogenetic control of drug response can be studied with an animal model system like the mouse and rat (Miles et al., 1991) or directly with humans. Types of mapping populations used to map genes for drug response are different between animals and humans because of their different biological properties and ethnic consideration. Below, we will discuss several commonly used mapping populations for animals and humans.

1.3.1   Experimental Crosses

Cross between two different mouse strains can widen genetic diversity because of segregation and recombination of genes (Darvasi, 1998). Consider two inbred strains in mouse which are homozygous for alternative alleles at each gene. These two strains as the parents are crossed to generate a heterozygous F_1 hybrid that contains two different alleles at each gene, each from a different parent. The F_1 is crossed to any one of the original parents to generate a backcross population in which there are two different genotypes at each gene (one heterozygote and one homozygote). Or, a cross between different F_1 individuals is made to form an F_2 population in which three different genotypes are generated for each gene (two homozygotes for each of the two alleles and one heterozygote). Some other mapping populations include recombinant inbred lines (RILs). An RIL population is generated by continuous inbreeding of the progeny, initiated with two heterozygous founders, for an adequately large number of generations that leads to the disappearance of any heterozygote (Broman, 2005). Thus, only homozygotes for the alternative alleles at each gene exist among RILs.

The backcross, F_2, and RILs provide excellent material for genetic mapping of quantitative traits. Genes that control a quantitative trait are called quantitative trait loci (QTLs). From a quantitative genetic viewpoint, an F_2 design is more informative than a backcross and RIL design because the former allows the estimation of additive and dominance genetic effects at a QTL whereas the latter can only provide the estimation of the additive genetic effect. The principle behind QTL mapping using an experimental cross is the occurrence of recombination events between genetic loci (measured by the recombination fraction) when gametes are formed and transmitted from parents to offspring. By estimating the recombination fractions between all pairs of markers, all markers can be grouped and ordered to construct a genetic linkage map that covers the entire genome or part of it. Functional QTLs for a quantitative trait, such as drug response, can then be located in terms of their recombination fractions with molecular markers with the linkage map and the genetic effects of these genes can be estimated and tested (Wu et al., 2007a). The greater coverage the linkage map has of the entire genome, the more likely it is that the map can detect a complete suite of QTLs for a trait.

1.3.2 Nuclear Families

In humans, neither adequate numbers of progeny can be generated from a single family nor can any controlled cross, like the backcross, F_2 or RILs, be made possible. For this species, a nuclear family with multiple successive generations is often used in order to accumulate a sufficient number of progeny for genetic mapping. For such a structured nuclear family, among-member relationships are defined by the probability of identical by descent (IBD) for observable marker alleles. Through the recombination fractions between the markers and a functional QTL, the IBD pattern of the QTL can be predicted with the IBD probability of the markers (Fulker and Cardon, 1994). A random-effect model can be incorporated into genetic mapping of a quantitative trait that segregates from family to family (Xu and Atchley, 1995), from which the genetic variance due to a QTL is estimated and tested. This design can be used to map QTLs for inherited diseases, such as diabetes or cancer.

1.3.3 Natural Populations with Unrelated Individuals

The genetic mapping of drug response can be conducted by sampling a collection of unrelated individuals at random from a natural population (Lou et al., 2003). In this case, mapping is based on (gametic) linkage disequilibrium (LD), i.e., a non-random association of alleles between different genes in a population. Because a particular allele at a marker locus tends to co-segregate with one allelic variant of the QTL of interest, provided the marker and QTL are very closely linked, LD mapping can potentially be used to map QTLs to very small regions (Wall and Pritchard, 2003). In order to perform efficient LD mapping, markers must be mapped at a density compatible with the distances in that LD extends in the population (Rafalski, 2002). Thus, an LD map that specified the decay of LD with genetic distance over the genome (Liu et al., 2006) is a prerequisite for genome-wide LD mapping. A high-density LD map is needed if LD decays dramatically with map distance (Fig. 1.4). Otherwise, a moderate-density map will be sufficient if LD extends over a wide distance. This design is appropriate for mapping QTLs that govern infectious diseases like HIV/AIDS.

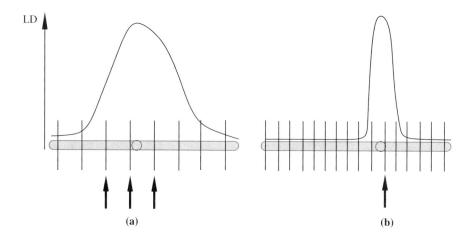

FIGURE 1.4

The extent of LD for appropriate resolution of association studies. (**a**) LD declines slowly with increasing distance from the gene (circle) responsible for the phenotype on a chromosome. In this case, even a low density of markers (shown as vertical lines) is sufficient to identify associated markers (thick arrows). (**b**) LD declines very rapidly around the causative gene (circle), and a much greater density of markers is required to identify an associated marker (thick arrow). Adapted from Rafalski (2002).

1.3.4   Natural Populations with Unrelated Families

Although LD mapping has tremendous potential to fine map functional QTLs for a quantitative trait, it may provide a spurious estimate of LD in practice when the association between a marker and QTL is due to evolutionary forces, such as mutation, drift, selection, and admixture (Lynch and Walsh, 1998). A mapping strategy that samples unrelated families (composed of parents and offspring) from a natural population (Wu and Zeng, 2001) is helpful for overcoming the limitation of LD mapping. At the upper hierarchy of a structured mapping strategy, associations among parents are specified by LD parameters, whereas the offspring within each family at the lower hierarchy of the strategy are related by the recombination fraction. Thus, the analysis of a structured natural population can not only simultaneously estimate the linkage and LD parameters, but also takes advantage of both linkage and LD mapping approaches in genome-wise scan and fine-structured estimates of QTL positions (Lou et al., 2005). This design is powerful for QTL identification of inherited diseases such as cancer or hypertension.

1.4   Strategies for Genomic Mapping

There are many different strategies used for genomic mapping. These strategies can be broadly classified into two categories, that is, linkage mapping based on the estimation of the recombination fraction and linkage disequilibrium mapping based on the estimation of LD between markers and functional genes. The determination of a proper mapping strategy depends on the genetic design developed for gene mapping. The characteristics of these two mapping strategies will be discussed as follows.

1.4.1   Linkage Mapping

If a genetic design contains the transmission of genes from parents to their offspring, linkage mapping can be used. Although the idea of linking a continuously varying phenotype with a discrete trait (marker) dates back to the work of Sax (1923), it was Lander and Botstein (1989) who first established an explicit principle for QTL mapping based on linkage analysis in a structured pedigree. They also provided a tractable statistical and computational algorithm for dissecting a quantitative trait into their individual QTL components. The basic idea of QTL mapping is to detect and test differences in the phenotypic value of a trait among different QTL genotypes, although these genotypes cannot be observed directly. Lander and Botstein (1989) used a pair of flanking markers to locate the position of a putative QTL on a marker interval by deriving the conditional probabilities of a QTL genotype given observed marker genotypes in terms of the marker-QTL recombination fractions. They constructed QTL mapping within a mixture model-based framework in which the QTL genotype that any individual carries can be inferred from its marker genotype and phenotypic observation.

Lander and Botstein's (1989) so-called *interval mapping* has proven powerful to map a QTL in simple situations in which a trait under study is strictly controlled by one QTL on a chromosome. But interval mapping may not produce precise results when more than one QTL are located on the same chromosome. To better separate the linkage among multiple QTLs, several researchers combined interval mapping based on two flanking markers and partial regression analysis on the other markers (excluding the two tested markers) (Jansen and Stam, 1994; Zeng, 1993, 1994). This approach, called *composite interval mapping* by Zeng (1994), can increase the precision of QTL detection by controlling the chromosomal region outside the marker interval under consideration. Composite interval mapping has been widely used in practical QTL mapping (Mackay, 2001), although the choice of suitable markers loci that serve as co-factors is still an open problem.

Both interval mapping and composite interval mapping model one QTL at a time, and fail to characterize the detailed genetic architecture of a quantitative trait. Because a complex trait may be controlled by a number of QTL, it is crucial to have a mapping approach that can model multiple QTL simultaneously and identify and locate all the QTLs responsible for quantitative variation. Such an approach, named multiple interval mapping (Kao and Zeng, 1997; Kao et al., 1999; Zeng et al., 2000), has been proposed by Zeng and colleagues who derived general formulae for obtaining maximum likelihood estimates for the positions and effects of multiple QTLs. Multiple interval mapping models multiple QTL in a way that QTLs can be directly controlled in the model through the simultaneous use of multiple marker intervals. It is supposed to have more power for estimating the positions and effects of QTLs than conventional interval mapping and composite interval mapping. In addition, by searching and mapping all possible QTLs in multiple marker intervals simultaneously, multiple interval mapping allows for the full estimation of the genetic architecture of a quantitative trait in terms of the number of the underlying QTLs, their genetic effects, pleiotropic effects, and epistatic network among different QTLs. The area of research that is open to multiple interval mapping is the procedure for the model selection of multiple QTLs – their genomic positions and effects – that collectively provide the best fit of the data observed.

Linkage mapping with an experimental cross or structured pedigree is a useful tool for detecting the existence and distribution of QTLs throughout the genome, but its mapping resolution is low unless a huge sample size is obtained. Given a currently available mapping population of 100–200 progeny for most genetic programs, linkage mapping only can map QTLs to 5–10 cM (Brown et al., 2003). This may translate into a physical distance of several megabases, which may contain several hundred genes. In order for linkage mapping to obtain the achievable mapping resolution for QTL positional cloning, the following measures can be taken (Georges, 2007):

(1) Increase the marker density. The more markers we have, the smaller the interval size, and therefore the higher the map resolution. This is technically the easiest approach for increasing the mapping resolution, although it is a labor-intensive and time-consuming "web lab" undertaking.

(2) Increase the crossover density by increasing current recombinants. By generating more progeny of a mapping population, one can increase the crossover density. However, this is an expensive way because the number of backcross or F_2 progeny required to map a QTL down to 5 cM or less is enormous (~ 5000) (Darvasi, 1998). One alternative for increasing the crossover density is to generate advanced intercross lines, such as F_3, F_4, ..., F_n.

(3) Increase the crossover density by utilizing historical recombinants. This approach capitalizes on the non-random allelic association, i.e, LD, between a

QTL and closely liked markers at the population level. In the following section, this approach will be discussed in detail.

1.4.2 Linkage Disequilibrium Mapping

Currently, population-based association studies or LD mapping have been proposed to map QTLs that are segregating in natural populations. Because any long range association between a marker and QTL has been broken due to recombinations that accumulate over many generations (Lynch and Walsh, 1998), this approach can only detect a marker-QTL association in a short stretch of the genome, and thus provides a vital tool for high-resolution mapping of QTLs. The theoretical basis of association studies is the LD at different loci in a population. LD-based mapping that directly capitalizes on existing natural populations is well suited to humans. Current LD mapping approaches can be used only for a population that is at Hardy–Weinberg equilibrium.

1.4.2.1 Hardy–Weinberg Equilibrium

A natural population sampled to perform LD-based QTL mapping consists of the offspring of a mixture of different mating types, and the ratios of the different genotypes in such a population are weighted averages of the segregation ratios of the different mating types, the weights being the relative frequencies of the different mating types. When the mating type frequencies arise from random mating, the ratios of the different genotypes follow a mathematical model established independently by the English mathematician Hardy (1908) and the German physician Weinberg (1908).

Consider a locus with alleles A in a frequency of p and a in a frequency of $1-p$. Let P_2, P_1, and P_0 be the frequencies of the three genotypes AA, Aa, and aa in a large population, respectively. Hardy and Weinberg's result was that, if individuals in the population mated with each other at random, these frequencies would satisfy the relationship

$$P_1^2 = 4P_2P_0.$$

When this relationship is held, the two theorems as follows can be proven:

Theorem 1 – In a Hardy–Weinberg equilibrium population, each of the genotype frequencies is expressed as the product of the corresponding allele frequencies, i.e.,

$$P_2 = p^2$$
$$P_1 = 2p(1-p) \tag{1.1}$$
$$P_0 = (1-p)^2$$

Theorem 2 – In a Hardy–Weinberg equilibrium population, each of the genotype frequencies is kept unchanged from generation to generation, i.e.,

$$P_2(t) = P_2(t+1) = \dots$$
$$P_1(t) = P_1(t+1) = \dots \tag{1.2}$$
$$P_0(t) = P_0(t+1) = \dots$$

where t denotes a generation of the population.

1.4.2.2   Principle of LD Mapping

The persistence of the equilibrium in genotype frequencies after one generation of random mating is true when all loci are considered separately. But it may not be true for the genotypes containing two or more loci considered jointly. Suppose there are two loci, one being a marker **M** with two alleles M, m and the second being a QTL **Q** with two alleles Q, q. The allelic frequencies of these two loci are denoted by p, $1 - p$ and q, $1 - q$, respectively. We denote the frequencies of four possible gametes MQ, Mq, mQ, and mq in the gametic output of a generation by p_{11}, p_{10}, p_{01}, and p_{00}.

The difference of the gamete frequency from the product of allele frequencies at two different loci is the amount of linkage disequilibrium denoted by D. Thus, the gametic frequencies of four gametes in the population are expressed as

$$
\begin{aligned}
p_{11} &= pq + D, \\
p_{10} &= p(1 - q) - D, \\
p_{01} &= (1 - p)q - D, \\
p_{00} &= (1 - p)(1 - q) + D.
\end{aligned}
\tag{1.3}
$$

From these expressions, it can be shown that $D = p_{11}p_{00} - p_{10}p_{01}$. Under linkage equilibrium, the gamete frequencies are expressed as $p_{11} = pq$, $p_{10} = p(1 - q)$, $p_{01} = (1 - p)q$, and $p_{00} = (1 - p)(1 - q)$.

The four gametes, whose frequencies in generation $t - 1$ are expressed as $p_{11}(t - 1)$, $p_{10}(t - 1)$, $p_{01}(t - 1)$, and $p_{00}(t - 1)$, randomly unite to form a zygote. The proportion $1 - r$ of the gametes produced by this zygote are parental (or non-recombinant) gametes and fraction r are nonparental (or recombinant) gametes. A particular gamete, say MQ, has a proportion $1 - r$ in the next generation t produced without recombination. The frequency with which this gamete is produced in this way is $(1 - r)p_{11}(t)$. This gamete can also be generated as a recombinant from the genotypes formed by the gametes containing marker allele M and the gametes containing QTL allele Q. The frequencies of the gametes containing alleles M or Q are $p(t - 1)$ and $q(t - 1)$, respectively. So the frequency with which MQ arises in this way is $rp(t - 1)q(t - 1)$. Therefore, the frequency of MQ in the generation t is

$$
p_{11}(t) = (1 - r)p_{11}(t - 1) + rp_1(t)q_1(t - 1)
$$

By subtracting $p(t - 1)q(t - 1)$ from both sides of the above equation, we have

$$
p_{11}(t) - p(t - 1)q(t - 1) = (1 - r)[p_{11}(t - 1) - p(t - 1)q(t - 1)].
\tag{1.4}
$$

If we assume that the allelic frequencies are constant from generation to generation, that is, that $p(t - 1)q(t - 1) = pq$, etc., and we define $D(t - 1) = p_{11}(t - 1) - pq$, then we can write (1.4) as

$$
D(t) = (1 - r)D(t - 1),
$$

which implies

$$D(t) = (1 - r)^t D(0), \tag{1.5}$$

or

$$\frac{D(t)}{D(0)} = (1 - r)^t,$$

where we define $D(t) = p_{11} - pq$.

We conclude that $D(t)$ converges to 0 at the geometric rate $1 - r$. Thus, *linkage equilibrium*, or gamete phase equilibrium, $p_{11} = pq$, is approached gradually and without oscillation. The larger is r, the faster is the rate of convergence, the most rapid being $(\frac{1}{2})^t$ for unlinked loci, in which case LD tends to disappear after 7–8 generations of mating (see Fig. 1.5).

The principle of LD decaying with generation builds up the LD-based mapping strategy (Weiss and Clark, 2002). The approach of using LD to construct a high-density linkage map allows us to increase the sample size by using all the recombination events occurring between different genes since the origin of the mutation, instead of only those events in the past few generations of a family. According to equation (1.5), a significant $D(t)$ value detected in the current generation implies that the decay rate after t generations, $(1 - r)^t$, should be significantly different from zero. As shown in Fig. 1.5, when t is very large, this rate $D(t)/D(0)$ will finally approach zero unless $r \to 0$. If r has an intermediate size, such as 0.1 or 0.2, $D(t)/D(0)$ will quickly approach zero. It can be seen from these analyses that under the assumption that the disequilibrium is generated a long time ago (t is large), the detection of significant LD implies high-resolution mapping of QTLs. LD-based mapping thus provides a powerful tool for fine mapping of genes affecting a quantitative trait.

1.4.3 Joint Linkage and Linkage Disequilibrium Mapping

A major problem with the LD mapping strategy is that it provides little insight into the mechanistic basis of LD detected in a natural population. Without such knowledge, however, the genomic localization and cloning of genes based on LD may not be successful, because a strong LD detected between two genetic loci may be due to the recent occurrence of disequilibrium rather than a close physical map distance of the two loci. The cause of LD can be revealed in a nuclear family through a combined linkage and LD analysis with a transmission/disequilibrium testing design (Allison, 1997; Rabinowitz, 1997; Camp, 1998). For a set of randomly sampled families from a natural population, linkage and LD mapping strategies can also be combined, taking advantage of each strategy to improve the power and resolution of QTL mapping (Wu et al., 2002b; Du et al., 2002). Obviously, simultaneous estimation of the recombination fraction and LD between the markers and QTL avoids false positive results (spurious LD) when LD is used to fine map genes for complex traits. Lou et al. (2005) developed a unifying model for integrating interval and LD mapping to fine map the locations of QTLs on a linkage map. As with interval mapping, this

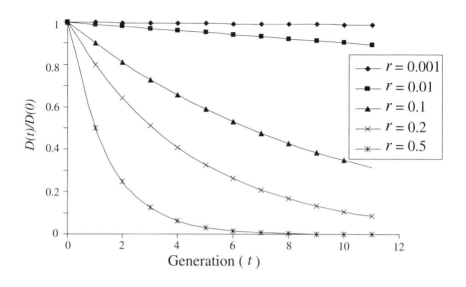

FIGURE 1.5

Plot of linkage disequilibrium decaying with generation (t) when a marker and QTL are linked to different degrees ($r = 0.001$–0.5).

model allows to scan for the existence and distribution of QTLs throughout the linkage map, and meanwhile provides the high-resolution estimation of QTL mapping by making use of recombinants that are generated over different generations.

1.5   From QTL to QTN

QTL mapping is robust and powerful for the detection of major QTLs and presents the most efficient way to utilize marker information when marker maps are sparse. However, this approach is limited in several aspects. First, because the markers and a bracketed QTL are located at different genomic positions, the significant linkage of a QTL detected with these markers cannot tell any information about the sequence structure and organization of the QTL. Second, the inference of the QTL position using the nearby markers is sensitive to marker informativeness, marker density, and mapping population type, which in turn affects the positional cloning of a QTL. Second, for outcrossing species, like humans, uncertainty about the linkage phase between genetic loci prevents the discovery of QTL alleles. Partly for these reasons, only a few QTLs detected from markers have been successfully cloned (Frary et al., 2000; Li et al., 2006a), despite a considerable number of QTLs reported in the literature.

A more accurate and useful approach for characterizing the genetic variants contributing to quantitative variation is to directly analyze DNA sequences that encode a particular trait with SNP data (Ron and Weller, 2007). If a string of DNA sequence or haplotype is known to increase disease risk, then a specialized drug can be designed to inhibit the expression of this DNA sequence. The control of this disease can be made more efficient if all of its underlying DNA sequences are identified over the entire genome. A new term, *quantitative trait nucleotides* or QTNs, has been coined to describe the sequence polymorphisms that cause phenotypic variation in a quantitative trait. The minimum nucleotide size of a QTN is a single SNP, whereas very often a QTN contains two or more closely linked SNPs. The association analysis between multi-SNP QTNs and phenotypes is statistically challenging because of the existence of unphased haplotypes. Haplotype-based QTN mapping of a complex disease and drug response will form the core of this book.

1.5.1 Genotype and Diplotype

A QTN is functional because its specific nucleotide or nucleotide sequence encodes a trait. The combination of nucleotides (alleles) at different SNPs on a single chromosome is called a *haplotype*. The pair of maternally- and paternally-derived haplotypes is called a *diplotype*. Unlike a genotype, a diplotype cannot always be observed although it has actual effects on a trait. Suppose there are two different SNPs on the same genomic region, one with two alleles A and a and the other with two alleles B and b, respectively. Allele A from SNP 1 and allele B from SNP 2 are located on the first homologous chromosome, whereas allele a from SNP 1 and allele b from SNP 2 are located on the second homologous chromosome. Thus, $[AB]$ is one haplotype and $[ab]$ is a second haplotype, and both constitute a diplotype $[AB][ab]$ (Fig. 1.6). Thus, for a double heterozygote Aa/Bb, there are two possible diplotypes $[AB][ab]$ and $[Ab][aB]$ (Fig. 1.7), which are unobservable.

While QTL mapping is to associate QTL genotypes with a phenotypic trait, QTN haplotyping aims to detect specific SNP haplotypes that encode the trait. With the QTL mapping standard, the double heterozygote Aa/Bb may be found to associate with a favorable phenotype. However, according to QTN haplotyping, Aa/Bb may have nothing to do with the favorable phenotype if its diplotype is $[Ab][aB]$ when haplotype $[AB]$ or $[ab]$ encodes the trait. Thus, if the genetic effect is expressed at the haplotype level, the same genotype Aa/Bb may perform differently, depending on what diplotype it carries. In practice, it is important to estimate haplotype effects on a quantitative trait based on the diplotypes and therefore genotypes. The statistical model being described in this book can be used to determine which diplotype is associated with a favorable phenotype in a natural population or experimental cross.

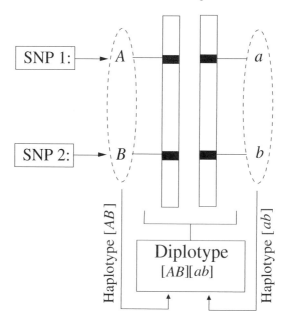

FIGURE 1.6
Haplotype configuration of a diplotype for two hypothesized SNPs.

1.5.2   Identification of QTNs

The completion of the human genome project has made it possible to construct genetic maps with the most abundant source of DNA variation – SNPs. The identification of QTNs in the human genome causally involved in the genetic etiology of disease risk and drug response in concordance with environmental influences can be based on two different approaches, candidate gene and genome-wide association studies. Candidate gene studies are usually hypothesis-driven, focusing on a particular gene or area of the genome, whereas genome-wide association studies are conducted without prior hypotheses by scanning for the entire genome.

The prerequisite of candidate gene studies is the generic nature of genes, whose functions can be potentially assigned to any genome. For example, miRNAs, discovered in *Caenorhabditis elegans*, are found to be functional in other organisms (Lee et al., 1993). For a small number of genes, prior knowledge about function suggests that they affect a complex phenotypic trait through particular pathways. Therefore, it is natural to observe associations between phenotypic variation and DNA sequence variation in these candidate genes. Yet, to date, the success rate of candidate gene studies has been very poor because their results could not have been replicated well. Although this may be partially due to the view that many complex traits are not caused by common genetic polymorphisms, many candidate gene studies were based on a poor experimental design (Zondervan and Cardon, 2007). Recent

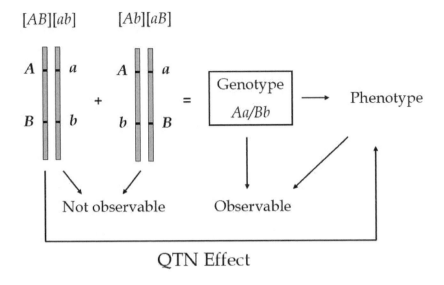

FIGURE 1.7
Diplotype configuration of a genotype for two hypothesized SNPs.

advances in molecular technologies have led to reductions in genotyping costs and more sophisticated specifications of the genotyping arrays in terms of SNP numbers and coverage. This has made it feasible to perform genome-wide association studies which allow the scan of functional or causal polymorphisms from 300,000–1 million SNPs. Many examples of replicated findings from genome-wide association studies have started to emerge (Klein et al., 2005; DeWan et al., 2006; Duerr et al., 2006; Cardon, 2006; Sladek et al., 2007; Frayling et al., 2007; Burton et al., 2007).

1.6   Functional Mapping of Drug Response

Since drug response is longitudinal in nature and should be measured repeatedly over different time points or dosage levels, conventional genomic mapping approaches cannot capture the dynamic and longitudinal change of genetic effects and therefore predict pharmacological reactions of a drug before it is administrated. Although the elucidation of the relationship between genetic control and drug response is statistically a pressing challenge, some of the key difficulties have been overcome by a statistical genetics group at the University of Florida (Ma et al., 2002; Wu et al., 2002a, 2003, 2004b,c,d; Wu and Lin, 2006). They have proposed a general statistical framework, i.e., *functional mapping*, to genome-wide map specific QTLs or QTNs that determine the developmental pattern of a complex trait.

The basic rationale of functional mapping lies in the connection between gene action/interaction effects or environmental effects and developmental change by parametric or nonparametric models. Functional mapping maps dynamic QTLs or QTNs that are responsible for a biological process that is measured at a finite number of time points. A number of mathematical models have been established to describe the developmental process of a biological phenotype. For example, a series of growth equations have been derived to describe growth in height, size, or weight (von Bertalanffy, 1957; Richards, 2002) that occur whenever the anabolic or metabolic rate exceeds the rate of catabolism. Based on fundamental principles behind biological or biochemical networks, West et al. (2001) have mathematically proved the universality of these growth equations. With mathematical functions incorporated into the QTL/QTN mapping framework, functional mapping estimates parameters that determine shapes and functions of a particular biological network, instead of directly estimating the gene effects at all possible time points. Because of such connections among these points through mathematical functions, functional mapping strikingly reduces the number of parameters to be estimated and, hence, displays increased statistical power.

From a statistical perspective, functional mapping is a problem of jointly modelling mean-covariance structures in longitudinal studies, an area that has recently received a considerable interest in the statistical literature (Pourahmadi, 1999, 2000; Pan and Mackenzie, 2003; Wu and Pourahmadi, 2003). However, different from general longitudinal modeling, functional mapping integrates the parameter estimation and test process within a biologically meaningful mixture-based likelihood framework. Functional mapping is thus advantageous in terms of biological relevance because biological principles are embedded into the estimation process of QTL/QTN parameters. The results derived from functional mapping will be closer to biological reality. In this book, we will demonstrate how functional mapping can be used to haplotype a QTN that contributes to inter-individual variability in drug response.

2

Genetic Haplotyping in Natural Populations

The association between haplotype diversity and phenotypic variation has been detected by several pharmacogenetic studies (Judson et al., 2000; Bader, 2001; Rha et al., 2007) and will shape our recognition of the genetic control of drug response. However, since haplotypes (comprising diplotypes) cannot be directly observed, the effects of different haplotypes on the phenotype need be postulated from observed zygotic genotypes. The inference of diplotypes for a particular genotype is statistically a missing data problem that can be formulated by a finite mixture model. Liu et al. (2004) proposed a statistical model for estimating and testing haplotype effects at a QTN in a random sample drawn from a natural population. This model is based on the population genetic properties of gene segregation. Through the implementation of the EM algorithm, population genetic parameters of SNPs, such as haplotype frequencies, allele frequencies, and linkage disequilibria, and quantitative genetic parameters, such as haplotype effects of a QTN, are estimated with closed forms.

In this chapter, we will describe the haplotyping method for the identification of DNA sequence variants that encode a quantitative trait in a natural population that is assumed at Hardy–Weinberg equilibrium. We will first introduce the basic concepts of QTN haplotyping and then provide a procedure for constructing and computing the likelihood of haplotype distribution and haplotype effects. We will use a published example to demonstrate the usefulness of genetic haplotyping to study the genetic architecture of a complex trait. The haplotyping theory described in this chapter will be integrated with functional mapping to understand the dynamic genetic control of QTNs for drug response.

2.1 Notation and Definitions

Suppose there is a QTN composed of only two SNPs each with two alleles designated as 1 and 0. These two SNPs segregating in the natural population form four haplotypes, [11], [10], [01], and [00], with frequencies arrayed in a population genetic parameter vector $\Omega_p = (p_{11}, p_{10}, p_{01}, p_{00})$. All the genotypes are consistent with diplotypes, except for the double heterozygote, 10/10, that contains two different diplotypes [11][00] with a frequency of $2p_{11}p_{00}$ and [10][01] with a frequency of $2p_{10}p_{01}$ (Table 2.1). The relative frequencies of different diplotypes for the double

TABLE 2.1

Diplotypes and their frequencies for each of nine genotypes at two SNPs within a QTN, haplotype composition frequencies for each genotype, and composite diplotypes for four possible risk haplotypes.

	Diplotype		Risk Haplotype			
Genotype	Configuration	Frequency	[11]	[10]	[01]	[00]
11/11	[11][11]	p_{11}^2	AA	$\bar{A}\bar{A}$	$\bar{A}\bar{A}$	$\bar{A}\bar{A}$
11/10	[11][10]	$2p_{11}p_{10}$	$A\bar{A}$	$A\bar{A}$	$\bar{A}\bar{A}$	$\bar{A}\bar{A}$
11/00	[10][10]	p_{10}^2	$\bar{A}\bar{A}$	AA	$\bar{A}\bar{A}$	$\bar{A}\bar{A}$
10/11	[11][01]	$2p_{11}p_{01}$	$A\bar{A}$	$\bar{A}\bar{A}$	$A\bar{A}$	$\bar{A}\bar{A}$
10/10	$\begin{cases} [11][00] \\ [10][01] \end{cases}$	$\begin{cases} 2p_{11}p_{00} \\ 2p_{10}p_{01} \end{cases}$	$\begin{cases} A\bar{A} \\ \bar{A}\bar{A} \end{cases}$	$\begin{cases} \bar{A}\bar{A} \\ A\bar{A} \end{cases}$	$\begin{cases} \bar{A}\bar{A} \\ A\bar{A} \end{cases}$	$\begin{cases} A\bar{A} \\ \bar{A}\bar{A} \end{cases}$
10/00	[10][00]	$2p_{10}p_{00}$	$\bar{A}\bar{A}$	$A\bar{A}$	$\bar{A}\bar{A}$	$A\bar{A}$
00/11	[01][01]	p_{01}^2	$\bar{A}\bar{A}$	$\bar{A}\bar{A}$	AA	$\bar{A}\bar{A}$
00/10	[01][00]	$2p_{01}p_{00}$	$\bar{A}\bar{A}$	$\bar{A}\bar{A}$	$A\bar{A}$	$A\bar{A}$
00/00	[00][00]	p_{00}^2	$\bar{A}\bar{A}$	$\bar{A}\bar{A}$	$\bar{A}\bar{A}$	AA

Two alleles for each of the two SNPs are denoted as 1 and 0, respectively. Genotypes at different SNPs are separated by a slash. Diplotypes are the combination of two bracketed maternally and paternally derived haplotypes. By assuming different haplotypes as a risk haplotype, composite diplotypes are accordingly defined and their genotypic values are given.

heterozygote are a function of haplotype frequencies.

A total of n subjects sampled are classified into nine genotypes for the two SNPs, each genotype with observation generally expressed as $n_{r_1 r_1'/r_2 r_2'}$ $(r_1 \geq r_1', r_2 \geq r_2' = 1, 0)$. The frequency of each genotype can be expressed in terms of haplotype frequencies (Table 2.1). Assume that this QTN triggers an effect on the trait because at least one haplotype is different from the remaining three. Without loss of generality, let [11] be such a distinct haplotype, called *risk haplotype*, designated as A. All the other non-risk haplotypes, [10], [01], and [00], are collectively expressed as $\bar{A}$. The risk and non-risk haplotypes form three *composite diplotypes* AA (symbolized as **2**), $A\bar{A}$ (symbolized as **1**), and $\bar{A}\bar{A}$ (symbolized as **0**). Let μ_2, μ_1, and μ_0 be the genotypic value of the three composite diplotypes, respectively (Table 2.1). The means for different composite diplotypes and residual variance are arrayed by a quantitative genetic parameter vector $\Omega_q = (\mu_2, \mu_1, \mu_0, \sigma^2)$.

We will use traditional quantitative genetic theory to characterize genetic effects of a QTN. Let a be the additive effect due to the substitution of a non-risk haplotype by a risk haplotype, and d be the dominance effect due to the interaction between the risk and non-risk haplotypes. The genotypic values of three composite diplotypes can now be expressed as

$$\begin{aligned}
\mu_2 &= \mu + a, \\
\mu_1 &= \mu + d, \\
\mu_0 &= \mu - a,
\end{aligned} \tag{2.1}$$

where μ is the overall population mean of the trait. Thus, if the genotypic values of composite diplotypes are estimated, then the additive and dominance genetic effects can be estimated by solving a system of equations (2.1).

2.2 Likelihoods

The log-likelihood of population and quantitative genetic parameters given the phenotypic (y) and SNP data ($\mathbf{S}$) is factorized into two parts, expressed as

$$\log L(\Omega_p, \Omega_q | y, \mathbf{S}) = \log L(\Omega_p | \mathbf{S}) + \log L(\Omega_q | y, \mathbf{S}, \Omega_p) \tag{2.2}$$

where

$$\begin{aligned}
\log L(\Omega_p | \mathbf{S}) &= \text{constant} \\
&+ 2n_{11/11} \log p_{11} \\
&+ n_{11/10} \log(2p_{11}p_{10}) \\
&+ 2n_{11/00} \log p_{10} \\
&+ n_{10/11} \log(2p_{11}p_{01}) \\
&+ n_{10/10} \log(2p_{11}p_{00} + 2p_{10}p_{01}) \\
&+ n_{10/00} \log(2p_{10}p_{00}) \\
&+ 2n_{00/11} \log p_{01} \\
&+ n_{00/10} \log(2p_{01}p_{00}) \\
&+ 2n_{00/00} \log p_{00}
\end{aligned}$$

$$\begin{aligned}
\log L(\Omega_q | y, \mathbf{S}, \Omega_p) &= \\
&\sum_{i=1}^{n_{11/11}} \log f_2(y_i) \\
&+ \sum_{i=1}^{n_{11/10}} \log f_1(y_i) \\
&+ \sum_{i=1}^{n_{11/00}} \log f_0(y_i) \\
&+ \sum_{i=1}^{n_{10/11}} \log f_1(y_i) \\
&+ \sum_{i=1}^{n_{10/10}} \log[\phi f_1(y_i) + (1 - \phi) f_0(y_i)] \\
&+ \sum_{i=1}^{n_{10/00}} \log f_0(y_i) \\
&+ \sum_{i=1}^{n_{00/11}} \log f_0(y_i) \\
&+ \sum_{i=1}^{n_{00/10}} \log f_0(y_i) \\
&+ \sum_{i=1}^{n_{00/00}} \log f_0(y_i)
\end{aligned}$$

$$\tag{2.3}$$

assuming that [11] is a risk haplotype with $f_j(y_i)$ being a normal distribution density function of composite diplotype j ($j = 2, 1, 0$).

It can be seen from the above likelihood functions that, although most zygote genotypes contain a single component (diplotype), the double heterozygote is the mixture of two possible diplotypes weighted by ϕ and $1 - \phi$, expressed as

$$\phi = \frac{p_{11}p_{00}}{p_{11}p_{00} + p_{10}p_{01}} \tag{2.4}$$

which represents the relative frequency of diplotype [11][00] for the double heterozygote.

It should be noted that $L(\Omega_q|y, S, \Omega_p)$ relies on the haplotype frequencies defined
in $L(\Omega_p|S)$ and, thus, the latter is thought to be nested within the former. The esti-
mates of parameters that maximize $L(\Omega_p|S)$ can also maximize the $L(\Omega_q|y, S, \Omega_p)$.

2.3   The EM Algorithm

The expectation-maximization (EM) algorithm, advocated originally by Dempster
et al. (1977), can be derived to estimate the unknown parameters that construct
the likelihoods (2.3). Liu et al. (2004) derived a closed-form solution for the EM
algorithm to estimate these parameters. The estimates of haplotype frequencies
are based on the log-likelihood function $L(\Omega_p|S)$, whereas the estimates of diplo-
type genotypic values and residual variance are based on the log-likelihood function
$L(\Omega_q|y, S, \Omega_p)$. These two different types of genetic parameters can be estimated
using a two-stage hierarchical EM algorithm.

At a higher hierarchy of the EM algorithm, the E step aims to calculate the relative frequency (ϕ) of diplotype [11][00] in the double heterozygote as calculated by
equation (2.4). The M step aims to estimate the haplotype frequencies based on the
probabilities calculated in the previous iteration using

$$
\begin{aligned}
\hat{p}_{11} &= \frac{1}{2n}(2n_{11/11} + \phi n_{10/10} + n_{11/10} + n_{10/11}) \\
\hat{p}_{10} &= \frac{1}{2n}[2n_{11/00} + n_{11/10} + (1-\phi)n_{10/10} + n_{10/00}] \\
\hat{p}_{01} &= \frac{1}{2n}[2n_{00/11} + n_{10/11} + (1-\phi)n_{10/10} + n_{00/10}] \\
\hat{p}_{00} &= \frac{1}{2n}(2n_{00/00} + \phi n_{10/10} + n_{01/00} + n_{10/00}).
\end{aligned}
\tag{2.5}
$$

At a lower hierarchy of the EM algorithm, the E step is derived to calculate the
posterior probability ($\Phi_{[11][00]i}$) with which a double heterozygote i carries diplotype
[11][00] by

$$
\Phi_{[11][00]|i} = \frac{\phi f_{[11][00]}(y_i)}{\phi f_{[11][00]}(y_i) + (1-\phi)f_{[10][01]}(y_i)}.
\tag{2.6}
$$

Note that for all the other genotypes, such posterior probabilities do not need to be
defined.

Under the assumption that [11] is the risk haplotype, the M step is derived to
estimate the genotypic values (μ_j) for each composite diplotype and the residual

variance based on the calculated posterior probabilities by

$$\hat{\mu}_2 = \frac{\sum_{i=1}^{n_{11/11}} y_i}{n_{11/11}},$$

$$\hat{\mu}_1 = \frac{\sum_{i=1}^{\dot{n}} y_i + \sum_{i=1}^{n_{10/10}} \Phi_{[11][00]|i} y_i}{\dot{n} + \sum_{i=1}^{n_{10/10}} \Phi_{[11][00]|i}},$$

$$\hat{\mu}_0 = \frac{\sum_{i=1}^{\ddot{n}} y_i + \sum_{i=1}^{n_{10/10}} (1 - \Phi_{[11][00]|i}) y_i}{\ddot{n} + \sum_{i=1}^{n_{10/10}} (1 - \Phi_{[11][00]|i})}, \tag{2.7}$$

$$\hat{\sigma}^2 = \frac{1}{n} \left\{ \sum_{i=1}^{n_{11/11}} (y_i - \hat{\mu}_2)^2 + \sum_{i=1}^{\dot{n}} (y_i - \hat{\mu}_1)^2 + \sum_{i=1}^{\ddot{n}} (y_i - \hat{\mu}_0)^2 \right.$$

$$\left. + \sum_{i=1}^{n_{10/10}} \left[\Phi_{[11][00]|i} (y_i - \hat{\mu}_1)^2 + (1 - \Phi_{[11][00]|ii})(y_i - \hat{\mu}_0)^2 \right] \right\},$$

where

$$\dot{n} = n_{11/10} + n_{10/11},$$

$$\ddot{n} = n_{11/00} + n_{10/00} + n_{01/01} + n_{01/00} + n_{00/00}.$$

Iterations including the E and M steps are repeated at the higher hierarchy between equations (2.4) and (2.5) and at the lower hierarchy among equations (2.6) and (2.7) until the estimates of the parameters converge to stable values. The estimates at convergence are the maximum likelihood estimates (MLEs) of the parameters.

2.4   Sampling Variances of Parameter Estimates

After the point estimates of parameters are obtained by the EM algorithm, it is necessary to derive the variance-covariance matrix and evaluate the standard errors of the estimates. Because the EM algorithm does not automatically provide the estimates of the asymptotic variance-covariance matrix for parameters, an additional procedure has been developed to estimate this matrix (and thereby standard errors) when the EM algorithm is used (Louis, 1982; Meng and Rubin, 1991). These techniques involve calculation of the incomplete-data information matrix which is the negative second-order derivative of the incomplete-data log-likelihood. Louis (1982) established an important relationship among the information for the complete ($\mathscr{I}_{com}$), incomplete ($\mathscr{I}_{incom}$), and missing data ($\mathscr{I}_{miss}$), expressed as

$$\mathscr{I}_{incom} = \mathscr{I}_{com} - \mathscr{I}_{miss}.$$

Since it is difficult to evaluate $\mathscr{I}_{incom}$ directly, Louis' relationship allows for an indirect evaluation based on $\mathscr{I}_{com}$ and $\mathscr{I}_{miss}$. Meng and Rubin (1991) proposed a so-called supplemented EM algorithm or SEM algorithm to estimate the asymptotic variance-covariance matrices, which can also be used for the calculations of the standard errors for the MLEs of the genetic parameters.

Below, the derivation procedures of $\mathscr{I}_{com}$ and $\mathscr{I}_{miss}$ are given. As seen in Louis (1982), the observed information ($\mathscr{I}_{incom}$) can be obtained by subtracting the missing information ($\mathscr{I}_{miss}$) from the complete information ($\mathscr{I}_{com}$). Using Louis' notation, we denote the complete data by $\mathbf{x} = (x_1, x_2, \cdots, x_n)'$, the observed data by $\mathbf{y} = (y_1, y_2, \cdots, y_n)'$ and the parameter vector by θ. The likelihood of the incomplete data is expressed as

$$f_Y(\mathbf{y}|\theta) = \int_{\mathbf{R}} f_X(\mathbf{x}|\theta) d\vartheta(\mathbf{x}), \tag{2.8}$$

where $\mathbf{R} = \{\mathbf{x} : y(\mathbf{x}) = \mathbf{y}\}$ and $\vartheta(\mathbf{x})$ is a dominating measure. For equation 2.8), we have the score,

$$\frac{\partial \ln f_Y(\mathbf{y}|\theta)}{\partial \theta} = \int_{\mathbf{R}} \frac{f_X(\mathbf{x}|\theta)}{\int_{\mathbf{R}} f_X(\mathbf{x}|\theta) d\vartheta(\mathbf{x})} \frac{\partial \ln f_X(\mathbf{x}|\theta)}{\partial \theta} d\vartheta(\mathbf{x}), \tag{2.9}$$

and the Hessian matrix,

$$\begin{aligned}
\frac{\partial^2 \ln f_Y(\mathbf{y}|\theta)}{\partial \theta \partial \theta'} &= \int_{\mathbf{R}} \frac{f_X(\mathbf{x}|\theta)}{\int_{\mathbf{R}} f_X(\mathbf{x}|\theta) d\vartheta(\mathbf{x})} \frac{\partial^2 \ln f_X(\mathbf{x}|\theta)}{\partial \theta \partial \theta'} d\vartheta(\mathbf{x}) \\
&+ \int_{\mathbf{R}} \frac{f_X(\mathbf{x}|\theta)}{\int_{\mathbf{R}} f_X(\mathbf{x}|\theta)} \frac{\partial \ln f_X(\mathbf{x}|\theta)}{\partial \theta} d\vartheta(\mathbf{x}) \frac{\partial \ln f_X(\mathbf{x}|\theta)}{\partial \theta'} d\vartheta(\mathbf{x}) \\
&- \int_{\mathbf{R}} \frac{f_X(\mathbf{x}|\theta)}{\int_{\mathbf{R}} f_X(\mathbf{x}|\theta) d\vartheta(\mathbf{x})} \frac{\partial \ln f_X(\mathbf{x}|\theta)}{\partial \theta} d\vartheta(\mathbf{x}) \\
&\int_{\mathbf{R}} \frac{f_X(\mathbf{x}|\theta)}{\int_{\mathbf{R}} f_X(\mathbf{x}|\theta) d\vartheta(\mathbf{x})} \frac{\partial \ln f_X(\mathbf{x}|\theta)}{\partial \theta'} d\vartheta(\mathbf{x}). \tag{2.10}
\end{aligned}$$

When $(x_1, x_2, \cdots, x_n)$ are independent but not necessarily identically distributed and $y_i(\mathbf{x}) = y_i(x_i)$, we have $\mathbf{R} = R_1 \times R_2 \times \cdots \times R_n$. To make the score (2.9) and the Hessian matrix (2.10) more tractable, we need to use Fubini's theorem. By incorporating Fubini's theorem, we can now express the score and the Hessian matrix as

$$\frac{\partial \ln f_Y(\mathbf{y}|\theta)}{\partial \theta} = \sum_{i=1}^n \int_{R_i} \frac{f_X(x_i|\theta)}{\int_{R_i} f_X(x_i|\theta) d\vartheta(x_i)} \frac{\partial \ln f_X(x_i|\theta)}{\partial \theta} d\vartheta(x_i) \tag{2.11}$$

and

$$\begin{aligned}
\frac{\partial^2 \ln f_Y(\mathbf{y}|\theta)}{\partial \theta \partial \theta'} &= \sum_{i=1}^n \int_{R_i} \frac{f_X(x_i|\theta)}{\int_{R_i} f_X(x_i|\theta) d\vartheta(x_i)} \frac{\partial^2 \ln f_X(x_i|\theta)}{\partial \theta \partial \theta'} d\vartheta(x_i) \\
&+ \sum_{i=1}^n \int_{R_i} \frac{f_X(x_i|\theta)}{\int_{R_i} f_X(x_i|\theta) d\vartheta(x_i)} \frac{\partial \ln f_X(x_i|\theta)}{\partial \theta} \frac{\partial \ln f_X(x_i|\theta)}{\partial \theta^{\mathbf{T}}} d\vartheta(x_i)
\end{aligned}$$

$$-\sum_{i=1}^{n}\left[\int_{R_i}\frac{f_X(x_i|\theta)}{\int_{R_i}f_X(x_i|\theta)d\vartheta(x_i)}\frac{\partial\ln f_X(x_i|\theta)}{\partial\theta}d\vartheta(x_i)\right.$$

$$\left.\int_{R_i}\frac{f_X(x_i|\theta)}{\int_{R_i}f_X(x_i|\theta)d\vartheta(x_i)}\frac{\partial\ln f_X(x_i|\theta)}{\partial\theta'}d\vartheta(x_i)\right]. \tag{2.12}$$

Based on (2.11) and (2.12), the complete $\mathscr{I}_{com}$ and missing information $\mathscr{I}_{miss}$ can be calculated using Louis' formulae, from which the observed information $\mathscr{I}_{incom}$ is estimated.

After these parameters are estimated, the observed information matrix can be readily estimated for haplotype frequencies, genotypic values of composite diplotypes, and the residual variance. A further step is needed to derive the variance-covariance matrices for allele frequencies and linkage disequilibria from the haplotype frequencies and for the additive and dominance genetic effects from the genotypic values. It is not difficult to obtain the variance-covariance matrices for allele frequencies and genetic effects because they are a linear function of haplotype frequencies and genotypic values, respectively.

Because the analytical expression for estimating the variances and covariances of linkage disequilibria does not exist, we derive an approximate expression for these variances and covariances using the *delta method* derived from Taylor's series expansions. If a parameter vector φ is a function of the characteristic parameter θ, i.e., $\varphi = f(\theta)$, then the approximate variance-covariance of $\hat{\varphi} = f(\hat{\theta})$ is,

$$\text{Var}(\hat{\varphi}) = \frac{\partial f(\theta)}{\partial\theta}\text{Var}(\hat{\theta})\frac{\partial f(\theta)}{\partial\theta'}. \tag{2.13}$$

Based on equation (2.13), the variance of the MLE of the linkage disequilibrium between the two SNPs can be derived as

$$\text{Var}(\hat{D}) = \begin{pmatrix} 1-2\hat{p}_{11}-\hat{p}_{10}-\hat{p}_{01} \\ -\hat{p}_{11}-\hat{p}_{01} \\ -\hat{p}_{11}-\hat{p}_{10} \end{pmatrix}' \times \widehat{\text{Var}}\begin{pmatrix} \hat{p}_{11} \\ \hat{p}_{10} \\ \hat{p}_{01} \end{pmatrix}$$

$$\times (1-2\hat{p}_{11}-\hat{p}_{10}-\hat{p}_{01} \quad -\hat{p}_{11}-\hat{p}_{01} \quad -\hat{p}_{11}-\hat{p}_{10}).$$

2.5 Model Selection

The likelihood $L(\Omega_q|y,\mathbf{S},\Omega_p)$ is formulated by assuming that haplotype [11][11] is a risk haplotype. However, a real risk haplotype is unknown from raw data $(y,\mathbf{S})$. An additional step for the choice of the most likely risk haplotype should be implemented. The simplest way to do so is to calculate the likelihood values by assuming that any one of the four haplotypes can be a risk haplotype (Table 2.1). Thus, we obtain four possible likelihood values as follows:

No.	Risk Haplotype	Likelihood
1	[11]	$L_1(\hat{\boldsymbol{\Omega}}_{1q}|y,\mathbf{S},\hat{\boldsymbol{\Omega}}_p)$
2	[10]	$L_2(\hat{\boldsymbol{\Omega}}_{2q}|y,\mathbf{S},\hat{\boldsymbol{\Omega}}_p)$
3	[01]	$L_3(\hat{\boldsymbol{\Omega}}_{3q}|y,\mathbf{S},\hat{\boldsymbol{\Omega}}_p)$
4	[00]	$L_4(\hat{\boldsymbol{\Omega}}_{4q}|y,\mathbf{S},\hat{\boldsymbol{\Omega}}_p)$

The largest likelihood value calculated is thought to correspond to the most likely risk haplotype. Under an optimal risk haplotype k, we estimate the quantitative genetic parameters $\hat{\boldsymbol{\Omega}}_{kq}$.

In practice, some haplotype my be missing in which case there may be fewer parameters to be estimated. If this happens, the likelihood cannot be used as a criterion to select the optimal risk and non-risk haplotype combination because the number of parameters is different. When the number of parameters is different, an AIC- or BIC-based model selection strategy (Burnham and Andersson 1998) should be used.

2.6 Hypothesis Tests

Two major hypotheses can be tested in the following sequence: (1) The two SNPs under study comprise a QTN by testing the significance of the linkage disequilibrium between these two SNPs, and (2) The QTN detected exerts significant additive and dominant effects on the trait by testing the difference of a given haplotype from the remaining haplotypes. The LD between two given SNPs can be tested using the following hypotheses:

$$H_0 : D = 0 \text{ vs. } H_1 : D \neq 0 \qquad (2.14)$$

The log-likelihood ratio test statistic for the significance of LD is calculated by comparing the likelihood values under the H_1 (full model) and H_0 (reduced model) using

$$\text{LR}_1 = -2[\log L_0(\tilde{p}_{11}\tilde{p}_{00} = \tilde{p}_{10}\tilde{p}_{01}|\mathbf{S}) - \log L_1(\hat{\boldsymbol{\Omega}}_p|\mathbf{S})], \qquad (2.15)$$

where the tildes and hats denote the MLEs of parameters under the null and alternative hypotheses of (2.14), respectively. The LR_1 is considered to asymptotically follow a χ^2 distribution with one degree of freedom.

Diplotype or haplotype effects on the trait, i.e., the existence of a significant QTN, can be tested using the following hypotheses expressed as

$$H_0 : \mu_j \equiv \mu \; (j = 2,1,0) \text{ vs. } H_1 : \text{at least one equality in } H_0 \text{ is not true,}$$

$$(2.16)$$

The log-likelihood ratio test statistic (LR_2) under these two hypotheses can be similarly calculated,

$$\text{LR}_2 = -2[\log L(\tilde{\mu},\tilde{\sigma}^2|y) - \log L(\hat{\boldsymbol{\Omega}}_q|y,\mathbf{S},\hat{\boldsymbol{\Omega}}_p)], \qquad (2.17)$$

where the tildes and hats denote the MLEs of parameters under the null and alternative hypotheses of (2.16), respectively. Although the critical threshold for determining the existence of a QTN can be based on empirical permutation tests, the LR_2 may asymptotically follow a χ^2 distribution with two degrees of freedom, so that the threshold can be obtained from the χ^2 distribution table.

The additive and dominance effects of the significant QTN can be tested separately by formulating the null hypothesis H_0: $a = 0$ or H_0: $d = 0$, respectively. The estimates of quantitative genetic parameters under these null hypotheses can be made with the same EM procedure as described in equations (2.6) and (2.7), except for a constraint expressed as

$$\tilde{\mu}_2 = \tilde{\mu}_0$$

for H_0: $a = 0$, and

$$\tilde{\mu}_1 = \tfrac{1}{2}(\tilde{\mu}_2 + \tilde{\mu}_0)$$

for H_0: $d = 0$. In both cases, the calculated log-likelihood values follow a χ^2-distribution with one degree of freedom.

2.7 Haplotyping with Multiple SNPs

It is possible that more than two SNPs form a QTN. We provide an estimation and test procedure for the genetic effect of a QTN composed of three different SNPs. Statistical methods for analyzing more than three QTNs can be derived in a similar way.

2.7.1 Haplotype Structure

Consider three SNPs, S_1, S_2, and S_3, for a QTN. Two alleles 1 and 0 at each SNP are symbolized by r_1, r_2, and r_3, respectively. The 1-allele frequencies at the three SNPs are denoted as p_1, p_2, and p_3, respectively, thus with the frequencies of the 0-alleles being $1 - p_1$, $1 - p_2$, and $1 - p_3$. Eight haplotypes, [111], [110], [101], [100], [011], [010], [001], and [000], formed by these three SNPs, have the frequencies arrayed in $\Omega_p = (p_{111}, p_{110}, p_{101}, p_{100}, p_{011}, p_{010}, p_{001}, p_{000})$. Some genotypes are consistent with diplotypes, whereas the others that are heterozygous at two or more SNPs are not. Each double heterozygote contains two different diplotypes. One triple heterozygote, i.e., 10/10/10, contains four different diplotypes, [111][000] (in a probability of $2p_{111}p_{000}$), [110][001] (in a probability of $2p_{110}p_{001}$), [101][010] (in a probability of $2p_{101}p_{010}$), and [100][011] (in a probability of $2p_{100}p_{011}$). The relative frequencies of different diplotypes for this double or triple heterozygote are a function of haplotype frequencies (Table 2.2).

In a natural population, there are 27 genotypes for the three SNPs. Let $n_{r_1 r_1'/r_2 r_2'/r_3 r_3'}$ $(r_1 \geq r_1', r_2 \geq r_2', r_3 \geq r_3' = 1, 0)$ be the number of offspring for a genotype. As seen

TABLE 2.2
Possible diplotypes and their frequencies for each of 27 genotypes at three SNPs within a QTN, and genotypic value vectors of composite diplotypes (assuming that [111] (A) is the risk haplotype and the others ($\bar{A}$) are the non-risk haplotype).

Genotype	Diplotype Configuration	Frequency	Composite Diplotype Symbol	Mean
11/11/11	[111][111]	p_{111}^2	AA	μ_2
11/11/10	[111][110]	$2p_{111}p_{110}$	$A\bar{A}$	μ_1
11/11/00	[110][110]	p_{110}^2	$\bar{A}\bar{A}$	μ_0
11/10/11	[111][101]	$2p_{111}p_{101}$	$A\bar{A}$	μ_1
11/10/10	$\begin{cases}[111][100]\\ [110][101]\end{cases}$	$\begin{cases}2p_{111}p_{100}\\ 2p_{110}p_{101}\end{cases}$	$\begin{cases}A\bar{A}\\ \bar{A}\bar{A}\end{cases}$	$\begin{cases}\mu_1\\ \mu_0\end{cases}$
11/10/00	[110][100]	$2p_{110}p_{100}$	$\bar{A}\bar{A}$	μ_0
11/00/11	[101][101]	p_{101}^2	$\bar{A}\bar{A}$	μ_0
11/00/10	[101][100]	$2p_{101}p_{100}$	$\bar{A}\bar{A}$	μ_0
11/00/00	[100][100]	p_{100}^2	$\bar{A}\bar{A}$	μ_0
10/11/11	[111][011]	$2p_{111}p_{011}$	$A\bar{A}$	μ_1
10/11/10	$\begin{cases}[111][010]\\ [110][011]\end{cases}$	$\begin{cases}2p_{111}p_{010}\\ 2p_{110}p_{011}\end{cases}$	$\begin{cases}A\bar{A}\\ \bar{A}\bar{A}\end{cases}$	$\begin{cases}\mu_1\\ \mu_0\end{cases}$
10/11/00	[110][010]	$2p_{110}p_{010}$	$\bar{A}\bar{A}$	μ_0
10/10/11	$\begin{cases}[111][001]\\ [101][011]\end{cases}$	$\begin{cases}2p_{111}p_{001}\\ 2p_{101}p_{011}\end{cases}$	$\begin{cases}A\bar{A}\\ \bar{A}\bar{A}\end{cases}$	$\begin{cases}\mu_1\\ \mu_0\end{cases}$
10/10/10	$\begin{cases}[111][000]\\ [110][001]\\ [100][011]\\ [101][010]\end{cases}$	$\begin{cases}2p_{111}p_{000}\\ 2p_{110}p_{001}\\ 2p_{100}p_{011}\\ 2p_{101}p_{010}\end{cases}$	$\begin{cases}A\bar{A}\\ \bar{A}\bar{A}\\ \bar{A}\bar{A}\\ \bar{A}\bar{A}\end{cases}$	$\begin{cases}\mu_1\\ \mu_0\\ \mu_0\\ \mu_0\end{cases}$
10/10/00	$\begin{cases}[110][000]\\ [100][010]\end{cases}$	$\begin{cases}2p_{110}p_{000}\\ 2p_{100}p_{010}\end{cases}$	$\begin{cases}\bar{A}\bar{A}\\ \bar{A}\bar{A}\end{cases}$	$\begin{cases}\mu_0\\ \mu_0\end{cases}$
10/00/11	[101][001]	$2p_{101}p_{001}$	$\bar{A}\bar{A}$	μ_0
10/00/10	$\begin{cases}[101][000]\\ [100][001]\end{cases}$	$\begin{cases}2p_{101}p_{000}\\ 2p_{100}p_{001}\end{cases}$	$\begin{cases}\bar{A}\bar{A}\\ \bar{A}\bar{A}\end{cases}$	$\begin{cases}\mu_0\\ \mu_0\end{cases}$
10/00/00	[100][000]	$2p_{100}p_{000}$	$\bar{A}\bar{A}$	μ_0
00/11/11	[011][011]	p_{011}^2	$\bar{A}\bar{A}$	μ_0
00/11/10	[011][010]	$2p_{011}p_{010}$	$\bar{A}\bar{A}$	μ_0
00/11/00	[010][010]	p_{010}^2	$\bar{A}\bar{A}$	μ_0
00/10/11	[011][001]	$2p_{011}p_{001}$	$\bar{A}\bar{A}$	μ_0
00/10/10	$\begin{cases}[011][000]\\ [010][001]\end{cases}$	$\begin{cases}2p_{011}p_{000}\\ 2p_{010}p_{001}\end{cases}$	$\begin{cases}\bar{A}\bar{A}\\ \bar{A}\bar{A}\end{cases}$	$\begin{cases}\mu_0\\ \mu_0\end{cases}$
00/10/00	[010][000]	$2p_{010}p_{000}$	$\bar{A}\bar{A}$	μ_0
00/00/11	[001][001]	p_{001}^2	$\bar{A}\bar{A}$	μ_0
00/00/10	[001][000]	$2p_{001}p_{000}$	$\bar{A}\bar{A}$	μ_0
00/00/00	[000][000]	p_{000}^2	$\bar{A}\bar{A}$	μ_0

in Table 2.2, the frequency of each genotype is expressed in terms of haplotype frequencies. By assuming [111] as a risk haplotype (labeled by A) and all the others as non-risk haplotypes (labelled by $\bar{A}$), Table 2.2 provides the formulation of genotypic values for three composite diplotypes, μ_2 for AA, μ_1 for $A\bar{A}$, and μ_0 for $\bar{A}\bar{A}$. The haplotype effect parameters and residual covariance matrix are arrayed by a quantitative genetic parameter vector $\Omega_q = (\mu_2, \mu_1, \mu_0, \sigma^2)$.

2.7.2 Likelihoods and Algorithms

With the above notation, we construct two likelihoods, one for haplotype frequencies (Ω_p) based on SNP data (**S**) and the other for quantitative genetic parameters (Ω_q) based on haplotype frequencies (Ω_p), phenotypic (y), and SNP data (**S**). They are, respectively, expressed as

$$\log L(\Omega_p|\mathbf{S}) = \text{constant}$$
$$+2n_{11/11/11}\log p_{111}$$
$$+n_{11/11/10}\log(2p_{111}p_{110})$$
$$+2n_{11/11/00}\log p_{110}$$
$$+n_{11/10/11}\log(2p_{111}p_{101})$$
$$+n_{11/10/10}\log(2p_{111}p_{100}+2p_{110}p_{101})$$
$$+n_{11/10/00}\log(2p_{110}p_{100})$$
$$+2n_{11/00/11}\log p_{101}$$
$$+n_{11/00/10}\log(2p_{101}p_{100})$$
$$+2n_{11/00/00}\log p_{100}$$
$$+n_{10/11/11}\log(2p_{111}p_{011})$$
$$+n_{10/11/10}\log(2p_{111}p_{010}+2p_{110}p_{011})$$
$$+n_{10/11/00}\log(2p_{110}p_{010})$$
$$+n_{10/10/11}\log(2p_{111}p_{001}+2p_{101}p_{011})$$
$$+n_{10/10/10}\log(2p_{111}p_{000}+2p_{101}p_{010}$$
$$\qquad\qquad +2p_{110}p_{001}+2p_{100}p_{011})$$
$$+n_{10/10/00}\log(2p_{110}p_{000}+2p_{100}p_{010})$$
$$+n_{10/00/11}\log(2p_{101}p_{001})$$
$$+n_{10/00/10}\log(2p_{101}p_{000}+2p_{100}p_{001})$$
$$+n_{10/00/00}\log(2p_{100}p_{000})$$
$$+2n_{00/11/11}\log p_{011}$$
$$+n_{00/11/10}\log(2p_{011}p_{010})$$
$$+2n_{00/11/00}\log p_{010}$$

$$\log L(\Omega_q|\Omega_p, y, \mathbf{S}) =$$
$$\sum_{i=1}^{n_{11/11/11}} \log f_2(y_i)$$
$$+\sum_{i=1}^{n_{11/11/10}} \log f_1(y_i)$$
$$+\sum_{i=1}^{n_{11/11/00}} \log f_0(y_i)$$
$$+\sum_{i=1}^{n_{11/10/11}} \log f_1(y_i)$$
$$+\sum_{i=1}^{n_{11/10/10}} \log[\phi_1 f_1(y_i) + \bar{\phi}_1 f_0(y_i)]$$
$$+\sum_{i=1}^{n_{11/10/00}} \log f_0(y_i)$$
$$+\sum_{i=1}^{n_{11/00/11}} \log f_0(y_i)$$
$$+\sum_{i=1}^{n_{11/00/10}} \log f_0(y_i)$$
$$+\sum_{i=1}^{n_{11/00/00}} \log f_0(y_i)$$
$$+\sum_{i=1}^{n_{10/11/11}} \log f_1(y_i)$$
$$+\sum_{i=1}^{n_{10/11/10}} \log[\phi_2 f_1(y_i) + \bar{\phi}_2 f_0(y_i)]$$
$$+\sum_{i=1}^{n_{10/11/00}} \log f_0(y_i)$$
$$+\sum_{i=1}^{n_{10/10/11}} \log[\phi_3 f_1(y_i) + \bar{\phi}_3 f_0(y_i)]$$
$$+\sum_{i=1}^{n_{10/10/10}} \log[\phi_4 f_1(y_i) + \bar{\phi}_4 f_0(y_i)]$$
$$+\sum_{i=1}^{n_{10/10/00}} \log[\phi_5 f_0(y_i) + \bar{\phi}_5 f_0(y_i)]$$
$$+\sum_{i=1}^{n_{10/00/11}} \log f_0(y_i)$$
$$+\sum_{i=1}^{n_{10/00/10}} \log[\phi_6 f_0(y_i) + \bar{\phi}_6 f_0(y_i)]$$
$$+\sum_{i=1}^{n_{10/00/00}} \log f_0(y_i)$$
$$+\sum_{i=1}^{n_{00/11/11}} \log f_0(y_i)$$
$$+\sum_{i=1}^{n_{00/11/10}} \log f_0(y_i)$$
$$+\sum_{i=1}^{n_{00/11/00}} \log f_0(y_i)$$

$$+n_{00/10/11}\log(2p_{011}p_{001}) \qquad\qquad +\sum_{i=1}^{n_{00/10/11}}\log f_{\mathbf{0}}(y_i)$$
$$+n_{00/10/10}\log(2p_{011}p_{000}+2p_{010}p_{001}) \quad +\sum_{i=1}^{n_{00/10/10}}\log[\phi_7 f_{\mathbf{0}}(y_i)+\bar{\phi}_7 f_{\mathbf{0}}(y_i)]$$
$$+n_{00/10/00}\log(2p_{010}p_{000}) \qquad\qquad +\sum_{i=1}^{n_{00/10/00}}\log f_{\mathbf{0}}(y_i)$$
$$+2n_{00/00/11}\log p_{001} \qquad\qquad\qquad +\sum_{i=1}^{n_{00/00/11}}\log f_{\mathbf{0}}(y_i)$$
$$+n_{00/00/10}\log(2p_{001}p_{000}) \qquad\qquad +\sum_{i=1}^{n_{00/00/10}}\log f_{\mathbf{0}}(y_i)$$
$$+2n_{00/00/00}\log p_{000} \qquad\qquad\qquad +\sum_{i=1}^{n_{00/00/00}}\log f_{\mathbf{0}}(y_i)$$

$$(2.18)$$

where ϕ.'s ($\bar{\phi}. = 1 - \phi.$) are defined below, and $f_j(y_i)$ ($j = \mathbf{2,1,0}$) is a normal distribution density function of composite diplotype j.

A two-stage hierarchical EM algorithm is derived to estimate haplotype frequencies and quantitative genetic parameters. At the higher hierarchy of the EM framework, we calculate the proportions of a particular diplotype within double or triple heterozygous genotypes (E step) by

$$\phi_1 = \frac{p_{111}p_{100}}{p_{111}p_{100}+p_{101}p_{110}}, \qquad\qquad \text{for genotype } 11/10/10$$

$$\phi_2 = \frac{p_{111}p_{010}}{p_{111}p_{010}+p_{011}p_{110}}, \qquad\qquad \text{for genotype } 10/11/10$$

$$\phi_3 = \frac{p_{111}p_{001}}{p_{111}p_{001}+p_{101}p_{011}}, \qquad\qquad \text{for genotype } 10/10/11$$

$$\phi_4 = \frac{p_{111}p_{000}}{p_{111}p_{000}+p_{101}p_{010}+p_{110}p_{001}+p_{100}p_{011}}, \text{ for genotype } 10/10/10$$

$$\phi_4' = \frac{p_{101}p_{010}}{p_{111}p_{000}+p_{101}p_{010}+p_{110}p_{001}+p_{100}p_{011}}, \text{ for genotype } 10/10/10$$

$$\phi_4'' = \frac{p_{110}p_{001}}{p_{111}p_{000}+p_{101}p_{010}+p_{110}p_{001}+p_{100}p_{011}}, \text{ for genotype } 10/10/10$$

$$\phi_4''' = \frac{p_{100}p_{011}}{p_{111}p_{000}+p_{101}p_{010}+p_{110}p_{001}+p_{100}p_{011}}, \text{ for genotype } 10/10/10$$

$$\phi_5 = \frac{p_{110}p_{000}}{p_{110}p_{000}+p_{100}p_{010}}, \qquad\qquad \text{for genotype } 10/10/00$$

$$\phi_6 = \frac{p_{101}p_{000}}{p_{101}p_{000}+p_{001}p_{100}}, \qquad\qquad \text{for genotype } 10/00/10$$

$$\phi_7 = \frac{p_{011}p_{000}}{p_{011}p_{000}+p_{001}p_{010}}, \qquad\qquad \text{for genotype } 00/10/10$$

$$(2.19)$$

The calculated relative proportions by equation (2.19) were used to estimate the haplotype frequencies with

$$\hat{p}_{111} = \frac{1}{2n}(2n_{11/11/11} + n_{11/11/10} + n_{11/10/11} + n_{10/11/11}$$
$$+ \phi_1 n_{11/10/10} + \phi_2 n_{10/11/10} + \phi_3 n_{10/10/11} + \phi_4 n_{10/10/10})$$

$$\hat{p}_{110} = \frac{1}{2n}(2n_{11/11/00} + n_{11/11/10} + n_{11/10/00} + n_{10/11/00}$$
$$+ \bar{\phi}_1 n_{11/10/10} + \bar{\phi}_2 n_{10/11/10} + \phi_4'' n_{10/10/10} + \phi_5 n_{10/10/00})$$

$$\hat{p}_{101} = \frac{1}{2n}(2n_{11/00/11} + n_{11/10/11} + n_{11/00/11} + n_{10/00/11}$$
$$+ \bar{\phi}_1 n_{11/10/10} + \bar{\phi}_3 n_{10/10/11} + \phi_4' n_{10/10/10} + \phi_6 n_{10/00/10})$$

$$\hat{p}_{100} = \frac{1}{2n}(2n_{11/00/00} + n_{11/10/00} + n_{11/00/10} + n_{10/00/00}$$
$$+ \phi_1 n_{11/10/10} + \phi_4''' n_{10/10/10} + \bar{\phi}_5 n_{10/10/00} + \bar{\phi}_6 n_{10/00/10})$$

$$\hat{p}_{011} = \frac{1}{2n}(2n_{00/11/11} + n_{10/11/11} + n_{00/10/11} + n_{00/11/10}$$
$$+ \bar{\phi}_2 n_{10/11/10} + \bar{\phi}_3 n_{10/10/11} + \phi_4''' n_{10/10/10} + \phi_7 n_{00/10/10})$$

$$\hat{p}_{010} = \frac{1}{2n}(2n_{00/11/00} + n_{10/11/00} + n_{00/11/10} + n_{00/10/00}$$
$$+ \phi_2 n_{10/11/10} + \phi_4' n_{10/10/10} + \bar{\phi}_5 n_{10/10/00} + \bar{\phi}_7 n_{00/10/10})$$

$$\hat{p}_{001} = \frac{1}{2n}(2n_{00/00/11} + n_{10/00/11} + n_{00/10/11} + n_{00/00/10}$$
$$+ \phi_3 n_{10/10/11} + \phi_4'' n_{10/10/10} + \bar{\phi}_6 n_{10/00/10} + \bar{\phi}_7 n_{00/10/10})$$

$$\hat{p}_{000} = \frac{1}{2n}(2n_{00/00/00} + n_{00/00/10} + n_{00/10/00} + n_{10/00/00}$$
$$+ \phi_5 n_{00/10/10} + \phi_6 n_{10/00/10} + \phi_7 n_{10/10/00} + \phi_4 n_{10/10/10}).$$

$$(2.20)$$

At the lower hierarchy of the EM framework, we calculate the posterior probabilities of a double or triple heterozygous individual i to be a particular diplotype ($A\bar{A}$) (E step), for which where [111] is assumed as the risk haplotype, expressed as

$$\Phi_{11|i} = \frac{\phi_1 f_1(y_i)}{\phi_1 f_1(y_i) + \bar{\phi}_1 f_0(y_i)}, \quad \overline{\Phi}_{11|i} = 1 - \Phi_{11|i} \quad \text{for genotype } 11/10/10$$

$$\Phi_{21|i} = \frac{\phi_2 f_1(y_i)}{\phi_2 f_1(y_i) + \bar{\phi}_2 f_0(y_i)}, \quad \overline{\Phi}_{12|i} = 1 - \Phi_{12|i} \quad \text{for genotype } 10/11/10$$

$$\Phi_{31|i} = \frac{\phi_3 f_1(y_i)}{\phi_3 f_1(y_i) + \bar{\phi}_3 f_0(y_i)}, \quad \overline{\Phi}_{31|i} = 1 - \Phi_{13|i} \quad \text{for genotype } 10/10/11$$

$$\Phi_{41|i} = \frac{\phi_4 f_1(y_i)}{\phi_4 f_1(y_i) + \bar{\phi}_4 f_0(y_i)}, \quad \overline{\Phi}_{41|i} = 1 - \Phi_{14|i} \quad \text{for genotype } 10/10/10$$

$$(2.21)$$

With the calculated posterior probabilities by the above equation (2.21), we then estimate the quantitative genetic parameters, Ω_q, based on the log-likelihood equations. These equations have similar, but more complicated, forms like (2.7). The sampling errors of the MLEs of the parameters can be estimated by calculating Louis' (1982) observed information matrix. For four different linkage disequilibria, D_{12} between SNPs $\mathbf{S}_1$ and $\mathbf{S}_2$, D_{13} between SNPs $\mathbf{S}_1$ and $\mathbf{S}_3$, D_{23} between SNPs $\mathbf{S}_2$ and $\mathbf{S}_3$, and D_{123} among SNPs $\mathbf{S}_1$, $\mathbf{S}_2$, and $\mathbf{S}_3$, their sampling variances can be estimated by

$$
\mathrm{Var}\begin{pmatrix}\hat{D}_{13}\\ \hat{D}_{12}\\ \hat{D}_{23}\\ \hat{D}_{123}\end{pmatrix} \hat{=}
$$

$$
\begin{pmatrix}
1-\hat{p}_1-\hat{p}_3 & 1-\hat{p}_1-\hat{p}_2 & 1-\hat{p}_3-\hat{p}_2 \\
-\hat{p}_3 & 1-\hat{p}_1-\hat{p}_2 & -\hat{p}_3 \\
1-\hat{p}_1-\hat{p}_3 & -\hat{p}_2 & -\hat{p}_2 \\
-\hat{p}_3 & -\hat{p}_2 & 0 \\
-\hat{p}_1 & -\hat{p}_1 & 1-\hat{p}_3-\hat{p}_2 \\
0 & -\hat{p}_1 & -\hat{p}_3 \\
-\hat{p}_1 & 0 & -\hat{p}_2 \\
\end{pmatrix}
$$

$$
\begin{pmatrix}
1-\hat{D}_{13}-\hat{D}_{12}-\hat{D}_{23}-\hat{p}_1-\hat{p}_3-\hat{p}_2+2\hat{p}_1\hat{p}_3+2\hat{p}_1\hat{p}_2+2\hat{p}_3\hat{p}_2\\
2\hat{p}_1\hat{p}_3+2\hat{p}_3\hat{p}_2-\hat{D}_{13}-\hat{D}_{23}-\hat{p}_3\\
2\hat{p}_1\hat{p}_2+2\hat{p}_3\hat{p}_2-\hat{D}_{12}-\hat{D}_{23}-\hat{p}_2\\
2\hat{p}_3\hat{p}_2-\hat{D}_{23}\\
2\hat{p}_1\hat{p}_3+2\hat{p}_1\hat{p}_2-\hat{D}_{13}-\hat{D}_{12}-\hat{p}_1\\
2\hat{p}_1\hat{p}_3-\hat{D}_{13}\\
2\hat{p}_1\hat{p}_2-\hat{D}_{12}\\
\end{pmatrix}'
$$

$$
\times\widehat{\mathrm{Var}}\begin{pmatrix}\hat{p}_{111}\\ \hat{p}_{110}\\ \hat{p}_{101}\\ \hat{p}_{100}\\ \hat{p}_{011}\\ \hat{p}_{010}\\ \hat{p}_{001}\end{pmatrix}\times
\begin{pmatrix}
1-\hat{p}_1-\hat{p}_3 & 1-\hat{p}_1-\hat{p}_2 & 1-\hat{p}_3-\hat{p}_2 \\
-\hat{p}_3 & 1-\hat{p}_1-\hat{p}_2 & -\hat{p}_3 \\
1-\hat{p}_1-\hat{p}_3 & -\hat{p}_2 & -\hat{p}_2 \\
-\hat{p}_3 & -\hat{p}_2 & 0 \\
-\hat{p}_1 & -\hat{p}_1 & 1-\hat{p}_3-\hat{p}_2 \\
0 & -\hat{p}_1 & -\hat{p}_3 \\
-\hat{p}_1 & 0 & -\hat{p}_2 \\
\end{pmatrix}
$$

$$
\begin{pmatrix}
1-\hat{D}_{13}-\hat{D}_{12}-\hat{D}_{23}-\hat{p}_1-\hat{p}_3-\hat{p}_2+2\hat{p}_1\hat{p}_3+2\hat{p}_1\hat{p}_2+2\hat{p}_3\hat{p}_2\\
2\hat{p}_1\hat{p}_3+2\hat{p}_3\hat{p}_2-\hat{D}_{13}-\hat{D}_{23}-\hat{p}_3\\
2\hat{p}_1\hat{p}_2+2\hat{p}_3\hat{p}_2-\hat{D}_{12}-\hat{D}_{23}-\hat{p}_2\\
2\hat{p}_3\hat{p}_2-\hat{D}_{23}\\
2\hat{p}_1\hat{p}_3+2\hat{p}_1\hat{p}_2-\hat{D}_{AB}-\hat{D}_{12}-\hat{p}_1\\
2\hat{p}_1\hat{p}_3-\hat{D}_{13}\\
2\hat{p}_1\hat{p}_2-\hat{D}_{12}\\
\end{pmatrix}.
$$

2.7.2.1   Selection of a Risk Haplotype and Hypothesis Testing

As in the 2-SNP case, an optimal risk haplotype should correspond to the largest likelihood when different risk haplotypes are assumed. When some haplotypes are missing whose possibility increases with the number of SNPs, the AIC or BIC criterion should be used for the determination of the optimal risk haplotype that best explains the given data set.

Hypothesis tests can be made for linkage disequilibria among three SNPs and haplotype effects. There are four different linkage disequilibria, D_{12}, D_{13}, D_{23}, and D_{123}, that describe non-random associations among three SNPs. These disequilibria can be estimated by the following equations:

$$
\begin{aligned}
D_{12} &= \tfrac{1}{4}[(p_{111} + p_{110} + p_{001} + p_{000}) - (p_{101} + p_{100} + p_{011} + p_{010})] \\
D_{23} &= \tfrac{1}{4}[(p_{111} + p_{011} + p_{100} + p_{000}) - (p_{110} + p_{010} + p_{101} + p_{001})] \\
D_{13} &= \tfrac{1}{4}[(p_{111} + p_{101} + p_{010} + p_{000}) - (p_{110} + p_{100} + p_{011} + p_{001})] \\
D_{123} &= \tfrac{1}{8}[(p_{111} + p_{100} + p_{010} + p_{001}) - (p_{110} + p_{101} + p_{011} + p_{000})].
\end{aligned}
\tag{2.22}
$$

Non-random associations of these SNPs can be detected by testing these disequilibria jointly or separately. Joint tests are based on the formulation of null hypotheses like H_0: $D_{12} = D_{23} = D_{13} = D_{123} = 0$, H_0: $D_{12} = D_{23} = D_{13} = 0$, H_0: $D_{12} = D_{13} = 0$, H_0: $D_{12} = D_{23} = 0$, or H_0: $D_{23} = D_{13} = 0$. The estimation of population genetic parameters under each of these null hypotheses needs the EM algorithm, as described by equations (2.19) and (2.20), implemented with respective constraints constructed by equation (3.6). The log-likelihood ratios for each hypothesis are thought to follow a χ^2 distribution with a specific degree of freedom.

After an optimal risk haplotype is determined, the significance of a QTN composed of three SNPs and its additive and dominance effects can be tested in a similar way. The critical values for these tests can be obtained from a χ^2 table.

2.8   R-SNP Model

The idea for haplotyping a quantitative trait is described for two- and three-SNP models. It is possible that these models are too simple to characterize genetic variants for quantitative variation. With the analytical line for the two- and three-SNP sequencing model, a model can be developed to include an arbitrary number of SNPs whose sequences are associated with the phenotypic variation. A key issue for the multi-SNP sequencing model is how to distinguish among 2^{r-1} different diplotypes for the same genotype heterozygous at r loci. The relative frequencies of these diplotypes can be expressed in terms of haplotype frequencies. The integrative EM algorithm can be employed to estimate the MLEs of haplotype frequencies. A general formula for estimating haplotype frequencies can be derived.

Example 2.1
The Pharmacogenomic Project: A Case Study: The haplotyping model was used to characterize the QTN variants that encode an obesity trait with the data from a pharmacogenomic project. Numerous genes have been investigated as potential

TABLE 2.3
The maximum likelihood estimates (MLE) and their standard errors (SE) of population and quantitative genetic parameters for DNA sequences associated with phenotypic variation in BMI for 155 patients.

Parameters		MLE	SE
Population genetic parameters			
Haplotype frequencies	p_{AC}	0.0250	0.0264
	p_{AG}	0.3686	0.0632
	p_{GC}	0.3589	0.0300
	p_{GG}	0.2476	0.0361
Allele frequencies and linkage disequilibrium	$p_G^{(16)}$	0.6065	0.0346
	$p_C^{(27)}$	0.3839	0.0346
	D	0.1261	0.0200
Quantitative genetic parameters			
Overall mean	μ	30.8269	1.0442
Additive effect	a	-1.7732	1.0442
Dominant effect	d	-3.0548	1.4758
Residual variance	σ^2	71.9070	20.4749

obesity-susceptibility genes (Mason et al., 1999; Chagnon et al., 2003). The $\beta 2$ adrenoceptor ($\beta 2$AR) is a major lipolytic receptor in human fat cells, whose function was found to be determined by known polymorphisms in codons 16 and 27 (Green et al., 1995). In a genetic study of obesity involving 149 women by Large et al. (1997), the Arg16Gly polymorphism at codon 16 was found to be associated with altered $\beta 2$AR function with Gly16 carriers showing a fivefold increased agonist sensitivity. The Gln27Glu polymorphism at codon 27 is markedly associated with obesity traits. The homozygotes for Glu27 display an average fat mass excess of 20 kg and 50% larger fat cells than controls. Although the Arg16Gly polymorphism is not significantly associated with obesity and the Gln27Glu polymorphism is not significantly associated with the $\beta 2$AR function, genetic variability in these two sites of the human $\beta 2$AR gene could be of major importance for obesity, energy expenditure, and lipolytic $\beta 2$AR function in adipose tissue, at least in women (Large et al., 1997).

To determine whether sequence variants at these two polymorphisms of $\beta 2$AR are associated with obesity phenotypes, a group of 155 men and women were investigated with ages from 32 to 86 years old with a large variation in body fat mass. Each of these patients was determined for their genotypes at codon 16 with two alleles Arg16 (A) and Gly16 (G) and codon 27 with two alleles Gln27 (C) and Glu27 (G), within the $\beta 2$AR gene and measured for body mass index (BMI). These two SNPs form four haplotypes designated as [AC], [AG], [GC], and [GG], which lead to nine genotypes, AA/CC, AA/CG, AA/GG, AG/CC, AG/CG, AG/GG, GG/CC, GG/CG, and GG/GG, and the 10 corresponding diplotypes, [AC][AC], [AC][AG], [AG][AG], [AC][GC], [AC][GG] or [AG][GC], [AG][GG], [GC][GC], [GC][GG], and [GG][GG]. Our model is used to associate diplotype differences with variation in BMI. The MLEs of the haplotype frequencies, allele frequencies, and linkage disequilibrium between the two SNPs were obtained (Table 2.3). These two SNPs are highly associated with each other based on hypothesis test (2.15), whose linkage disequilibrium is estimated as 0.1261. The allele frequencies are about 0.61 for allele G at codon 16 and 0.38 for allele G at codon 27, respectively, suggesting that both of them have fairly high heterozygosity.

The haplotyping model is based on the assumption that one haplotype is different from the rest of the haplotypes. In a practical situation, as in our example used here, the issue of how these haplotypes differ in their effects is not known. By assuming that any one haplotype is different from the rest of the haplotypes, we can find a best difference pattern based on the estimates of the log-likelohood ratio test statistics (LR_2) using equation (2.17). When haplotype [GG], [GC], [AG], or [AC] is assumed to differ in BMI from the other haplotypes, the LR_2 values were estimated as 10.35, 3.11, 1.52, and 2.32, respectively. Obviously, haplotype GG is considered as a reference haplotype that has an effect on BMI in a comparison with the other haplotypes. The difference between haplotype GG and the rest of the haplotypes can significantly explain some variation in BMI. The resultant LR_2 for testing the association between DNA sequence and BMI phenotype (10.35) is well beyond the critical threshold value at the significance level of 1% (6.07) estimated from 1000 simulation replicates. Haplotype GG displays large additive and dominant effects. Patients with homozygous genotype GG/GG have significantly lower BMI than those with homozygous genotype AA/CC, as indicated by a large additive effect ($a = -1.77$). For double heterozygote AG/GC, diplotype [GG][AC] reduces BMI by 3.05 (the dominant effect) compared to diplotype [AG][GC]. The difference between haplotype [GG] and the rest of the haplotypes can explain 6.3% of the observed variation in BMI. We estimate the standard errors of the MLEs of the population and quantitative genetic parameters based on Louis' (1982) observed information matrix, suggesting that all MLEs have reasonable estimation precision although the estimates of quantitative genetic parameters are not as precise as those of population genetic parameters due to a small sample size used Liu et al. (2004). ⊔

3

Genetic Haplotyping in Experimental Crosses

Drug metabolism, drug disposition, and drug-body interactions are complex traits determined by many genetic and environmental factors. The genetic factors are usually studied in an experimental cross (backcross, F_2, or advanced intercross lines) of animal models, such as the mouse, through quantitative trait locus (QTL) analysis. Although QTL mapping shows an advantage of detecting novel key genes involved in various metabolic pathways of drug interactions, this approach has proven to have low-resolution characterization of genes that underlie a complex trait. Recently, it is suggested that haplotype analysis with genome-wide typed single nucleotide polymorphisms (SNPs) in controlled crosses may be useful for fine mapping of complex traits including those related to drug response (Park et al., 2003; Wiltshire et al., 2003; Wang et al., 2004; Yalcin et al., 2004, 2005; Cuppen, 2005; Payseur et al., 2007).

High-density genotyping of SNPs in some particular genomic regions suggests that genetic variation in the mouse may follow a similar pattern detected in the human genome, that is, the distribution of SNP diversity is not random but appears to be organized in discrete haplotype blocks (Wiltshire et al., 2003; Guryev et al., 2006). But different from the variation between individuals in humans, relatively little variation exists among the commonly used inbred mice strains. Also, the variation in the mouse is found to be organized in a mosaic structure of regions with either extremely low or high levels of polymorphisms. Such regions are assumed to be the result of a recent genetic bottleneck (reviewed in Cuppen (2005)). These characteristics of genetic variation in the mouse genome and their orthologous relationship in humans suggest that results from genetic haplotyping of the mouse will have significant implications for understanding the genetic architecture of complex traits in humans.

Unlike a natural population in which gene co-segregation analysis is based on linkage disequilibria (LD) (Lou et al., 2003), an experimental cross is usually analyzed in terms of the linkage between different genes to locate the position of a functional QTL on a linkage map, which is different from the objective of genetic haplotyping, aimed to detect the association between haplotypes and complex phenotypes. Hou et al. (2007) developed a general statistical model for integrating the concepts of linkage and LD in a cross mapping population and provided the design and interpretation of genetic experiments for a joint linkage and LD analysis. By estimating the linkage between different SNPs, this model is able to test haplotype effects within the context of linkage disequilibrium analysis. A variety of genetic parameters that build the model are estimated through the derivations of closed forms for the EM algorithm.

In this chapter, we will describe population genetic theory that connects linkage and LD in an experimental cross, and then introduce the statistical model for haplotype discovery responsible for quantitative variation. The relationship between linkage analysis and linkage disequilibrium analysis will be explored in different types of crosses. An example from a mouse mapping project will show the utilization of the haplotyping model and result interpretation.

3.1 LD Analysis in the F_1's Gamete Population

3.1.1 A General Model

Consider two genetically different inbred lines that are crossed to generate a heterozygous F_1 population. During meioses, the F_1 produces segregating gametes for its next generation, such as the backcross or F_2. Suppose there is a QTN (composed of L SNPs) responsible for a quantitative trait. The two alleles, 1 and 0, at each of these SNPs are symbolized by $r_1, \ldots, r_L$, respectively. According to Mendel's first law, allele frequencies for each SNP should be 1/2 in the F_1's gamete population. The frequency of a haplotype constituted by the L SNPs, denoted as $p_{r_1 r_2 \cdots r_L}$, will depend on the linkage among these SNPs, which can be decomposed into the following components:

$$
\begin{aligned}
p_{r_1 r_2 \ldots r_L} = & \\
& \left(\tfrac{1}{2}\right)^L & \text{No LD} \\
& + (-1)^{r_1+r_2}\left(\tfrac{1}{2}\right)^{L-2} D_{12} + \ldots + (-1)^{r_{L-1}+r_L}\left(\tfrac{1}{2}\right)^{L-2} D_{(L-1)L} & \text{Digenic LD} \\
& + (-1)^{r_1+r_2+r_3}\left(\tfrac{1}{2}\right)^{L-3} D_{123} + \ldots & \\
& \qquad + (-1)^{r_{L-2}+r_{L-1}+r_L}\left(\tfrac{1}{2}\right)^{L-3} D_{(L-2)(L-1)L} & \text{Trigenic LD} \\
& + \ldots & \\
& + (-1)^L (-1)^{r_1+\ldots+r_L} D_{1\ldots L} & \text{L-genic LD}
\end{aligned}
$$

$$(3.1)$$

where D's are the linkage disequilibria of different orders among particular SNPs.

Totally, L SNPs form 2^L haplotypes expressed as

$$[r_1 \ldots r_L],$$

$2^{L-1}(2^L + 1)$ diplotypes, i.e., a pair of maternally- (m) and paternally-derived haplotypes (p), expressed as

$$[r_1^m \ldots r_L^m][r_1^p \ldots r_L^p] \quad (r_1^m, r_1^p; \ldots; r_L^m, r_L^p = 1, 0),$$

and 3^L genotypes expressed as

$$r_1 r_1' / \ldots / r_L r_L' \quad (r_1 \geq r_1', \ldots, r_L \geq r_L' = 1, 0).$$

Only genotypes can be observed. The number of diplotypes is smaller than the number of genotypes because the genotypes that are heterozygous at two or more SNPs contain multiple different diplotypes. Diplotype (and therefore genotype) frequencies can be expressed in terms of haplotype frequencies. Let $P_{[r_1^m \ldots r_L^m][r_1^p \ldots r_L^p]}$ and $P_{r_1 r_1' / \ldots / r_L r_L'}$ denote the diplotype and genotype frequencies, respectively, and $n_{r_1 r_1' / \ldots / r_L r_L'}$ denote genotype observation.

3.1.2 A Special Case: Two-Point LD

For two given SNPs (S_1 and S_2), there are four different haplotypes in a cross population. According to the definition given above, these four haplotypes are denoted as [11], [10], [01], and [00], whose frequencies in a cross population are, respectively, expressed as

$$
\begin{aligned}
p_{11} &= \tfrac{1}{4} + D \\
p_{10} &= \tfrac{1}{4} - D \\
p_{01} &= \tfrac{1}{4} - D \\
p_{00} &= \tfrac{1}{4} + D.
\end{aligned}
\tag{3.2}
$$

Assume that the two SNPs are linked with the recombination fraction of r. The haplotype frequencies can be expressed in terms of r, i.e., $p_{11} = \tfrac{1}{2}(1-r)$, $p_{10} = \tfrac{1}{2}r$, $p_{01} = \tfrac{1}{2}r$, and $p_{00} = \tfrac{1}{2}(1-r)$. Combining equation (3.2), this establishes the relation between the linkage disequilibrium and recombination fraction as

$$D = \tfrac{1}{4}(1 - 2r), \tag{3.3}$$

or

$$r = \tfrac{1}{2}(1 - 4D). \tag{3.4}$$

3.1.3 A Special Case: Three-Point LD

For three linked SNPs (S_1, S_2, and S_3), there are eight different haplotypes, i.e., [111], [110], [101], [100], [011], [010], [001], and [000]. Let D_{12}, D_{23}, and D_{13} be the LD between SNP S_1 and S_2, between S_2 and S_3, and between S_1 and S_2, respectively, and D_{123} be the LD among the three SNPs. Since allele frequencies at each SNP are 1/2, assuming no segregation distortion, the haplotype frequencies in a gametic population are expressed as

$$p_{111} = \tfrac{1}{8} + \tfrac{1}{2}D_{12} + \tfrac{1}{2}D_{13} + \tfrac{1}{2}D_{23} + D_{123}$$
$$p_{110} = \tfrac{1}{8} + \tfrac{1}{2}D_{12} - \tfrac{1}{2}D_{13} - \tfrac{1}{2}D_{23} - D_{123}$$
$$p_{101} = \tfrac{1}{8} - \tfrac{1}{2}D_{12} + \tfrac{1}{2}D_{13} - \tfrac{1}{2}D_{23} - D_{123}$$
$$p_{100} = \tfrac{1}{8} - \tfrac{1}{2}D_{12} - \tfrac{1}{2}D_{13} + \tfrac{1}{2}D_{23} + D_{123}$$
$$p_{011} = \tfrac{1}{8} - \tfrac{1}{2}D_{12} - \tfrac{1}{2}D_{13} + \tfrac{1}{2}D_{23} - D_{123} \tag{3.5}$$
$$p_{010} = \tfrac{1}{8} - \tfrac{1}{2}D_{12} + \tfrac{1}{2}D_{13} - \tfrac{1}{2}D_{23} + D_{123}$$
$$p_{001} = \tfrac{1}{8} + \tfrac{1}{2}D_{12} - \tfrac{1}{2}D_{13} - \tfrac{1}{2}D_{23} + D_{123}$$
$$p_{000} = \tfrac{1}{8} + \tfrac{1}{2}D_{12} + \tfrac{1}{2}D_{13} + \tfrac{1}{2}D_{23} - D_{123}$$

By solving equations (3.5), the four disequilibrium coefficients can be estimated as

$$D_{12} = \tfrac{1}{4}[(p_{111} + p_{110} + p_{001} + p_{000}) - (p_{101} + p_{100} + p_{011} + p_{010})]$$
$$D_{23} = \tfrac{1}{4}[(p_{111} + p_{011} + p_{100} + p_{000}) - (p_{110} + p_{010} + p_{101} + p_{001})]$$
$$D_{13} = \tfrac{1}{4}[(p_{111} + p_{101} + p_{010} + p_{000}) - (p_{110} + p_{100} + p_{011} + p_{001})] \tag{3.6}$$
$$D_{123} = \tfrac{1}{8}[(p_{111} + p_{100} + p_{010} + p_{001}) - (p_{110} + p_{101} + p_{011} + p_{000})]$$

The first three first-order LD can be used to describe the linkage between different SNPs and crossover interference, whereas the last second-order LD is thought to be associated with chromatid interference.

Theorem: *The first-order linkage disequilibria, D_{12}, D_{23} and D_{13}, estimated from a three-point linkage disequilibrium analysis are only related to the degree of linkage between the corresponding pair of SNPs.*

Proof: Let r_{12}, r_{23}, and r_{13} be the recombination fractions between SNP S_1 and S_2, between S_2 and S_3 and between S_1 and S_2, respectively. According to the definition of the recombination fraction, r_{12} should include all possible gametes that have one crossover between SNP S_1 and S_2, irrespective of whether or not there is a crossover between the other SNP pairs. These gametes are [101], [100], [011], and [010], whose frequencies sum to r_{12}, i.e.,

$$r_{12} = p_{101} + p_{100} + p_{011} + p_{010}$$
$$= \tfrac{1}{8} \times 4 - \tfrac{1}{2}D_{12} \times 4$$
$$= \tfrac{1}{2}(1 - 4D_{12}).$$

The other two recombination fractions r_{23} and r_{13} can be proven similarly.

During crossover two types of interference may affect this process: crossover interference (I_r), the non-random placement of chiasma along a chromosome, and chromatid interference (I_h), the non-random choice of chromatids involved in adjacent chiasma. With the estimation of the three recombination fractions between three SNPs in an assumed order S_1-S_2-S_3, we can estimate and test the crossover interference expressed as

$$I_r = 1 - \frac{r_{12} + r_{23} - r_{13}}{2r_{12}r_{23}}$$

$$= 1 - \frac{1 - D_{12} - 4D_{23} + 4r_{13}}{(1 - 4D_{12})(1 - 4D_{23})}. \tag{3.7}$$

Rearrange the expression of D_{123} in equation (3.6) as

$$D_{123} = \tfrac{1}{8}[(p_{111} - p_{000}) + (p_{100} - p_{011}) + (p_{010} - p_{101}) + (p_{001} - p_{110}))]. \tag{3.8}$$

It can be seen that D_{123} reflects the accumulative difference between the frequencies of gametes that carry alternative alleles. Thus, D_{123} can be thought to be associated with the non-random choice of chromatids involved in adjacent chiasma (I_h).

3.2 LD Analysis in the Backcross

3.2.1 Design

Haplotype analysis in the backcross is straightforward because the diplotype can be determined uniquely for all the backcross genotype. Simple t-test or analysis of variance can be used to detect haplotype effects on a quantitative trait. Consider a pair of linked SNPs, each with two alleles symbolized as 1 and 0. A homozygous parent for the 1-allele (P_1) is crossed with a homozygous parent for the 0-allele (P_2) to generate a double heterozygote F_1. If the F_1 is backcrossed to parent P_2, then the four diplotypes (i.e., genotypes) generated in the backcross are [11][00], [10][00], [01][00], and [00][00]. Among these diplotypes, [11][00] and [00][00] are derived from non-recombinant types, whereas [10][00] and [01][00] derived from recombinant types. All the backcross progeny are measured for a quantitative trait. The sample sizes, trait means, and sampling variances of these diplotypes are tabulated in Table 3.1.

The maximum likelihood estimates (MLE) of the recombination fraction (r) and LD (D) between the two SNPs are given by

$$\hat{r} = \frac{n_{10} + n_{01}}{n_{11} + n_{10} + n_{01} + n_{00}},$$

and

$$\hat{D} = \frac{n_{11} + n_{00} - n_{10} - n_{01}}{4(n_{11} + n_{10} + n_{01} + n_{00})}.$$

The significance for the linkage or LD can be tested by calculating a test statistic,

$$\chi^2 = \frac{(n_{11} + n_{00} - n_{10} - n_{01})^2}{n_{11} + n_{10} + n_{01} + n_{00}},$$

which follows a χ^2-distribution with one degree of freedom.

TABLE 3.1
Basic statistics of two SNPs in a backcross population.

Diplotype	Sample Size	Mean	Sampling Variance	Composite Diplotype [11]	[10]	[01]	[00]
[11][00]	n_{11}	μ_{11}	s_{11}^2	$A\bar{A}$	$\bar{A}\bar{A}$	$\bar{A}\bar{A}$	$\bar{A}A$
[10][00]	n_{10}	μ_{10}	s_{10}^2	$\bar{A}\bar{A}$	$A\bar{A}$	$\bar{A}\bar{A}$	$\bar{A}A$
[01][00]	n_{01}	μ_{01}	s_{01}^2	$\bar{A}\bar{A}$	$\bar{A}\bar{A}$	$A\bar{A}$	$\bar{A}A$
[00][00]	n_{00}	μ_{00}	s_{00}^2	$\bar{A}\bar{A}$	$\bar{A}\bar{A}$	$\bar{A}\bar{A}$	AA
Total	n		t-value	t_{11}	t_{10}	t_{01}	t_{00}

TABLE 3.2
ANOVA for haplotype analysis in a backcross population.

Source	df	Mean Square	Expected Mean Square	F-value
Among Diplotypes	3	MS_1	$\sigma_e^2 + H\sigma_h^2$	MS_1/MS_2
Within Diplotypes	$n-4$	MS_2	σ_e^2	

H is the harmonic mean expressed as $\dfrac{n_{11}+n_{10}+n_{01}+n_{00}}{\frac{1}{n_{11}}+\frac{1}{n_{10}}+\frac{1}{n_{01}}+\frac{1}{n_{00}}}$.

3.2.2 Analysis of Variance

To investigate whether there is a haplotype effect for this pair of SNPs on a complex trait, we can perform an analysis of variance in which the sources of variance are due to the differences among and within diplotypes (Table 3.2). The F-value is calculated to test the significance of the difference among the four diplotypes by comparing this value with the critical $F_{(3,n-4;0.05)}$ value. If a significant difference is detected, this indicates that haplotypes trigger significant effects on phenotypic variation. The proportion of the total variance explained by diplotype difference is calculated as

$$R = \frac{\sigma_h^2}{\sigma_h^2 + \sigma_e^2} = \frac{MS_2 - MS_1}{MS_2 + (k-1)MS_1},$$

where σ_h^2 is the genetic variance due to the differences among diplotypes and σ_e^2 is the residual variance.

3.2.3 *t*-Test

In some cases, we are interested in finding the genetic effect of a specific haplotype on a trait. This can be done by assuming a risk haplotype among four haplotypes for the two SNPs studied, [11], [10], [01], and [00]. Let haplotype [11] be the risk haplotype (labeled as A) and the other three be the non-risk haplotype (labeled as $\bar{A}$). The assumed risk and non-risk haplotypes comprise composite diplotypes shown in display (3.1). Then, we use a t-test approach to detect whether haplotype [11] exerts a significant effect on the phenotypic value of a trait measured in the backcross. The t-test statistic is calculated as

$$t_{11} = \frac{\mu_{11} - \frac{1}{3}(\mu_{10} + \mu_{01} + \mu_{00})}{s\sqrt{\frac{1}{n_{11}} + \frac{1}{9n_{10}} + \frac{1}{9n_{01}} + \frac{1}{9n_{00}}}}, \tag{3.9}$$

where

$$s^2 = \frac{(n_{11} - 1)s_{11}^2 + (n_{10} - 1)s_{10}^2 + (n_{01} - 1)s_{01}^2 + (n_{00} - 1)s_{00}^2}{n_{11} + n_{10} + n_{01} + n_{00} - 4}.$$

If the calculated t value is greater than the critical t threshold with $n_{11} + n_{10} + n_{01} + n_{00} - 4$ degrees of freedom at the 5% significance level, this means that [11] is a significant risk haplotype. Otherwise, it is not a risk haplotype. An optimal risk haplotype can be selected on the basis of t values calculated when different risk haplotypes are assumed, i.e.,

$$t_{10} = \frac{\mu_{10} - \frac{1}{3}(\mu_{11} + \mu_{01} + \mu_{00})}{s\sqrt{\frac{1}{n_{10}} + \frac{1}{9n_{11}} + \frac{1}{9n_{01}} + \frac{1}{9n_{00}}}},$$

for risk haplotype [10],

$$t_{01} = \frac{\mu_{01} - \frac{1}{3}(\mu_{11} + \mu_{10} + \mu_{00})}{s\sqrt{\frac{1}{n_{01}} + \frac{1}{9n_{11}} + \frac{1}{9n_{10}} + \frac{1}{9n_{00}}}},$$

for risk haplotype [01], or

$$t_{00} = \frac{\mu_{00} - \frac{1}{3}(\mu_{11} + \mu_{10} + \mu_{01})}{s\sqrt{\frac{1}{n_{00}} + \frac{1}{9n_{11}} + \frac{1}{9n_{10}} + \frac{1}{9n_{01}}}},$$

for risk haplotype [10].

In practice, two-point analysis based on the t-test can be used to scan for every two adjacent SNPs over a linkage map, from which t values are then plotted against map position. Genomic regions that carry significant haplotypes can be detected from the t profile.

The procedure for analyzing two SNPs with the t test can be extended to consider a QTN composed of three or more SNPs. Multi-point analysis may not gain more information about the test of haplotype effects compared with two-point analysis, but the former helps to increase the linkage estimation through considering interference at adjacent intervals. As shown by equations (3.7) and (3.8), three-point analysis can test the significance of two types of interference during meioses.

TABLE 3.3
Observations, genotypic means, and sampling
variances for body mass in a mouse backcross.

Diplotype	Sample Size	Mean	Sampling Variance
[11][00]	49	30	4
[10][00]	3	28	6
[01][00]	2	43	5
[00][00]	51	29	5

Example 3.1

In a backcross experiment with the mouse, two SNPs, each with two alleles (labeled as 1, inherited from one parental inbred strain, and 0, inherited from the other parental inbred strain), were found to be linked on a mouse chromosome. Observations for four possible groups of genotypes (and therefore diplotypes) at the two SNPs in the backcross were tabulated in Table 3.3. Each backcross progeny was measured for body mass at age 10 weeks, with the means and sampling variances for each genotype group also given in Table 3.3. Note that the body mass data given have been adjusted for the size of litter, body mass at birth, birth time, and sex.

Based on the genotypes of two parental inbred strain, diplotypes [10][00] and [01][00] should be derived from the recombinant haplotype of the F_1, whereas diplotypes [11][00] and [00][00], derived from the non-recombinant haplotype. Thus, with the data given in Table 3.3, the recombination fraction between the two SNPs is estimated as

$$\hat{r} = \frac{3+2}{49+3+2+51} = 0.05.$$

The pooled sampling variance is estimated as

$$s^2 = \frac{(49-1) \times 4 + (3-1) \times 6 + (2-1) \times 5 + (51-1) \times 5}{49+3+2+51-4} = 4.54.$$

By assuming that [11] is a risk haplotype, the t-value is calculated as

$$t_{11} = \frac{30 - \frac{1}{3}(28+43+29)}{\sqrt{4.54} \times \sqrt{\frac{1}{49} + \frac{1}{9 \times 3} + \frac{1}{9 \times 2} + \frac{1}{9 \times 51}}} = -6.40.$$

Similarly, we calculate the t values as $t_{10} = -4.49$, $t_{01} = 8.93$, and $t_{00} = -6.47$, when the risk haplotype is assumed to be [10], [01], and [00], respectively. By comparison, we conclude that haplotype [01] is the most likely to be a risk haplotype. Haplotype [01] as a risk haplotype is statistically significant because its t value (8.93) is greater than the critical value $t_{\mathrm{df}=49+3+2+51-4, 0.05} = 1.66$. □

3.3   LD Analysis in the F_2

In the F_2 derived from the mating between two F_1 individuals, a simple t-test is not adequate for haplotyping because there exists a double heterozygote that is a mixture of two possible diplotypes. Suppose a panel of SNPs are typed, each of which is segregating in a 1:2:1 Mendelian ratio in the F_2 population. As assumed above, we use 1 and 0 to denote alternative alleles at each SNP. The EM algorithm will be needed to estimate genetic effects of haplotypes in the F_2.

3.3.1   Mixture Model

The statistical method for the genome-wide scan of functional haplotypes in the F_2 is formulated on the basis of a finite mixture model. The mixture model assumes that each observation comes from one of an assumed set of distributions. The mixture model derived to detect haplotype effects on a quantitative trait based on SNP genotype data contains three major parts:

(1) The mixture proportions of each distribution, denoted as the relative frequencies of different diplotypes for the same double-heterozygous genotype,

(2) The mean for each diplotype in the density function,

(3) The residual variance common to all diplotypes.

Let us first consider a pair of SNPs. For these two SNPs, there are four haplotypes [11], [10], [01], and [00], generated from the heterozygous F_1. The haplotype frequencies in segregating in the F_1 gametic population are arrayed in vector $\Omega_p = (p_{11}, p_{10}, p_{01}, p_{00})$. In the F_2, all the genotypes are consistent with diplotypes, except for the double heterozygote, 10/10, that contains two different diplotypes [11][00] (with a frequency of $2p_{11}p_{00}$) and [10][01] (with a frequency of $2p_{10}p_{01}$) (Table 2.1).

The F_2 population of size n contains nine genotypes for the two SNPs, whose genotypic observations are generally expressed as $n_{r_1 r'_1/r_2 r'_2}$ ($r_1 \geq r'_1, r_2 \geq r'_2 = 1, 0$). Considering a quantitative trait controlled by diplotype (rather than genotype) diversity, the phenotypic value of the trait (y_i) for individual i is expressed by a linear model, i.e.,

$$y_i = \sum_{r_1^m=0}^{1} \sum_{r_1^p=0}^{1} \sum_{r_2^m=0}^{1} \sum_{r_2^p=0}^{1} \xi_{j|i} u_{[r_1^m r_2^m][r_1^p r_2^p]} + e_i, \qquad (3.10)$$

where $\xi_{j|i}$ is the indicator variable defined as 1 if a diplotype considered is compatible with individual i and as 0 otherwise, $u_{[r_1^m r_2^m][r_1^p r_2^p]} = u_{[r_1^p r_2^p][r_1^m r_2^m]}$ is the genotypic value for diplotype $[r_1^m r_2^m][r_1^p r_2^p]$ derived from the combination between maternally

(M)- and paternally (P)-derived haplotypes, and e_i is the residual error distributed as $N(0, \sigma^2)$.

If haplotypes are composed of three SNPs, there are 27 genotypes for the three SNPs in the F_2 population. Let $n_{r_1r_1'/r_2r_2'/r_3r_3'}$ $(r_1 \geq r_1', r_2 \geq r_2', r_3 \geq r_3' = 1, 0)$ be the number of progeny for a genotype. As seen in Table 2.2, the frequency of each genotype is expressed in terms of haplotype frequencies. Similar to equation (8.16), the phenotypic value of the trait for individual i is expressed, at the diplotype level, as

$$y_i = \sum_{r_1^m=0}^{1} \sum_{r_1^P=0}^{1} \sum_{r_2^m=0}^{1} \sum_{r_2^P=0}^{1} \sum_{r_3^m=0}^{1} \sum_{r_3^P=0}^{1} \xi_{j|i} u_{[r_1^m r_2^m r_3^m][r_1^P r_2^P r_3^P]} + e_i, \qquad (3.11)$$

where $\xi_{j|i}$ is the indicator variable defined as 1 if a diplotype considered is compatible with individual i and as 0 otherwise, $u_{[r_1^m r_2^m r_3^m][r_1^P r_2^P r_3^P]} = u_{[r_1^P r_2^P r_3^P][r_1^m r_2^m r_3^m]}$ is the genotypic value for diplotype $[r_1^m r_2^m r_3^m][r_1^P r_2^P r_3^P]$, and e_i is the residual error distributed as $N(0, \sigma^2)$.

Regardless of two- or three-SNP analysis, we assume that one haplotype is the risk haplotype (designated as A), and thus all the others are the non-risk haplotypes (designated as $\bar{A}$). In the F_2, the risk and non-risk haplotypes will form three composite diplotypes AA (**2**), $A\bar{A}$ (**1**), and $\bar{A}\bar{A}$ (**0**), whose genotypic values are expressed as μ_2, μ_1, and μ_0, respectively. We array the genotypic values of different composite diplotypes and residual variance in a quantitative genetic parameter vector $\Omega_q = (\mu_2, \mu_1, \mu_0, \sigma^2)$.

3.3.2 Likelihoods, Estimation, and Hypothesis Tests

As shown in Sections (2.2) and (2.7), we can construct two likelihoods to estimate haplotype frequencies with $\log L(\Omega_p | S)$ and quantitative genetic parameters with $\log L(\Omega_q | y, S, \hat{\Omega}_p))$ by implementing with the EM algorithms. An optimal risk haplotype is determined on the basis of the likelihood values calculated when different risk haplotypes are assigned. The hypothesis tests of interest include the linkage (or linkage disequilibrium) between two SNPs, and overall genetic effects, and additive (a) and dominance genetic effects (d) of composite diplotypes constructed by the two SNPs. All these procedures are similar to those described in Sections (2.2) and (2.7), respectively, for two- and three-point analysis in a natural population. Readers are referred to these sections for any detail.

3.3.3 Model Selection: Two- vs. Three-Point LD Analysis

For a given set of three SNPs, S_1, S_2, and S_3, two-point analysis based on every possible pair or three-point analysis combining all the SNPs can be performed. An optimal analysis with these three SNPs should lead to the following results: (1) an optimal risk haplotype has been detected, and (2) such a risk haplotype is constituted by a minimum number of SNPs. For this reason, we need to perform all possible two- and three-point analyses. In each analysis, we calculate the likelihoods and AIC or BIC values when a particular risk haplotype (k) is assumed tabulated below:

	SNP Combination	Likelihood	AIC	BIC	Risk Haplotype (k)
Double	S_1-S_2	$L_{k\|12}$	$A_{k\|12}$	$B_{k\|12}$	$1,\cdots,4$
	S_1-S_3	$L_{k\|13}$	$A_{k\|13}$	$B_{k\|13}$	$1,\cdots,4$
	S_2-S_3	$L_{k\|23}$	$A_{k\|23}$	$B_{k\|23}$	$1,\cdots,4$
Triple	S_1-S_2-S_3	$L_{k\|123}$	$A_{k\|123}$	$B_{k\|123}$	$1,\cdots,8$

The likelihood values calculated can be compared among different risk haplotype assignments within the same pair of SNPs. However, an AIC or BIC value will be needed if the comparison is made between different pairs. The largest likelihood and/or minimum AIC or BIC value corresponds to an optimal choice of risk haplotype.

Example 3.2

The statistical model described above is used to map and identify risk haplotypes for a quantitative trait in an F_2 population. Because the marker data obtained for the mouse are microsatellites rather than SNPs, Hou et al. (2007) used these microsatellite markers as a surrogate of SNPs to demonstrate the utility of the model. The marker data were from Vaughn et al.'s (1999) study in which a linkage map composed of 19 chromosomes was constructed with 96 microsatellite markers for 502 F_2 mice (259 males and 243 females) derived from two strains, the Large (LG/J) and Small (SM/J). This map has a total map distance of $\sim$1780 cM (in Haldane's units) and an average interval length of $\sim$23 cM. The F_2 progeny was measured for their body mass at 10 weekly intervals starting at age 7 days. The raw weights were corrected for the effects of each covariate due to dam, litter size at birth, parity, and sex (Vaughn et al., 1999). Here, only adult body weights at week 10 are used for diplotype analysis.

For each F_2 mouse, the parental origin of alleles at each marker can be discerned in molecular studies. Let L and S be the alleles inherited from the Large (LG/J) and Small (SM/J) strains, respectively. For any pair of markers, there are four different haplotypes, LL, LS, SL, and SS, whose frequencies are accordingly denoted as

$$p_{LL} = p_{SS} = p$$

and

$$p_{LS} = p_{SL} = \frac{1}{2} - p.$$

By assigning each of the four haplotypes as a risk haplotype, respectively, the above model allows the estimates of haplotype frequencies by the EM iteration at the higher hierarchy and of composite genotypic values by the EM iteration at the lower hierarchy. The estimated haplotype frequencies are used to estimate linkage disequilibrium based on equation (3.2) and the recombination fraction (r) based on

TABLE 3.4

The MLEs of haplotype frequencies and significant log-likelihood ratios (LR) by assuming different risk haplotypes in the F_2 population of mice. The results were obtained by using a two-SNP QTN model.

Marker Pair	Association $\hat{D}$	$\hat{r}$	Selection of Risk Haplotype Risk Haplotype	Fre-quency	LR_2	Effect $\hat{a}$	$\hat{d}$
D4Mit16−D4Mit13	0.16	0.17	LL	0.36	157.59	0.53	0.46
			LS	0.08	152.57	0.60	-0.91
			SL	0.10	155.55	-1.46	0.90
			SS	0.47	153.26	-0.35	0.18
D6Mit9−D6Nds5	0.18	0.14	LL	0.42	19.95	1.17	0.24
D6Nds5−D6Mit15	0.14	0.22	LL	0.38	25.14	1.25	0.44
			SS	0.41	37.98	-1.69	0.51
D7Mit21−D7Nds1	0.09	0.32	LL	0.34	30.84	0.93	1.41
			SS	0.34	36.54	-1.70	-0.07
D7Nds1−D7Mit17	0.19	0.12	LL	0.44	46.87	1.66	0.59
			SS	0.45	43.31	-1.75	0.50
D7Mit17−D7Mit9	0.19	0.12	LL	0.43	33.41	1.42	0.57
			SS	0.45	34.35	-1.47	0.99
D7Mit9−D7Nds4	0.12	0.26	SS	0.38	19.84	-1.15	1.09
D14Mit5−D14Mit7	0.17	0.16	LL	0.43	19.35	1.10	0.33

equation (3.4). This estimation process is moved from the first ($\mathbf{M}_1$-$\mathbf{M}_2$) to last pair of markers ($\mathbf{M}_6$-$\mathbf{M}_7$) on chromosome 1 and then from chromosome 1 to 19.

Table 3.4 tabulates the results of the MLEs of haplotype frequencies and log-likelihoods under the assumptions of different risk haplotypes. A total of 96 markers are sparsely located on 19 mouse chromosomes, with the estimated recombination fractions from the linkage disequilibrium model (13.5) consistent with those obtained from the linkage model (Vaughn et al., 1999). Significant likelihood ratios for testing haplotype effects were determined by critical values obtained from the χ^2-square distribution with two degrees of freedom with a Bonferroni adjustment to the type I error. The adjusted critical values for the two- and three-marker models are 18.20 and 18.76, respectively, at the 5% significance level. Significant haplotype effects are detected for a total of eight marker pairs (Table 3.4), which include one pair on chromosome 4, two consecutive pairs on chromosome 6, four consecutive pairs on chromosome 7, and one pair on chromosome 14. For some pairs, multiple significant risk haplotypes were detected. Risk haplotypes purely composed of alleles inherited from the LG/J or SM/J parent exert a positive or negative additive effect on body weight, respectively. Based on the relative values of estimated additive and dominant effects, the significant marker pairs detected display partial dominant effects (Table 3.4).

The results from the three-marker model are basically consistent with those from the two-marker model (Table 3.5). The advantage of the three-marker model is that

TABLE 3.5

The MLEs of haplotype frequencies and significant log-likelihood ratios (LR) by assuming different risk haplotypes in the F_2 population of mice. The results were obtained by using a three-SNP QTN model.

Marker Pair	Selection of Risk Haplotype			Effect	
	Risk Haplotype	Fre-quency	LR_2	$\hat{a}$	$\hat{d}$
D4Mit45−D4Mit16−D4Mit13	LLL	0.29	124.70	0.40	0.61
	LLS	0.07	121.34	1.08	-1.00
	LSL	0.01	122.58	-1.88	-2.56
	LSS	0.07	122.08	-0.44	1.40
	SLL	0.07	122.14	0.80	0.12
	SLS	0.01	132.86	–	–
	SSL	0.09	122.33	-1.32	1.09
	SSS	0.40	123.65	-0.55	0.28
D6Mit9−D6Nds5−D6Mit15	SSS	0.34	22.65	-1.51	0.39
D7Mit21−D7Nds1−D7Mit17	LLL	0.29	38.28	0.81	1.85
	SSS	0.30	33.74	-1.80	0.09
D7Nds1−D7Mit17−D7Mit9	LLL	0.38	34.39	1.48	0.47
	SSS	0.40	32.18	-1.61	0.61
D7Mit17−D7Mit9−D7Nds4	LLL	0.33	21.74	1.20	0.45
	SSS	0.33	29.41	-1.60	1.36
D14Nds1−D14Mit5−D14Mit7	LLL	0.30	19.55	1.44	-0.50

it incorporates the interferences between adjacent marker intervals into the estimation process and, thus, can potentially increase the estimation precision of haplotype effects. ☐

3.4 LD Analysis in a Full-Sib Family

3.4.1 Introduction

With the completion of the mouse genome sequence, the mouse as a model organism widely used in genetic research will be more useful to define the amount, type, and organization of genetic variation in a complex trait related to drug response. Two types of strains have been used in the genetic study of mice: classical laboratory inbred lines and wide-derived inbred lines (Ideraabdullah et al., 2004). The first analyses of genome-wide patterns of genetic variation in the mouse have used inbred mouse strains (Lindblad-Toh et al., 2000; Wade et al., 2002; Wiltshire et al., 2003; Yalcin et al., 2004). But these studies suggest that classical inbred strains have limited levels of genetic variation as compared with humans, and that there exists a mosaic structure in the genome of classical inbred strains, i.e., some regions show low levels of polymorphism but others show high levels of polymorphism (Lindblad-Toh et al., 2000; Wade et al., 2002; Wiltshire et al., 2003; Yalcin et al., 2004).

More recently, the approach for hybridization between wild-derived and classical strains has been used to increase levels of genetic diversity in the mouse, leading to the high-resolution linkage map of the mouse for the study of genome imprinting, X-inactivation, and complex traits (Guénet and Bonhomme, 2003). For this reason, understanding the structure and patterns of sequence variation in the controlled cross of wild mice has become an essential task for genetic haplotyping of complex traits (Ideraabdullah et al., 2004). In this section, we will describe a statistical model for characterizing haplotype diversity and its association with phenotypic variation in a full-sib family of wild mice.

3.4.2   A General Model

Consider a pair of SNPs each with alleles 1 and 0. These two SNPs are linked with a recombination fraction of r. Suppose there are two heterozygous parents, both of which have genotypes 10/10 for the two SNPs. For an outcrossing double heterozygote, there is uncertainty on its diplotype, which can be either [11][00] or [10][01]. Assume that the probability with which P is diplotype [11][00] is p, and then the probability with which this parent is diplotype [10][01] is $1 - p$. Similar denotation for the diplotype probability of Q is q and $1 - q$, respectively. We array the unknown parameters in $\Omega_p = (r, p, q)$.

Two different diplotypes [11][00] and [10][01] will produce the same group of haplotypes (gametes), i.e., [11], [10], [01], and [00], but with haplotype frequencies depending on diplotype (Table 3.6). Let p_{11}^P, p_{01}^P, p_{10}^P, and p_{00}^P be the haplotype frequencies produced by parent P, and p_{11}^Q, p_{01}^Q, p_{10}^Q, and p_{00}^Q be the haplotype frequencies produced by parent Q. Considering the two possible diplotypes for each parent, these parent-specific haplotype frequencies are expressed as

$$
\begin{aligned}
p_{11}^P &= \tfrac{1}{2}(1-r)p + \tfrac{1}{2}r(1-p),\\
p_{10}^P &= \tfrac{1}{2}rp + \tfrac{1}{2}(1-r)(1-p),\\
p_{01}^P &= \tfrac{1}{2}rp + \tfrac{1}{2}(1-r)(1-p),\\
p_{00}^P &= \tfrac{1}{2}(1-r)p + \tfrac{1}{2}r(1-p),\\
p_{11}^Q &= \tfrac{1}{2}(1-r)q + \tfrac{1}{2}r(1-q),\\
p_{10}^Q &= \tfrac{1}{2}rq + \tfrac{1}{2}(1-r)(1-q),\\
p_{01}^Q &= \tfrac{1}{2}rq + \tfrac{1}{2}(1-r)(1-q),\\
p_{00}^Q &= \tfrac{1}{2}(1-r)q + \tfrac{1}{2}r(1-q),
\end{aligned}
\tag{3.12}
$$

which will be used to derive the diplotype and genotype frequencies under the assumption of Hardy–Weinberg equilibrium (see Table 3.7).

TABLE 3.6

The frequencies of four haplotypes derived from each of two diplotypes for a double heterozygote.

Diplotype	Diplotype Probability		Haplotype Frequency			
	P	Q	[11]	[10]	[01]	[00]
[11][00]	p	q	$\frac{1}{2}(1-r)$	$\frac{1}{2}r$	$\frac{1}{2}r$	$\frac{1}{2}(1-r)$
[10][01]	$1-p$	$1-q$	$\frac{1}{2}r$	$\frac{1}{2}(1-r)$	$\frac{1}{2}(1-r)$	$\frac{1}{2}r$

3.4.3 Estimation

Assume that the full-sib family produced by parents P_1 and P_2 contains n progeny with nine possible genotypes for the two SNPs considered. We use $n_{r_1r_1'/r_2r_2'}$ ($r_1 \geq r_1', r_2 \geq r_2' = 1, 0$) to denote the number of progeny for a genotype. The diplotypes and their frequencies for the nine genotypes are given in Table 3.7. From these observations and diplotype frequencies, a multinomial likelihood is constructed as

$$
\begin{aligned}
\log L(\Omega_p|S) &= \text{constant} \\
&+ n_{11/11}\log(p_{11}^P p_{11}^Q) + n_{11/10}\log(p_{11}^P p_{10}^Q + p_{10}^P p_{11}^Q) \\
&+ n_{11/00}\log(p_{11}^P p_{11}^Q) + n_{10/11}\log(p_{11}^P p_{01}^Q + (p_{01}^P p_{11}^Q) \\
&+ n_{10/10}\log(p_{11}^P p_{00}^Q + p_{00}^P p_{11}^Q + p_{10}^P p_{01}^Q + p_{01}^P p_{10}^Q) \\
&+ n_{10/00}\log(p_{10}^P p_{00}^Q + p_{00}^P p_{10}^Q) + n_{00/11}\log(p_{01}^P p_{01}^Q) \\
&+ n_{00/10}\log(p_{01}^P p_{00}^Q + p_{00}^P p_{01}^Q) + n_{00/00}\log(p_{00}^P p_{00}^Q).
\end{aligned}
\tag{3.13}
$$

From likelihood (3.13), the EM algorithm can be derived to obtain the maximum likelihood estimates (MLEs) of the recombination fraction (r) and parental diplotype probabilities (p and q). Each of these parameters has a closed form for their estimation. Below, the EM algorithm is described for the parameter estimation.

In the E step, calculate the expected numbers of recombinants within each genotype by

$$
\phi_1 = \frac{r(1-r)[(1-p)q + p(1-q)] + 2r^2(1-p)(1-q)}{\omega_1^P \omega_1^Q},
$$

$$
\phi_2 = \frac{2r(1-r)[pq + (1-p)(1-q)] + 2r^2[(1-p)q + p(1-q)]}{\omega_1^P \omega_2^Q + \omega_2^P \omega_1^Q},
$$

$$
\phi_3 = \frac{r(1-r)[p(1-q) + (1-p)q] + 2r^2 pq}{\omega_2^P \omega_2^Q},
$$

TABLE 3.7
Diplotypes and their frequencies for each of nine genotypes at two SNPs within a QTN, haplotype composition frequencies for each genotype, and composite diplotypes for four possible risk haplotypes.

	Diplotype		Risk Haplotype			
Genotype	Configuration	Frequency	[11]	[10]	[01]	[00]
11/11	[11][11]	$p_{11}^P p_{11}^Q$	AA	$\bar{A}\bar{A}$	$\bar{A}\bar{A}$	$\bar{A}\bar{A}$
11/10	[11][10]	$p_{11}^P p_{10}^Q + p_{10}^P p_{11}^Q$	$A\bar{A}$	$A\bar{A}$	$\bar{A}\bar{A}$	$\bar{A}\bar{A}$
11/00	[10][10]	$p_{10}^P p_{10}^Q$	$\bar{A}\bar{A}$	AA	$\bar{A}\bar{A}$	$\bar{A}\bar{A}$
10/11	[11][01]	$p_{11}^P p_{01}^Q + p_{01}^P p_{11}^Q$	$A\bar{A}$	$\bar{A}\bar{A}$	$A\bar{A}$	$\bar{A}\bar{A}$
10/10	$\begin{cases} [11][00] \\ [10][01] \end{cases}$	$\begin{cases} p_{11}^P p_{00}^Q + p_{00}^P p_{11}^Q \\ p_{10}^P p_{01}^Q + p_{01}^P p_{10}^Q \end{cases}$	$\begin{cases} A\bar{A} \\ \bar{A}\bar{A} \end{cases}$	$\begin{cases} \bar{A}\bar{A} \\ A\bar{A} \end{cases}$	$\begin{cases} \bar{A}\bar{A} \\ A\bar{A} \end{cases}$	$\begin{cases} A\bar{A} \\ \bar{A}\bar{A} \end{cases}$
10/00	[10][00]	$p_{10}^P p_{00}^Q + p_{00}^P p_{10}^Q$	$\bar{A}\bar{A}$	$A\bar{A}$	$\bar{A}\bar{A}$	$A\bar{A}$
00/11	[01][01]	$p_{01}^P p_{01}^Q$	$\bar{A}\bar{A}$	$\bar{A}\bar{A}$	AA	$\bar{A}\bar{A}$
00/10	[01][00]	$p_{01}^P p_{00}^Q + p_{00}^P p_{01}^Q$	$\bar{A}\bar{A}$	$\bar{A}\bar{A}$	$A\bar{A}$	$A\bar{A}$
00/00	[00][00]	$p_{00}^P p_{00}^Q$	$\bar{A}\bar{A}$	$\bar{A}\bar{A}$	$\bar{A}\bar{A}$	AA

$$\phi_4 = \frac{2r(1-r)[p(1-q)+(1-p)q]+2r^2[pq+(1-p)(1-q)]}{\omega_1^P \omega_1^Q + \omega_2^P \omega_2^Q},$$

$$\phi_5 = \frac{2r(1-r)[pq+(1-p)(1-q)]+2r^2[p(1-q)+(1-p)q]}{\omega_2^P \omega_1^Q + \omega_1^P \omega_2^Q},$$

$$(3.14)$$

the expected numbers of p within each genotype by

$$\psi_1^P = \frac{(1-r)p}{\omega_1^P},$$

$$\psi_2^P = \frac{(1-r)p\omega_2^Q + rp\omega_1^Q}{\omega_1^P \omega_2^Q + \omega_2^P \omega_1^Q},$$

$$\psi_3^P = \frac{rp}{\omega_2^P},$$

$$\psi_4^P = \frac{(1-r)p\omega_1^Q + rp\omega_2^Q}{\omega_1^P \omega_1^Q + \omega_2^P \omega_2^Q},$$

$$\psi_5^P = \frac{rp\omega_1^Q + (1-r)p\omega_2^Q}{\omega_2^P\omega_1^Q + \omega_1^P\omega_2^Q},$$

$$(3.15)$$

and the expected numbers of q within each genotype by

$$\psi_1^Q = \frac{(1-r)q}{\omega_1^Q},$$

$$\psi_2^Q = \frac{(1-r)q\omega_2^P + rq\omega_1^P}{\omega_1^P\omega_2^Q + \omega_2^P\omega_1^Q},$$

$$\psi_3^Q = \frac{rq}{\omega_2^Q},$$

$$(3.16)$$

$$\psi_4^Q = \frac{(1-r)q\omega_1^P + rq\omega_2^P}{\omega_1^P\omega_1^Q + \omega_2^P\omega_2^Q},$$

$$\psi_5^Q = \frac{rq\omega_1^P + (1-r)q\omega_2^P}{\omega_2^P\omega_1^Q + \omega_1^P\omega_2^Q},$$

where

$$\omega_1^P = (1-r)p + r(1-p),$$
$$\omega_1^Q = (1-r)q + r(1-q),$$
$$\omega_2^P = (1-r)(1-p) + rp),$$
$$\omega_2^Q = (1-r)(1-q) + rq.$$

In the M step, estimate the recombination fraction and diplotype probabilities with the numbers estimated from the E step by

$$\hat{r} = \frac{1}{2n}[\phi_1(n_{11/11} + n_{00/00}) + \phi_2(n_{11/10} + n_{10/11})$$
$$+ \phi_3(n_{11/00} + n_{00/11}) + \phi_4 n_{10/10} + \phi_5(n_{10/00} + n_{00/10})], \qquad (3.17)$$

$$\hat{p} = \frac{1}{2n}[\psi_1^P(n_{11/11} + n_{00/00}) + \psi_2^P(n_{11/10} + n_{10/11})$$
$$+ \psi_3^P(n_{11/00} + n_{00/11}) + \psi_4^P n_{10/10} + \psi_5^P(n_{10/00} + n_{00/10})], \qquad (3.18)$$

and

$$\hat{q} = \frac{1}{2n}[\psi_1^Q(n_{11/11} + n_{00/00}) + \psi_2^Q(n_{11/10} + n_{10/11})$$
$$+ \psi_3^Q(n_{11/00} + n_{00/11}) + \psi_4^Q n_{10/10} + \psi_5^Q(n_{10/00} + n_{00/10})]. \qquad (3.19)$$

Iterations between the E and M steps are repeated with equations (3.14), (3.15), and (3.16) and equations (3.17), (3.18), and (3.19) until the estimates of parameters converge. The estimates at the convergence are the MLEs of r, p, and q. Louis' (1982) approach can be used to derive and estimate the sampling variances of these MLEs.

The MLEs of r, p, and q will be used to estimate haplotype effects in the next step. To estimate genotypic values of the three composite diplotypes by assuming a risk haplotype (Table 3.7) and residual variance, arrayed in $\Omega_q = (\mu_2, \mu_1, \mu_0, \sigma^2)$, we formulate a likelihood, $\log L(\Omega_q | \mathbf{y}, \mathbf{S}, \hat{\Omega}_p)$, based on a mixture model. The EM algorithm is derived to estimate Ω_q with the same procedure, as described in Section 2.3, except for ϕ of equation (2.4) replaced by

$$\phi = \frac{p_{11}^P p_{00}^Q + p_{00}^P p_{11}^Q}{p_{11}^P p_{00}^Q + p_{00}^P p_{11}^Q + p_{10}^P p_{01}^Q + p_{01}^P p_{10}^Q}.$$

An optimal risk haplotype is chosen on the basis of the likelihoods calculated by assuming all possible risk haplotypes. After an optimal risk haplotype is determined, hypothesis tests can be performed about the existence of significant risk haplotypes as well as the additive and dominance genetic effects on the trait triggered by these haplotypes.

3.4.4   Multiple Segregating Types of Markers

For a given outcrossing line, its genotype for each SNP can be either homozygous (11 or 00), or heterozygous (10). For any pair of SNPs, its genotype can be 11/11 (1), 11/10 (2), 11/00 (3), 10/11 (4), 10/10 (5), 10/00 (6), 00/11 (7), 00/10 (8), or 00/00 (9). If both outcrossing lines (P and Q) are a double homozygote, their cross will not produce any segregation. Thus, such a cross type will not be useful for haplotyping analysis. Risk haplotypes can be discerned when parent cross types P $\times$ Q = (1,2,3,4,5,6,7,8,9) $\times$ (2,4,5,6,8) or (2,4,5,6,8) $\times$ (1,2,3,4,5,6,7,8,9). When a double heterozygote (5) is crossed with any genotype (2,4,5,6,8), the EM algorithm is needed to estimate the recombination fraction, diplotype probabilities, and quantitative genetic parameters. In the progeny of these cross types, there is a double heterozygote that has diplotypes [11][00] or [10][01]. For cross types $(1,2,3,4,6,7,8,9) \times (2,4,6,8)$ or $(2,4,6,8) \times (1,2,3,4,6,7,8,9)$, a simple t-test will be sufficient to detect a risk haplotype (see Section 3.2) because no double heterozygote is formed from these cross types.

It can be seen that genetic haplotyping is more informative than genetic mapping for a full-sib family. First, QTL mapping in a single full-sib family cannot estimate the linkage phase between markers and a QTL bracketed by the markers, although the genetic effect of the QTL can be estimated. Second, QTL mapping can use a very limited number of markers segregating in a full-sib family. For example, cross types $(1,2,3,4,6,7,8,9) \times (2,4,6,8)$ or $(2,4,6,8) \times (1,2,3,4,6,7,8,9)$ do not provide any information for linkage analysis. Of course, genetic haplotyping largely relies on the availability of a high-density SNP-based linkage map.

3.4.5 Three-Point Haplotyping

The model can be extended to model three SNPs (in order S_1-S_2-S_3) simultaneously. Consider two outcrossing parents P and Q whose three-SNP genotypes are 10/10/10. It is possible that such a triple heterozygote has four diplotypes, [111][000], [110][001], [101][010], and [100][011]. Let p_1, p_2, p_3, and $1 - p_1 - p_2 - p_3$ be the diplotype probabilities with which parent P is these four diplotypes, respectively. The diplotype probabilities of parent Q are q_1, q_2, q_3, and $1 - q_1 - q_2 - q_3$. For any one of the diplotypes, there are eight possible haplotypes [111], [110], [101], [100], [011], [010], [001], and [000]. Let g_{11}, g_{10}, g_{01}, and g_{00} be the probabilities with which there is one recombinant at each of intervals S_1-S_2 and S_2-S_3, there is one recombinant at interval S_1-S_2 but no recombinant at interval S_2-S_3, there is no recombinant at interval S_1-S_2 but one recombinant at interval S_2-S_3, and there is no recombinant at both intervals S_1-S_2 and S_2-S_3, respectively. Thus, haplotype frequencies can be expressed in terms of these g-probabilities, with the form depending on the diplotype of a given parent (see Table 3.8).

Based on the information provided by Table 3.8, we can derive the frequency expressions of 36 three-SNP diplotypes and 27 three-SNP genotypes, separately for different parents. A multinomial likelihood is formulated to estimate the g probabilities and parental diplotype probabilities (p_1, p_2, p_3 and q_1, q_2, q_3). Let r_{12}, r_{23}, and r_{13} be the recombination fractions between SNPs S_1 and S_2, between SNPs S_2 and S_3, and between SNPs S_1 and S_3. Thus, we have

$$r_{12} = g_{11} + r_{10}$$
$$r_{23} = g_{11} + r_{01}$$
$$r_{13} = g_{10} + r_{01}.$$

Also, by assigning one haplotype as a risk haplotype, a mixture-based likelihood can be formulated to estimate genotypic values of the three composite diplotypes and residual variance. An optimal risk haplotype is determined by choosing one haplotype that gives the largest likelihood if it is assumed as a risk haplotype. After the parameters are estimated, hypothesis tests can be made, including the linkage between each pair of SNPs, and overall effects of composite diplotypes and their additive and dominance effects.

As shown in Wu et al. (2007a), three-point analysis allows the characterization of an optimal order among the three SNPs by estimating the likelihood under each possible order. The second advantage of three-point analysis is that the degree of interference (I) between two adjacent intervals can be tested (Wu et al., 2007a). Assuming that the three SNPs are arranged in order S_1-S_2-S_3, the interference between intervals S_1-S_2 and S_2-S_3 can be estimated as

$$I = \frac{r_{12} + r_{23} - r_{13}}{2r_{12}r_{23}}.$$

If $I = 1$, this indicates that there is no interference. If $I = 0$, this indicates that there is no double recombination. A log-likelihood approach can be used to test whether

TABLE 3.8
The frequencies of eight haplotypes derived from each of four diplotypes for a triple heterozygote.

	Diplotype			
	$[111][000]$	$[110][001]$	$[101][010]$	$[011][100]$
	Parental Diplotype Probability			
P	p_1	p_2	p_3	$1 - p_1 - p_2 - p_3$
Q	q_1	q_2	q_3	$1 - q_1 - q_2 - q_3$
	Haplotype Frequency			
$[111]$	$\frac{1}{2}g_{00}$	$\frac{1}{2}g_{01}$	$\frac{1}{2}g_{11}$	$\frac{1}{2}g_{10}$
$[110]$	$\frac{1}{2}g_{01}$	$\frac{1}{2}g_{00}$	$\frac{1}{2}g_{10}$	$\frac{1}{2}g_{11}$
$[101]$	$\frac{1}{2}g_{11}$	$\frac{1}{2}g_{10}$	$\frac{1}{2}g_{00}$	$\frac{1}{2}g_{01}$
$[100]$	$\frac{1}{2}g_{10}$	$\frac{1}{2}g_{11}$	$\frac{1}{2}g_{01}$	$\frac{1}{2}g_{00}$
$[011]$	$\frac{1}{2}g_{10}$	$\frac{1}{2}g_{11}$	$\frac{1}{2}g_{01}$	$\frac{1}{2}g_{00}$
$[010]$	$\frac{1}{2}g_{11}$	$\frac{1}{2}g_{10}$	$\frac{1}{2}g_{00}$	$\frac{1}{2}g_{01}$
$[001]$	$\frac{1}{2}g_{01}$	$\frac{1}{2}g_{00}$	$\frac{1}{2}g_{10}$	$\frac{1}{2}g_{11}$
$[000]$	$\frac{1}{2}g_{00}$	$\frac{1}{2}g_{01}$	$\frac{1}{2}g_{11}$	$\frac{1}{2}g_{10}$

I is equal to 1 or 0, thus providing a procedure for studying the behavior of meioses among these three SNPs.

3.5 Prospects

Quantitative trait locus (QTL) mapping aims to identify narrow chromosomal segments for a quantitative trait by using a statistical method, and has proven its value to study the genetic architecture of the trait in a variety of species (Frary et al., 2000; Mackay, 2001; Li et al., 2006a). The limitations of this technique lie in its inability to characterize the structure and organization of DNA sequences and statistical difficulty in deriving the distribution of test statistics under the null hypothesis of no QTL (Lander and Schork, 1994). At least partly for these reasons, despite thousands of QTL reported for different traits and populations, a very small portion of them have been cloned (Flint et al., 2005). With the completion of the genome projects for several important organisms, a new line of thought in the post genomic era has begun to emerge for the identification of specific combinations of nucleotides or haplotypes that contribute to a complex quantitative trait (Liu et al., 2004; Lin and Wu, 2006b).

The haplotyping model offers a powerful tool for positional cloning of QTLs that affect a complex trait. Flint et al. (2005) reviewed the potential of currently available cloning strategies, such as probabilistic ancestral haplotype reconstruction, Yin-Yang crosses, and in silico analysis of sequence variants, to identify genes that underlie QTL in rodents. Our model, in conjunction with these strategies, may open a new gateway for the illustration of a detailed picture of the genetic architecture for a complex trait.

4

A General Quantitative Model for Genetic Haplotyping

The models for genetic haplotyping described in Chapters 2 and 3 are based on two critical assumptions. One is that the population chosen to haplotype a complex trait should be at Hardy–Weinberg equilibrium (HWE). As a fundamental assumption in general population genetic studies (Lynch and Walsh, 1998), this allows the diplotype frequency to be expressed as the product of the frequencies of two pairing haplotypes. The second is that one haplotype composed of alleles at multiple SNPs is different from the remaining haplotypes in terms of genetic effects on a trait. This assumption allows the direct use of a traditional biallelic quantitative genetic model (Lynch and Walsh, 1998) and facilitates the definition and estimation of genetic effects triggered by different haplotypes.

These two assumptions may not be realistic in practice. First, a population of interest to be mapped may be in a non-equilibrium state, in which non-random associations of alleles at different loci may occur both within and between haplotypes. A non-equilibrium population may be due to non-random mating and various evolutionary forces, such as selection, mutation, genetic drift and population structure (Falconer and Mackay, 1996). Liu et al. (2006) showed that ignoring non-random associations between different haplotypes, as assumed for a HWE population, will lead to a misleading inference of haplotype diversity and distribution in the population. Second, when there is substantial variation among the non-risk haplotypes, the estimation of haplotype effects will be misleading. Wu et al. (2007b) developed a multi-allelic quantitative genetic model to detect multiple risk haplotypes and estimate additive and dominance genetic effects of different kinds due to these risk haplotypes.

In this chapter, we will describe a general genetic model for relaxing these two assumptions. First, a statistical procedure will be introduced for characterizing multiple risk haplotypes in a natural population. We will focus on the selection of the optimal number and combination of risk haplotypes responsible for a quantitative trait through conventional model selection criteria. The closed forms for the EM algorithm will be shown to estimate a variety of genetic parameters including haplotype frequencies and haplotype effects. Second, we will show the limitation of the HWE assumption in analyzing linkage disequilibria for a non-equilibrium population through simulation studies. The investigation of the statistical properties of parameter estimation in a non-equilibrium population is still a open question in the current genetic literature.

4.1   Quantitative Genetic Models

4.1.1   Population Structure

In this section, we will describe a series of multiple risk haplotype models for genetic haplotyping. When relaxing the assumption of a single risk haplotype, the HWE assumption of a population being mapped will be used. Suppose there are genetically associated SNPs each with two alleles designated as 1 and 0. Let p and q be the 1-allele frequencies for the first and second SNP, respectively. Thus, the 0-allele frequencies at different SNPs will be $1 - p$ and $1 - q$. These two SNPs segregating in a natural population form four haplotypes, [11], [10], [01], and [00], whose frequencies are constructed by allele frequencies and linkage disequilibrium (D) between the two SNPs, i.e.,

$$
\begin{aligned}
p_{11} &= pq + D, \\
p_{10} &= p(1 - q) - D, \\
p_{01} &= (1 - p)q - D, \\
p_{00} &= (1 - p)(1 - q) - D.
\end{aligned}
$$

We use $\Omega_p = (p_{11}, p_{10}, p_{01}, p_{00})$ to denote the haplotype frequency vector. These haplotypes unite randomly to generate 10 distinct diplotypes and 9 distinct genotypes. If the population is at Hardy–Weinberg equilibrium, the frequency of a diplotype is expressed as the product of the frequencies of the two haplotypes that constitute the diplotype (see Table 2.1). The frequency of the double zygotic genotype is the summation of the frequencies of its two possible diplotypes.

4.1.2   Biallelic Model

Liu et al. (2004) assumed that all haplotypes are sorted into two groups, risk and non-risk, and defined the combination of risk and non-risk haplotypes as a composite diplotype. Let A_1 and A_0 be the risk and non-risk haplotypes, respectively, which are equivalent to two alternative alleles if the two associated SNPs considered are viewed as a "locus." Thus, for such a "biallelic locus," we have three possible composite diplotypes. Let μ be the overall mean, a be the additive effect due to the substitution of the risk haplotype by the non-risk haplotype, and d be the dominance effect due to the interaction between the risk and non-risk haplotypes. Table 4.1 shows the genotypic values of three composite diplotypes and their additive and dominant components. We use $\Omega_{qB} = (\mu, a, d)$ to denote the genetic parameters for the biallelic model.

In this model, all the haplotypes, [11], [10], [01], and [00], are sorted into two categories, risk haplotype and non-risk haplotype. This includes two cases. In the first case, because any one of the haplotypes can be risk, there are four choices for determining the risk haplotype. In the second case, any two haplotypes can be

TABLE 4.1

Genetic values and their partitioning components for three different composite diplotypes.

Composite Diplotype	Genotypic Value	Components
A_1A_1	μ_1	$\mu + a$
A_1A_0	μ_2	$\mu + d$
A_0A_0	μ_3	$\mu - a$

TABLE 4.2

Seven options to choose a risk haplotype for a biallelic model.

Option	Risk Haplotype	Non-risk Haplotype
B_1	11	10,01,00
B_2	10	11,01,00
B_3	01	11,10,00
B_4	00	11,10,01
B_5	11,10	01,00
B_6	11,01	10,00
B_7	11,00	10,01

together a risk haplotype, and the other two are collectively a non-risk haplotype. These two cases lead to a total of seven options to choose the risk haplotype which is tabulated in Table 4.2. The optimal choice of a risk haplotype for the biallelic model is based on the maximum of the likelihoods calculated for each of the seven options described above.

4.1.3 Triallelic Model

It is possible that there are two distinct risk haplotypes which are each different from non-risk haplotypes. This case is regarded as a "triallelic locus." Let A_1 and A_2 be the first and second risk haplotypes, and A_0 be the non-risk haplotype, which form six composite diplotypes with genotypic values tabulated in Table 4.3. In the table, μ is the overall mean, a_1 and a_2 are the additive effects due to the substitution of the first and second risk haplotype by the non-risk haplotype, and d_{12}, d_{10}, and d_{20} are the dominance effects due to the interaction between the first and second risk haplotype, between the first risk haplotype and the non-risk haplotype, and between the second risk haplotype and non-risk haplotype, respectively. These parameters are arrayed in $\Omega_{qr} = (\mu, a_1, a_2, d_{12}, d_{10}, d_{20})$.

The triallelic model may include a total of six haplotype combinations (Table 4.4). The optimal combination of risk haplotypes for the triallelic model corresponds to the maximum of the likelihoods calculated for each of the six possibilities.

TABLE 4.3

Genetic values and their partitioning components for six
different composite diplotypes.

Composite Diplotype	Genotypic Value	Component
A_1A_1	μ_1	$\mu + a_1$
A_2A_2	μ_2	$\mu + a_2$
A_0A_0	μ_3	$\mu - a_1 - a_2$
A_1A_2	μ_4	$\mu + \frac{1}{2}(a_1 + a_2) + d_{12}$
A_1A_0	μ_5	$\mu - \frac{1}{2}a_2 + d_{10}$
A_2A_0	μ_6	$\mu - \frac{1}{2}a_1 + d_{20}$

Note: The additive effect (a_1 or a_2) due to a given haplotype sums to zero across all
the composite diplotypes that contain this haplotype, and the dominance effects (d_{12},
d_{10} and d_{20} are the interactions between different haplotypes.

TABLE 4.4

Seven options to choose a risk haplotype for
a triallelic model.

Option	Risk Haplotype 1	2	Non-risk Haplotype
T_1	11	10	01,00
T_2	11	01	10,00
T_3	11	00	10,01
T_4	10	01	11,00
T_5	10	00	11,01
T_6	01	00	11,10

4.1.4 Quadriallelic Model

If there are three distinct risk haplotypes, we need a quadriallelic genetic model to
specify haplotype effects. Let A_1, A_2, and A_3 be the first, second, and third risk haplotypes, and A_0 be the non-risk haplotype, which form 10 composite diplotypes with
genotypic values expressed in Table 4.5. In the table, μ is the overall mean, a_1, a_2,
and a_3 are the additive effects due to the substitution of the first, second, and third
risk haplotype by the non-risk haplotype, and d_{12}, d_{13}, d_{23}, d_{10}, d_{20}, and d_{30} are the
dominance effects due to the interaction between the first and second risk haplotype,
between the first and third risk haplotype, between the second and third risk haplotype, between the first risk and non-risk haplotype, between the second risk and
non-risk haplotype, and between the third risk and non-risk haplotype, respectively.
These parameters are arrayed in $\Omega_{qQ} = (\mu, a_1, a_2, a_3, d_{12}, d_{13}, d_{23}, d_{10}, d_{20}, d_{30})$.

TABLE 4.5

Genetic values and their partitioning components for 10 different composite diplotypes.

Composite Diplotype	Genotypic Value	Component
A_1A_1	μ_1	$\mu + a_1$
A_2A_2	μ_2	$\mu + a_2$
A_3A_3	μ_3	$\mu + a_3$
A_0A_0	μ_4	$\mu - (a_1 + a_2 + a_3)$
A_1A_2	μ_5	$\mu + \frac{1}{2}(a_1 + a_2) + d_{12}$
A_1A_3	μ_6	$\mu + \frac{1}{2}(a_1 + a_3) + d_{13}$
A_2A_3	μ_7	$\mu + \frac{1}{2}(a_2 + a_3) + d_{23}$
A_1A_0	μ_8	$\mu - \frac{1}{2}(a_2 + a_3) + d_{10}$
A_2A_0	μ_9	$\mu - \frac{1}{2}(a_1 + a_3) + d_{20}$
A_3A_0	μ_{10}	$\mu - \frac{1}{2}(a_1 + a_2) + d_{30}$

4.2 Likelihood

Assume that a total of n subjects are sampled from a Hardy–Weinberg equilibrium population and that each subject is genotyped for many SNPs and phenotyped for a quantitative trait. Consider two of the SNPs that form nine genotypes with observed numbers generally expressed as $n_{r_1 r_1'/r_2 r_2'}$ $(r_1 \geq r_1', r_2 \geq r_2' = 0, 1)$. The phenotypic value of the trait for subject i is expressed in terms of the two-SNP haplotypes as

$$y_i = \sum_{j=1}^{J} \xi_i \mu_J + e_i, \tag{4.1}$$

where ξ_i is the indicator variable defined as 1 if subject i has a composite diplotype j and 0 otherwise, e_i is the residual error, normally distributed as $N(0, \sigma^2)$, and J is the number of composite diplotypes expressed as

$$J = \begin{cases} 3 & \text{for the biallelic model} \\ 6 & \text{for the triallelic model} \\ 10 & \text{for the quadriallelic model.} \end{cases} \tag{4.2}$$

The genotypic values of composite diplotypes and variance are arrayed by a quantitative genetic parameter vector $\Omega_q = (\Omega_{q_B}, \sigma^2)$ for the biallelic model, (Ω_{q_T}, σ^2) for the triallelic model, and (Ω_{q_Q}, σ^2) for the quadriallelic model.

The log-likelihood of haplotype frequencies, genotypic values of the diplotypes, and residual variance given the phenotypic ($\mathbf{y}$) and SNP data ($\mathbf{S}$) is factorized into two parts, expressed as

$$\log L(\Omega_p, \Omega_q | \mathbf{y}, \mathbf{S}) = \log L(\Omega_p | \mathbf{S}) + \log L(\Omega_q | \mathbf{y}, \mathbf{S}, \Omega_p) \tag{4.3}$$

where

$$\log L(\boldsymbol{\Omega}_p|\mathbf{S}) = \text{constant}$$
$$+ n_{11/11}\log(2p_{11}) + n_{11/10}\log(2p_{11}p_{10}) + n_{11/00}\log(2p_{10})$$
$$+ n_{10/11}\log(2p_{11}p_{01}) + n_{10/10}\log(2p_{11}p_{00} + 2p_{10}p_{01})$$
$$+ n_{10/00}\log(2p_{10}p_{00}) + n_{00/11}\log(2p_{01})$$
$$+ n_{00/10}\log(2p_{01}p_{00}) + n_{00/00}\log(2p_{00}), \tag{4.4}$$

$$\log L(\boldsymbol{\Omega}_{q_B}|\mathbf{y},\mathbf{S},\hat{\boldsymbol{\Omega}}_p)$$
$$= \sum_{i=1}^{n_{11/11}} \log f_1(y_i) + \sum_{i=1}^{n_{11/10}} \log f_2(y_i) + \sum_{i=1}^{n_{11/00}} \log f_3(y_i)$$
$$+ \sum_{i=1}^{n_{10/11}} \log f_2(y_i) + \sum_{i=1}^{n_{10/10}} \log[\phi f_2(y_i) + (1-\phi)f_3(y_i)] + \sum_{i=1}^{n_{10/00}} \log f_3(y_i)$$
$$+ \sum_{i=1}^{n_{00/11}} \log f_3(y_i) + \sum_{i=1}^{n_{00/10}} \log f_3(y_i) + \sum_{i=1}^{n_{00/00}} \log f_3(y_i) \tag{4.5}$$

for the biallelic model assuming that haplotype [11] is a risk haplotype,

$$\log L(\boldsymbol{\Omega}_{q_T}|\mathbf{y},\mathbf{S},\hat{\boldsymbol{\Omega}}_p)$$
$$= \sum_{i=1}^{n_{11/11}} \log f_1(y_i) + \sum_{i=1}^{n_{11/10}} \log f_4(y_i) + \sum_{i=1}^{n_{11/00}} \log f_2(y_i)$$
$$+ \sum_{i=1}^{n_{10/11}} \log f_5(y_i) + \sum_{i=1}^{n_{10/10}} \log[\phi f_5(y_i) + (1-\phi)f_6(y_i)] + \sum_{i=1}^{n_{10/00}} \log f_6(y_i)$$
$$+ \sum_{i=1}^{n_{00/11}} \log f_3(y_i) + \sum_{i=1}^{n_{00/10}} \log f_3(y_i) + \sum_{i=1}^{n_{00/00}} \log f_3(y_i) \tag{4.6}$$

for the triallelic model assuming that haplotypes [11] and [10] are the first and second risk haplotypes, respectively,

$$\log L(\boldsymbol{\Omega}_{q_Q}|\mathbf{y},\mathbf{S},\hat{\boldsymbol{\Omega}}_p)$$
$$= \sum_{i=1}^{n_{11/11}} \log f_1(y_i) + \sum_{i=1}^{n_{11/10}} \log f_5(y_i) + \sum_{i=1}^{n_{11/00}} \log f_2(y_i)$$
$$+ \sum_{i=1}^{n_{10/11}} \log f_6(y_i) + \sum_{i=1}^{n_{10/10}} \log[\phi f_8(y_i) + (1-\phi)f_7(y_i)] + \sum_{i=1}^{n_{10/00}} \log f_9(y_i)$$
$$+ \sum_{i=1}^{n_{00/11}} \log f_3(y_i) + \sum_{i=1}^{n_{00/10}} \log f_{10}(y_i) + \sum_{i=1}^{n_{00/00}} \log f_4(y_i) \tag{4.7}$$

for the quadriallelic model assuming that haplotypes [11], [10], and [01] are the first, second and third risk haplotypes, respectively, with $f_j(y_i)$ being a normal distribution density function of composite diplotype j with mean μ_j and variance σ^2.

We have shown that maximizing $L(\Omega_p, \Omega_q | \mathbf{y}, \mathbf{S})$ in equation (4.3) is equivalent to individually maximizing $\log L(\Omega_p | \mathbf{S})$ in equation (4.3) and $\log L(\Omega_q | \mathbf{y}, \mathbf{S}, \hat{\Omega}_p)$ in equation (4.5), (4.6), or (4.7).

4.2.1   The EM Algorithm

A closed-form solution for the EM algorithm has been derived to estimate the unknown parameters that maximize the likelihoods. The estimates of haplotype frequencies are based on the log-likelihood function $L(\Omega_p | \mathbf{S})$, whereas the estimates of genotypic values of composite diplotypes and the residual variance are based on the log-likelihood function $L(\Omega_q | \mathbf{y}, \mathbf{S}, \hat{\Omega}_p)$. These two different types of parameters can be estimated using a two-stage hierarchical EM algorithm. Liu et al. (2004) provided a detailed implementation of the algorithm.

4.2.2   Model Selection

The formulation of likelihoods (4.5), (4.6), and (4.7) is based on the assumption that one or more haplotypes are risk haplotypes for the biallelic, triallelic, and quadriallelic model. However, a real risk haplotype under each of these models is unknown from raw data $(\mathbf{y}, \mathbf{S})$. Also, we are uncertain about the optimal number of risk haplotypes. An additional step for the choice of the most likely risk haplotypes and their number should be implemented. The simplest way to do so is to calculate and compare the likelihood values within the model by assuming that any one or more of the four haplotypes can be a risk haplotype, and AIC or BIC among the models by assuming different numbers of risk haplotypes (Burnham and Anderson, 2002). Thus, we obtain possible likelihood values and AIC or BIC as follows:

Model	No.	Likelihood	AIC	BIC
Biallelic	B_l	$\log L_{B_l}(\hat{\Omega}_p, \hat{\Omega}_{qB} | \mathbf{y}, \mathbf{S})$	A_{B_l}	B_{B_l}
Triallelic	T_l	$\log L_{T_l}(\hat{\Omega}_p, \hat{\Omega}_{qT} | \mathbf{y}, \mathbf{S})$	A_{T_l}	B_{T_l}
Quadriallelic	Q	$\log L_Q(\hat{\Omega}_p, \hat{\Omega}_{qQ} | \mathbf{y}, \mathbf{S})$	A_Q	B_Q

The largest likelihood and the smallest AIC or BIC value calculated is thought to correspond to the most likely risk haplotypes and their optimal number.

4.2.3   Hypothesis Tests

The genetic architecture of a quantitative trait is characterized by quantitative genetic parameters (including haplotype effects and the mode of their inheritance). The model proposed provides a meaningful way for estimating the genetic architecture of a trait. The estimated genotypic values for the composite diplotypes can be used to estimate additive and dominance genetic effects of haplotypes by using the formulas shown in Table 4.6.

TABLE 4.6
Additive and dominance effects under different models.

Model	Additive	Dominance
Biallelic	$a = (\mu_1 - \mu_3)/2$	$d = \mu_2 - (\mu_1 + \mu_3)/2$
Triallelic	$a_1 = [2\mu_2 - (\mu_1 + \mu_3)]/3$ $a_2 = [2\mu_1 - (\mu_2 + \mu_3)]/3$	$d_{12} = \mu_4 - (\mu_1 + \mu_2)/2$ $d_{10} = \mu_5 - (\mu_1 + \mu_3)/2$ $d_{20} = \mu_6 - (\mu_2 + \mu_3)/2$
Quadriallelic	$a_1 = [3\mu_1 - (\mu_2 + \mu_3 + \mu_4)]/4$ $a_2 = [3\mu_2 - (\mu_1 + \mu_3 + \mu_4)]/4$ $a_3 = [3\mu_3 - (\mu_1 + \mu_2 + \mu_4)]/4$	$d_{12} = \mu_5 - (\mu_1 + \mu_2)/2$ $d_{13} = \mu_6 - (\mu_1 + \mu_3)/2$ $d_{23} = \mu_7 - (\mu_2 + \mu_3)/2$ $d_{10} = \mu_8 - (\mu_1 + \mu_4)/2$ $d_{20} = \mu_9 - (\mu_2 + \mu_4)/2$ $d_{30} = \mu_{10} - (\mu_3 + \mu_4)/2$

The additive and dominance effects under different models can be tested by formulating the null hypothesis that the effect being tested is equal. The estimates of the parameters under the null hypotheses can be obtained with the same EM algorithm derived for the alternative hypotheses but with a constraint of the tested effect equal to zero. The log-likelihood ratio test statistics for each hypothesis is thought to asymptotically follow a χ^2-distributed with the degree of freedom equal to the difference of the numbers of the parameters being tested under the null and alternative hypotheses.

4.3 Three-SNP Haplotyping

Li et al. (2006b) constructed a conceptual framework and statistical algorithm for haplotyping a quantitative trait with three SNPs. For a set of three SNPs, there are eight different haplotypes, among which it is possible to have one to seven risk haplotypes. The biallelic model specifies one risk haplotype which may be composed of one (8 cases), two (24 cases), three (56), or four haplotypes (170). The triallelic, quadrialleli, pentaallelic, hexaallelic, septemallelic, and octoallelic models contain 28, 56, 170, 56, 24, and 8 cases, respectively. It can be seen that the model selection procedure to determine the optimal number and combination of risk haplotypes will become exponentially more complicated when the number of SNPs increases.

Example 4.1
Wu et al. (2007b) performed a simulation study to show how quantitative genetic parameters can well be estimated under different sample sizes and heritabilities when different numbers of risk haplotypes are assumed. Because the number and combina-

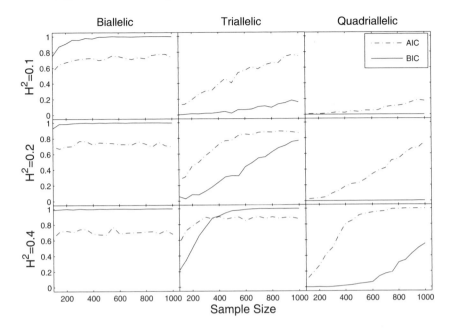

FIGURE 4.1
Power to detect correct risk haplotypes from the data simulated by a biallelic, trial-lelic, and quadriallelic model, respectively, under different heritabilities and sample sizes. Model selection criteria are based on AIC and BIC. Adapted from Wu et al. (2007b).

tion of risk haplotypes that govern a phenotypic trait is unknown for a practical data set, their simulation also includes power analysis for determining risk haplotypes correctly (see Fig. 4.1). Basic conclusions are summarized below.

(1) For the data set simulated with the biallelic model, the additive and dominance effects can be precisely estimated even with a heritability of 0.1 and a sample size of 100.

(2) The data set simulated under the triallelic model contains two additive effects and three dominance effects. A sample size of 100 is adequate for precise esti-mates of the additive effects even for a low heritability (0.1), but the reasonable estimates of the dominance effects need increasing sample size (400 or more) if the heritability is 0.1 (Table 3). For a high heritability (0.4), a small sample size (100) can provide relatively precise estimates of the dominance effects.

(3) For the data set simulated with the quadriallelic model, three additive effects and six dominance effects are included. Still, a low sample size (100) can provide very good estimates of the additive effects even for a low heritability.

To reasonably estimate the dominance effects, we need a large sample size (1000) for the heritability of 0.1 or a moderately large sample size (400) for the heritability of 0.4 (Table 4).

(4) For the data simulated under the biallelic model, a correct risk haplotype can well be determined with a sample size of 200 even when the heritability of the trait is modest (0.1). In this case in which a small number of genetic parameters are included, the BIC performs better than the AIC. For a data set simulated under the triallelic model, the power of haplotype detection reduces considerably, compared with the data set simulated by the biallelic model. If the heritability of a trait is as low as 0.1, about 1000 subjects are needed to achieve the power of 0.8. With the heritability increasing to 0.2 or 0.4, the same power needs about 600 or 300 subjects, respectively. It is interesting to note that the AIC performs better than the BIC when the heritability is low (0.1 or 0.2), whereas the two criteria perform similarly when the heritability is high (0.4).

⬜

4.4   Haplotyping in a Non-Equilibrium Population

Current population genetic studies consider that a population is at Hardy–Weinberg equilibrium (HWE) if individual genes studied are thought to be at HWE through a statistical test. However, this is basically not true because the HWE for individual genes is not a representation of the HWE for haplotypes. In this section, we will show how the results from a HWE analysis are misleading in the inference of haplotype distribution when the population violates this assumption at the haplotype level.

Consider a pair of SNPs each with alleles 1 and 0. These SNPs form 4 haplotypes, [11], [10], [01], and [00], 10 diplotypes and 9 observed genotypes. Let p_{11}, p_{10}, p_{01}, and p_{00}, be the haplotype frequencies in a natural population from which a random sample of size n is drawn. The diplotype frequencies are expressed as a function of haplotype frequencies by assuming that the population is at Hardy–Weinberg disequilibrium (HWD) or HWE, respectively (Table 4.7). When HWD is assumed, we will define six linkage disequilibria to describe the diplotype frequencies. These include the linkage disequilibria D_1, D_2, D_3, D_4, D_5, and D_6 due to non-random associations between haplotypes [11] and [10], between haplotypes [11] and [01], between haplotypes [11] and [00], between haplotypes [10] and [01], between haplotypes [10] and [00], and between haplotypes [01] and [00], respectively. The HWD defined here is thought to happen at the haplotype level.

At HWD, we have three independent haplotype frequencies and six linkage disequilibria, summing to nine parameters. However, 10 diplotypes cannot be totally distinguished from each other with 9 observed genotypes because the double het-

TABLE 4.7

Genotypes, diplotypes, and diplotype frequencies for two SNPs under Hardy–Weinberg disequilibrium (HWD) and Hardy–Weinberg equilibrium (HWE).

| Genotype | Diplotype | Diplotype Frequency | | Size |
		HWD	HWE	
11/11	[11][11]	$p_{11}^2 + D_1 + D_2 + D_3$	p_{11}^2	$n_{11/11}$
11/10	[11][10]	$2p_{11}p_{10} - 2D_1$	$2p_{11}p_{10}$	$n_{11/10}$
11/00	[10][10]	$p_{10}^2 + D_1 + D_4 + D_5$	p_{10}^2	$n_{11/00}$
10/11	[11][01]	$2p_{11}p_{01} - 2D_2$	$2p_{11}p_{01}$	$n_{10/11}$
10/10	[11][00]	$2p_{11}p_{00} - 2D_3$	$2p_{11}p_{00}$	$n_{1|10/10}$
	[10][01]	$2p_{10}p_{01} - 2D_4$	$2p_{10}p_{01}$	$n_{2|10/10}$
10/00	[10][00]	$2p_{10}p_{00} - 2D_5$	$2p_{10}p_{00}$	$n_{10/00}$
00/11	[01][01]	$p_{01}^2 + D_2 + D_4 + D_6$	p_{01}^2	$n_{00/11}$
00/10	[01][00]	$2p_{01}p_{00} - 2D_6$	$2p_{01}p_{00}$	$n_{00/10}$
00/00	[00][00]	$p_{00}^2 + D_3 + D_5 + D_6$	p_{00}^2	$n_{00/00}$

Note: A non-random association (D_1, D_2, D_3, D_4, D_5, and D_6) between any two different haplotypes sums to zero across the two homozygous diplotypes composed of each of the two haplotypes and the heterozygous diplotype composed of these two different diplotypes.

erozygote 10/10 contains 2 possible diplotypes [11][00] and [10][01]. Thus, we will have only eight degrees of freedom which is inadequate to solve nine parameters. A simulation study will be performed to study the results from an HWD-based simulated data set analyzed by the currently used HWE model.

Given certain values of the haplotype frequencies and six linkage disequilibria as well as a sample size of n, a multinomial likelihood is used to simulate the SNP data based on diplotype frequencies. The simulated sizes of each diplotype are listed in Table 4.7. We simulate two diplotypes, [11][00] and [10][01], as if they can be distinguished from each other. The simulated data set is analyzed by the HWE model. But this analysis is based on a summed size of two diplotypes, [11][00] and [10][01], which mimics a real situation. The EM algorithm can be used to estimate four haplotype frequencies (see Section 2.3). Table 4.8 gives the results from the simulation strategy described above under different schemes. Schemes 1 to 5 assume different haplotype frequencies, with the HWD among haplotypes and linkage disequilibrium within haplotypes occurring at different rates. Scheme 6 assumes the same haplotype frequency, with a full occurrence of HWD and linkage disequilibrium. The results are summarized as follows:

(1) If a population deviates from the HWE equilibrium at the haplotype level, currently available models that were derived under the HWE assumption provide

TABLE 4.8

MLEs of population genetic parameters and their standard errors (SE) from HWD-based simulated data estimated with 200 simulation replicates by using currently available models.

Scheme		p_{11}	p_{10}	p_{01}	p_{00}	D_1	D_2	D_3	D_4	D_5	D_6
1	True	0.350	0.150	0.150	0.350	0.05	0.05	0.05	0.05	0.05	0.05
	MLE	0.309	0.190	0.190	0.311						
	SE	0.033	0.028	0.029	0.033						
2	True	0.350	0.150	0.150	0.350	0.05	0.05	0.05	0.05	0.05	0
	MLE	0.310	0.190	0.190	0.311						
	SE	0.033	0.029	0.026	0.031						
3	True	0.350	0.150	0.150	0.350	0.05	0.05	0.05	0	0	0
	MLE	0.360	0.141	0.139	0.360						
	SE	0.033	0.023	0.023	0.028						
4	True	0.350	0.150	0.150	0.350	0.05	0	0	0	0	0
	MLE	0.352	0.150	0.150	0.348						
	SE	0.028	0.024	0.020	0.026						
5	True	0.350	0.150	0.150	0.350	0	0	0	0	0	0
	MLE	0.350	0.151	0.149	0.350						
	SE	0.025	0.020	0.020	0.025						
6	True	0.250	0.250	0.250	0.250	0.05	0.05	0.05	0.05	0.05	0.05
	MLE	0.250	0.249	0.251	0.250						
	SE	0.030	0.030	0.031	0.030						

misleading estimates of haplotype frequencies for the population;

(2) More misleading estimates of the parameters are yielded when non-random associations among different haplotypes are more common and/or stronger (comparing schemes 1 to 5). With more among-haplotype disequilibria, the precision and accuracy of parameter estimation reduce;

(3) Parameter estimation is also affected by the occurrence and size linkage disequilibrium within haplotypes. When there is a linkage disequilibrium within haplotypes (scheme 1), the estimation of haplotype frequencies is more biased and imprecise as compared to the situation in which such a disequilibrium does not exist (scheme 6).

In a zygotic linkage disequilibrium analysis with a complex canine pedigree, Liu et al. (2006) showed that the models that consider the non-equilibrium nature of a population cover those derived from the HWE assumption. The results from such general models will provide more meaningful scientific guidance about candidate gene or whole-genome association studies. For a practical data set in which the

HWE assumption may, or may not, be a case, the zygotic model will always provide a reasonable estimation and test of linkage disequilibria. But this will not be true for the gametic model when the HWE assumption is violated.

4.5 Prospects

To deal with multiple risk haplotypes, an issue arises naturally about the selection of most likely risk haplotypes from a pool of haplotypes. This will include the optimal number of risk haplotypes and their combination that provide a best fit to the given data. We implemented model selection procedures into the test process of haplotype diversity and effects with two commonly used criteria, AIC and BIC. Extensive simulation studies were performed to investigate the statistical properties of the model and its utilization. Given a real data set, we do not know about the type and number of risk haplotypes. But these can be estimated with model selection by assuming different types of genetic models, biallelic (one risk haplotype), triallelic (two risk haplotypes), and quadriallelic (three risk haplotypes). Simulation studies with two-SNP haplotypes provide a table of model selection approaches (Tables 2–4) to detect most likely risk haplotypes hidden in a genetic association data set based on a range of sample size and heritability as well as the types of genetic models.

The human genome contains millions of SNPs distributed over 23 pairs of chromosomes (Altshuler et al., 2005). However, these SNPs were observed to locate in different haplotype blocks of the human genome (Patil et al., 2001; Terwilliger and Hiekkalinna, 2006). For a given block, there are a particular number of representative SNPs or htSNPs that uniquely identify the common haplotypes in this block or QTN. Several algorithms have been developed to identify a minimal subset of htSNPs that can characterize the most common haplotypes (Zhang et al., 2002; Sebastiani et al., 2003; Eyheramendy et al., 2007). The idea given in this article can be used to find risk haplotypes of these htSNPs by modelling an arbitrary number of SNPs (Li et al., 2006b), and extended to detect haplotype-haplotype interactions (Lin and Wu, 2006b), haplotype-environment interactions, parent-of-origin effects of haplotypes in genetic association studies, and haplotypes regulating pharmacodynamic reactions of drugs (Lin et al., 2005). Although these works will be computationally expensive, it should not be computationally prohibitive if combinatorial mathematics, graphical models, and machine learning are incorporated into closed forms of parameter estimation. With detailed extensions that take into account more realistic biological and genetic problems, our model may provide an efficient solution to the growing need for haplotype data collection and association studies.

5

Basic Principle of Functional Mapping

A real biological existence includes two aspects, static and dynamic. From a philosophical viewpoint, static existence is merely a phase of dynamic existence. On the other hand, dynamic existence is composed of infinite static existence. Dynamic existence becomes static upon its termination, while static existence will be dynamic when it is propelled by a new force. From a philosophical standpoint, dynamics is absolute, whereas statics is relative. Genetic studies of static biological existence are capable of centering on its specific features, but, in most cases, they are powerless to elucidate an entire process of a biological entity. Thus, while we have focused on genetic models and computational algorithms for haplotyping DNA sequence variants that control a static trait in the preceding chapters, it is now a time to relate genetic haplotyping with developmental and dynamic processes of trait formation.

In the real world, many biological entities, such as growth, HIV load trajectories, biological clock, or drug response, can be better described by a dynamic process. In statistics, they often arise as curves. More recently, a collection of statistical methods implemented with growth model theories have been proposed to map quantitative trait loci (QTLs) that govern the pattern and shape of growth curves using molecular linkage maps (Ma et al., 2002; Wu and Lin, 2006). These models, called *functional mapping*, can be used to study the genetic architecture of other dynamic processes in various mapping populations. While mathematical models are increasingly used in modern biology to understand how phenotypes emerge from the collective interactions of molecular and cellular components, functional mapping integrated by these mathematical models will find its immediate application to quantify the dynamic change of the genetic control of a complex biological system. Dynamic QTLs for growth forms have been detected with functional mapping in poplar trees (Ma et al., 2002; Wu et al., 2004b,c,d), rice (Zhao et al., 2004c), mice (Zhao et al., 2004a; Wu et al., 2005), and soybeans (Li et al., 2007).

In this chapter, we will describe the basic statistical and biological principle of functional mapping for growth curves, outline the algorithmic procedure for computing functional mapping, and pinpoint the advantages of functional mapping over conventional interval mapping in biological relevance and statistical power through the analysis of live examples. We show how a variety of biologically meaningful hypotheses can be asked and tested within the framework of functional mapping. Functional mapping is a new area in which there are still a number of open biological, statistical and algorithmic questions that deserve further investigations. We will provide some future directions for functional mapping.

5.1 Dynamic Genetic Control

Many traits, such as growth, disease (e.g., AIDS) progression, and drug response, are dynamic in nature and should be measured in a longitudinal way. To obtain a clear picture of the genetic control of these traits, it is crucial to characterize the change pattern of genetic expression during development. Figure 5.1 illustrates four representative patterns of time-dependent genetic effects triggered by different genes, explained as follows:

(1) Long genes. These genes are permanently expressed, which gives rise to two parallel growth curves in which one genotype is consistently better than the other throughout the entire growth process (A). The expression of this gene is not affected by development, and therefore shows no interaction with age.

(2) Early genes. These genes are expressed at early developmental stages and are switched off after a particular age. The two genotypes have different growth at early stages, but tend to converge at later stages (B). This gene displays interactions with age because there is a change of variance of the genetic effects during development.

(3) Late genes. These genes remain silent during early stages and are expressed only after a particular age. The two genotypes display similar growth at early stages, but tend to diverge at later stages (C). Analogous to Pattern 2, there is a genotype $\times$ age interaction in this case, operating with the conditional neutrality mechanism.

(4) Inverse genes. One genotype performs better than the other at the early stage of growth, but this changes at the later stage (D). This gene displays inverse effects at a particular age. Genotype $\times$ age interactions occur because there is a change of the direction of the genetic effects during development.

For each of these patterns (Fig. 5.1), curve parameters for developmental trajectories can be estimated and tested for individual genotypes with a genetic mapping approach. If different genotypes at a given gene correspond to different trajectories, this gene must affect differentiation of this trait. Thus, by estimating the curve parameters that define the trait trajectory of each genotype and testing the differences in these parameters among genotypes, we can determine whether a dynamic gene exists and how it affects the formation and expression of a trait during development.

Elucidating the relationship between genetic control and development for dynamic traits is statistically challenging in terms of the high-dimensionality and complexity of its underlying structure. The statistical genetics group at the University of Florida has for the first time proposed a series of statistical models, called functional mapping, for characterizing developmental and dynamic QTLs in time course (Ma et al., 2002; Wu and Lin, 2006). Functional mapping based on genetic linkage maps has

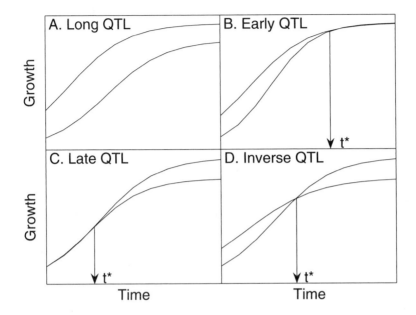

FIGURE 5.1
Four representative patterns for the genetic control of growth curves by a dynamic gene. Each curve is presented by a different genotype.

been proposed to genome-wide map specific QTLs that determine the developmental pattern of a complex trait. Functional mapping shows remarkable advantages in asking fundamental biological questions about developmental mechanisms and addressing these questions at interplay between genetic action and developmental pathways (Wu et al., 2004b). More specifically, functional mapping can determine when a QTL turns on and then turns off, how long this QTL is expressed in a time course, and how this QTL for growth and development also affects other biological events and processes.

5.2 Structure of Functional Mapping

The basic principle of functional mapping is to express the expected values of a genotype for a QTL at different time points as a continuous growth function with respect to time t. Under this principle, the parameters that describe the shape of growth curves, rather than the genotypic means at individual time points as determined in traditional mapping strategies, are estimated within a maximum likelihood framework. Also unlike traditional mapping strategies, functional mapping estimates the parameters that model the structure of a longitudinal covariance among multiple different time points and, therefore, largely reduces the number of parameters being estimated for variances and covariances, especially when the number of time points is large.

5.2.1   Mixture Model

Functional mapping is, in spirit, a statistical problem of jointly modeling mean-covariance structures in longitudinal studies, an area that has recently received considerable interest in the statistical literature (Pourahmadi, 1999, 2000; Pan and Mackenzie, 2003; Wu and Pourahmadi, 2003; Diggle et al., 2002; Ramsay and Silverman, 2002; Ferraty and Vieu, 2006). However, in contrast to general longitudinal modeling, functional mapping integrates the estimation and test process of its underlying parameters within a mixture-based likelihood framework. Each mixture component in the likelihood model is given a particular biological rationale. For a finite mixture model, each observation is assumed to have arisen from one of a known or unknown number of components, each component being modeled by a density from the parametric family. Assuming that there are J QTL genotypes contributing to a longitudinal trait measured at T time points (denoted by $\mathbf{y}$), this mixture model is expressed as

$$\mathbf{y} \sim p(\mathbf{y}|\boldsymbol{\omega}, \boldsymbol{\Omega}_u, \boldsymbol{\Omega}_v) = \omega_{1|i} f_1(\mathbf{y}; \boldsymbol{\Omega}_{u_1}, \boldsymbol{\Omega}_v) + \cdots + \omega_{J|i} f_J(\mathbf{y}; \boldsymbol{\Omega}_{u_J}, \boldsymbol{\Omega}_v), \quad (5.1)$$

where $\omega = (\omega_{1|i}, \cdots, \omega_{J|i})$ are the proportions of mixture components (defined by QTL genotypes) which are constrained to be non-negative and sum to unity, $\boldsymbol{\Omega}_u = (\boldsymbol{\Omega}_{u_1}, \cdots, \boldsymbol{\Omega}_{u_J})$ is a vector that contains the parameters specific to component (i.e., QTL genotype) j, and $\boldsymbol{\Omega}_v$ includes the parameters common to all components.

We use the multivariate normal distribution to model the density of each component, and for individual i the density is expressed as

$$f(\mathbf{y}_i; \boldsymbol{\Omega}_{u_j}, \boldsymbol{\Omega}_v) = \frac{1}{(2\pi)^{T/2}|\boldsymbol{\Sigma}|^{1/2}} \exp\left[-\frac{1}{2}(\mathbf{y}_i - \mathbf{u}_j)\boldsymbol{\Sigma}^{-1}(\mathbf{y}_i - \mathbf{u}_j)'\right], \quad (5.2)$$

where $\mathbf{y}_i = [y_i(1), \cdots, y_i(T)]$ is a vector of observation measured at T time points and $\mathbf{u}_j = [u_j(1), \cdots, u_j(T)]$ is a vector of expected values for QTL genotype j at different points. Ignoring the effects of covariates, the relationship between the observation and expected value at a particular time t can be described by a regression model, that is

$$y_i(t) = \sum_{j=1}^{J} \xi_{ij} u_j(t) + e_i(t), \quad (5.3)$$

where ξ_{ij} is the indicator variable denoted as 1 if a QTL genotype j is considered for individual i and 0 otherwise; $e_i(t)$ is the residual error (i.e., the accumulative effect of polygenes and errors) that is iid (independently and identically distributed) normal with mean zero and variance $\sigma^2(t)$. The errors at two different time points, t_1 and t_2, are correlated with covariance $\sigma(t_1, t_2)$.

The mixture proportions $(\omega_{1|i}, \cdots, \omega_{J|i})$ in (5.1) can be viewed as the prior probabilities of QTL genotypes in the mapping population. These proportions can be derived as the conditional probabilities of a QTL genotype given the marker genotype of individual i. The conditional probabilities are expressed in terms of the cosegregation between the markers and QTL. For a structured pedigree, they are expressed in terms of the recombination fractions, whereas for a natural population, they can be expressed in terms of linkage disequilibria.

To estimate the parameters contained within the mixture model (5.1) constructed by repeated measurements at multiple time points, one can extend the traditional interval mapping approach (Lander and Botstein, 1989) to accommodate the multivariate nature of time-dependent traits (Jiang and Zeng, 1995). However, this extension is limited in three aspects:

(1) Expected values of different QTL genotypes at all time points and all elements in the matrix Σ need to be estimated, resulting in substantial computational difficulties when the vector and matrix dimensions are large;

(2) The result from this approach may not be biologically meaningful because the underlying biological principle for growth is not incorporated;

(3) This approach cannot be well deployed in a practical scheme because of (2). Thus, some biologically interesting questions cannot be asked and answered.

5.2.2 Modeling the Mean-Covariance Structure

Functional mapping will model the multivariate normal distribution (5.2). This includes two tasks, one being the approximation of the time-dependent mean vector for each QTL genotype ($\mathbf{u}_j$) using the growth equation and the second being the structuring of the residual covariance matrix (Σ) according to the pattern of time series autocorrelation. The first task includes the selection of biologically sensible growth equations, which allows us to estimate the parameters that define the shape of growth curves and further make any relevant biological hypothesis tests. The second task needs appropriate statistical processes to model the structure and pattern of the covariance matrix, making functional mapping efficient and powerful for QTL detection.

5.2.2.1 Modeling the Mean Vector

The time-dependent expected values of QTL genotype j can be modeled by a growth equation. Consider a simple logistic growth equation. We then have

$$g_j(t) = \frac{a_j}{1 + b_j e^{-r_j t}} \qquad (5.4)$$

which is specified by a set of curve parameters arrayed in $\Omega_{u_j} = (a_j, b_j, r_j)$. The overall form of the growth curve of QTL genotype j is determined by the curve parameters contained in Ω_{u_j}. If different genotypes at a putative QTL have different combinations of these parameters, this implies that this QTL plays a role in governing the differentiation of growth trajectories. Thus, by testing for the difference of Ω_{u_j} among different genotypes, we can determine whether there exists a specific QTL that confers an effect on growth curves.

With equation (5.4), the genotypic mean vector for QTL genotype j from time 1 to T in the multivariate normal density function is specified by

$$\mathbf{u}_j = [u_j(1), u_j(2), \cdots, u_j(T)]$$
$$= \left[\frac{a_j}{1 + b_j e^{-r_j}}, \frac{a_j}{1 + b_j e^{-2r_j}}, \cdots, \frac{a_j}{1 + b_j e^{-r_j T}} \right]. \qquad (5.5)$$

The time-dependent growth curves for different genotypes, $u_j(t)$, can be used to estimate the dynamic changes of various genetic effects, including the additive, dominant, and epistatic effects as well as the interaction between these effects and environmental factors. For example, an F_2 population containing three genotypes at a QTL, coded as 2 for QQ, 1 for Qq, and 0 for qq, allow for the estimates of the time-dependent additive effect,

$$a(t) = \frac{1}{2}[u_2(t) - u_0(t)],$$

and dominant effects,

$$d(t) = \mu_1(t) - \frac{1}{2}[u_2(t) + u_0(t)],$$

during growth trajectories. The time-dependent epistatic and genotype × environment effects can be characterized in a similar way if a multi-QTL model is assumed or if an experimental design with multiple environments is used.

As discussed above, modeling the mean vector by the growth curve has biological merits in terms of biological significance of the growth equation and the implementation of biologically meaningful hypotheses (see below). The modeling of the mean vector is also statistically advantageous. As compared to conventional multivariate interval mapping for studying multi-point growth data, functional mapping estimates much fewer parameters. Below is the comparison of the number of parameters being estimated between the two approaches:

Number of time points	Backcross		F_2	
	Conventional	Funtional	Conventional	Funtional
3	$2 \times 3 = 6$	$2 \times 3 = 6$	$3 \times 3 = 9$	$3 \times 3 = 9$
5	$2 \times 5 = 10$	$2 \times 3 = 6$	$3 \times 5 = 10$	$3 \times 3 = 9$
10	$2 \times 10 = 20$	$2 \times 3 = 6$	$3 \times 10 = 30$	$3 \times 3 = 9$
20	$2 \times 20 = 40$	$2 \times 3 = 6$	$3 \times 20 = 60$	$3 \times 3 = 9$
50	$2 \times 50 = 100$	$2 \times 3 = 6$	$3 \times 50 = 150$	$3 \times 3 = 9$

It can be seen that the conventional approach will quickly become powerless when the number of time points increases. However, an increasing number of time points does not affect the estimation of parameters for functional mapping.

5.2.2.2 Modeling the Covariance Matrix

In statistics, theories and methods have been available to model the structure of co-variances between measurements repeatedly made at a series of time points (Diggle et al., 2002). Because of elegant mathematical and statistical properties, the autore-gressive process has been widely used for studies of longitudinal data measurements. The first-order autoregressive (AR(1)) model has been successfully applied to model the structure of the within-subject covariance matrix for functional mapping. The AR(1) model is based on two simplified assumptions, i.e., variance stationarity–the residual variance (σ^2) is unchanged over time points, and covariance stationarity-the correlation between different measurements decreases proportionally (in ρ) with increased time interval. Mathematically, the AR(1) is described as

$$\sigma^2(1) = \cdots = \sigma^2(T) = \sigma^2$$

for the variance, and

$$\sigma(t_1, t_2) = \sigma^2 \rho^{|t_2 - t_1|}$$

for the covariance between any two time intervals t_1 and t_2, where $0 < \rho < 1$ is the proportion parameter with which the correlation decays with time lag. The parameters that model the structure of the covariance matrix are arrayed in $\Omega_v = (\rho, \sigma^2)$.

When the residual covariance matrix (Σ) is modelled by the AR(1) model, the closed forms can be derived for its inverse and determinant, which facilitate model computing and parameter estimating. The inverse Σ^{-1} is a tridiagonal symmetric matrix, whose diagonal elements are

$$\frac{(1, 1+\rho^2, 1+\rho^2, \ldots, 1+\rho^2, 1+\rho^2, 1)^{\mathrm{T}}}{\sigma^2(1-\rho^2)}$$

and second diagonal elements are all

$$\frac{-\rho}{\sigma^2(1-\rho^2)}.$$

The determinant of the matrix is derived as

$$|\Sigma| = [\sigma^2(1-\rho^2)]^{T-1}\sigma^2.$$

Let $\mathbf{z}_{i|j} = [z_i(1), \ldots, z_i(T)] = \mathbf{y}_i - \mathbf{u}_j, (j = 0, \ldots, J)$, then we have

$$\mathbf{z}_i \Sigma^{-1} \mathbf{z}_i = \frac{\sum_{t=1}^{T-1}(z_{i|j}(t) - \rho z_{i|j}(t+1))^2 + (1-\rho^2)z_{i|j}^2(T)}{\sigma^2(1-\rho^2)}.$$

5.3   Estimation of Functional Mapping

5.3.1   Likelihood

Assume that the mapping population used is derived from an experimental cross, so that the mixture proportions (i.e., the conditional probabilities of QTL genotypes) are specified by a marker-QTL linkage parameter (θ). Let $\Omega = (\theta, \Omega_q)$ with $\Omega_q = (\{\Omega_{u_j}\}_{j=1}^{J}, \Omega_v)$. The likelihood of the unknown vector for the mapping population of size n with T-dimensional measurements can be constructed by a multivariate mixture model, i.e.,

$$L(\Omega) = \prod_{i=1}^{n} \left[\sum_{j=0}^{J} \omega_{j|i} f_j(\mathbf{y}_i) \right], \tag{5.6}$$

with $J = 1$ for the backcross, or 2 for the F_2, The maximum likelihood estimates (MLEs) of the unknown parameters for a QTL can be computed by implementing the EM algorithm.

The log-likelihood is given by

$$\log L(\Omega) = \sum_{i=1}^{n} \log \left[\sum_{j=0}^{J} \omega_{j|i} f_j(\mathbf{y}_i) \right], \tag{5.7}$$

with derivative with respect to an element Ω_ς in the unknown vector

$$\frac{\partial}{\partial \Omega_\varsigma} \log L(\Omega) = \sum_{i=1}^{n} \sum_{j=0}^{J} \frac{\frac{\partial \omega_{j|i}}{\partial \theta} f_j(\mathbf{y}_i) + \omega_{j|i} \frac{\partial}{\partial \Omega_q} f_j(\mathbf{y}_i)}{\sum_{j'=0}^{J} \omega_{j'|i} f_{j'}(\mathbf{y}_i)}$$

$$= \sum_{i=1}^{n} \sum_{j=0}^{J} \frac{\omega_{j|i} f_j(\mathbf{y}_i)}{\sum_{j'=0}^{J} \omega_{j'|i} f_{j'}(\mathbf{y}_i)} \left[\frac{\partial \omega_{j|i}}{\partial \theta} \frac{1}{\omega_{j|i}} + \frac{\partial}{\partial \Omega_q} \log f_j(\mathbf{y}_i) \right]$$

$$= \sum_{i=1}^{n} \sum_{j=0}^{J} \omega_{j|i} \left[\frac{\partial \omega_{j|i}}{\partial \theta} \frac{1}{\omega_{j|i}} + \frac{\partial}{\partial \Omega_q} \log f_j(\mathbf{y}_i) \right]$$

where we define

$$P_{j|i} = \frac{\omega_{j|i} f_j(\mathbf{y}_i)}{\sum_{j'=0}^{J} \omega_{j'|i} f_{j'}(\mathbf{y}_i)}, \tag{5.8}$$

which could be thought of as a posterior probability that progeny i has QTL genotype j. Conditional on $\mathbf{P} = \{P_{j|i}; i = 1, \ldots, n; j = 0, \ldots, J\}$, we solve for

$$\frac{\partial}{\partial \Omega_\varsigma} \log L(\Omega) = 0. \tag{5.9}$$

Below, we derived a series of log-likelihood equations to solve the parameters that model the mean-covariance structure.

5.3.2 Algorithm

If a simple growth equation (5.4) and the AR(1) model are used, we have $\Omega_{u_j} = (a_j, b_j, r_j)$ and $\Omega_v = (\rho, \sigma^2)$. In this case, the general form of the log-likelihood equations (5.9) is derived as follows:

$$a'_j = \frac{\sum_{i=1}^{n} P_{j|i} \left[\sum_{t=1}^{T-1} \delta_i^y(t) \left(\frac{1}{g_j(t)} - \frac{\rho}{g_j(t+1)} \right) + \frac{(1-\rho^2) y_i(T)}{g_j(T)} \right]}{\sum_{i=1}^{n} P_{j|i} \left[\sum_{t=1}^{T-1} \left(\frac{1}{g_j(t)} - \frac{\rho}{g_j(t+1)} \right)^2 + \frac{1-\rho^2}{g_j^2(T)} \right]},$$

$$b'_j = \frac{b_j \sum_{i=1}^{n} P_{j|i} \left[\sum_{t=1}^{T-1} \delta_j^u(t) \left(\frac{e^{-r_j t}}{g_j^2(t)} - \frac{\rho e^{-r_j(t+1)}}{g_j^2(t+1)} \right) + \frac{(1-\rho^2) a_j e^{-r_j T}}{g_j^2(T)} \right]}{\sum_{i=1}^{n} P_{j|i} \sum_{t=1}^{T-1} \left[\delta_i^y(t) \left(\frac{e^{-r_j t}}{g_j^{-2}(t)} - \frac{\rho e^{-r_j(t+1)}}{g_j^2(t+1)} \right) + \frac{(1-\rho^2) y_i(T) e^{-r_j T}}{g_j^2(T)} \right]},$$

$$r'_j =$$

$$\frac{r_j \sum_{i=1}^{n} P_{j|i} \left[\sum_{t=1}^{T-1} \delta_i^y(t) \left(\frac{t e^{-r_j t}}{g_j^2(t)} - \frac{(t+1)\rho e^{-r_j(t+1)}}{g_j^2(t+1)} \right) + \frac{(1-\rho^2) T y_i(T) e^{-r_j T}}{g_j^2(T)} \right]}{\sum_{i=1}^{n} P_{j|i} \left[\sum_{t=1}^{T-1} \delta_j^u(t) \left(\frac{t e^{-r_j t}}{g_j^2(t)} - \frac{(t+1)\rho e^{-r_j(t+1)}}{g_j^2(t+1)} \right) + \frac{(1-\rho^2) T a_j e^{-r_j T}}{(g_j^2(T)} \right]},$$

$$\sigma^{2'} = \frac{1}{nT(1-\rho^2)} \sum_{j=1}^{2} \sum_{i=1}^{n} P_{j|i} \left\{ \sum_{t=1}^{T-1} \Delta_{ij}^2(t) + (1-\rho^2)(y_i(T) - u_j(T))^2 \right\},$$

$$\rho' = \left[1 - \frac{nT - n - B}{A} \rho^3 - \frac{B + C - nT + n}{A} \rho \right]^{1/2},$$

$$(5.10)$$

where symbol $'$ denotes the estimates of parameters from the previous step, and

$$g_j(t) = 1 + b_j e^{-r_j t},$$
$$\delta_i^y(t) = y_i(t) - \rho y_i(t+1)$$
$$\delta_j^u(t) = u_j(t) - \rho u_j(t+1)$$
$$\Delta_{ij}(t) = (y_i(t) - u_j(t)) - \rho(y_i(t+1) - u_j(t+1))$$
$$A = \sum_{j=1}^{2} \sum_{i=1}^{n} P_{j|i} \sum_{t=1}^{T-1} (y_i(t) - u_j(t))(y_i(t+1) - u_j(t+1)),$$
$$B = \sum_{j=1}^{2} \sum_{i=1}^{n} P_{j|i} \left[\sum_{t=2}^{T} (y_i(t) - u_j(t))^2 - (y_i(T) - u_j(T))^2 \right],$$

$$C = \frac{1}{\sigma^2} \sum_{j=1}^{2} \sum_{i=1}^{n} P_{j|i} \left\{ \sum_{t=1}^{T-1} \Delta_{ij}^2(t) + (1-\rho^2)(y_i(T) - u_j(T))^2 \right\}.$$

Equations (5.8) and (5.10) form a loop of iterations for the EM algorithm. In the E step, given initial values for the unknown parameters, the posterior probability with which progeny i carries QTL genotype j is calculated with equation (5.8). In the M step, a series of equations (5.10) are solved with the estimated values from the E step. The two steps are repeated until the convergence of estimates. The values of $(a_j', b_j', r_j', \rho', \sigma^{2'})$ estimated from the above equations will be used to provide new estimators of Ω in the next step.

In practical computations, the QTL position parameter (ω) can be viewed as a fixed parameter because a putative QTL can be searched at every 1 or 2 cM on a map interval bracketed by two markers throughout the entire linkage map. The amount of support for a QTL at a particular map position is often displayed graphically through the use of likelihood maps or profiles, which plot the likelihood ratio test statistic as a function of map position of the putative QTL.

5.4   Hypothesis Tests of Functional Mapping

After the MLEs of the parameters are obtained, the existence of a QTL affecting an overall growth curve should be tested by formulating the following hypotheses,

$$\begin{cases} H_0 : \ \Omega_{u_j} \equiv \Omega_u, \ \ j = 1, ..., J \\ H_1 : \ \text{at least one of the equalities in } H_0 \text{ does not hold,} \end{cases} \tag{5.11}$$

where H_0 corresponds to the reduced model, in which the data can be fit by a single growth curve, and H_1 corresponds to the full model, in which there exist different growth curves to fit the data. The test statistic for testing the hypotheses in (5.11) is calculated as the ratio of the log-likelihoods (LR) under the full and reduced models:

$$\mathrm{LR} = -2[\log L(\tilde{\Omega}|\mathbf{y}) - \log L(\hat{\Omega}|\mathbf{y}, \mathbf{M})], \tag{5.12}$$

where $\tilde{\Omega}$ and $\hat{\Omega}$ denote the MLEs of the unknown parameters under the H_0 and H_1, respectively. Note that the estimation of $\hat{\Omega}$ depends on both phenotypic values ($\mathbf{y}$) and marker data ($\mathbf{M}$), whereas the estimation of $\tilde{\Omega}$ only depends on $\mathbf{y}$. The critical threshold for the declaration of a QTL can be determined from permutation tests.

After the existence of QTL is tested, a number of biologically meaningful hypotheses regarding the interplay between gene action and development can be formulated (Wu et al., 2004c). The hypothesis test can be performed on the time at which the detected QTL starts to exert or ceases an effect on growth trajectories, by comparing the difference of the expected means between different genotypes at various time points. At a given time t^*, the hypothesis is

$$\begin{cases} H_0 : u_j(t^*) \equiv u(t^*) \\ H_1 : \text{at least one of the equalities in } H_0 \text{ does not hold} \end{cases} \tag{5.13}$$

If H_0 is rejected, this means that the QTL has a significant effect on variation in growth at time t^*. By scanning time points from 1 to T, one can find the time point at which the QTL starts or ceases to exert an effect on growth.

The effect of QTL $\times$ age interaction on the growth at any two different time points t_1 and t_2 can be tested with the restriction

$$u_{j_1}(t_1) - u_{j_2}(t_1) = u_{j_1}(t_2) - u_{j_2}(t_2), \tag{5.14}$$

where j_1 and j_2 $(j_1, j_2 = 0, \ldots, J)$ are two different QTL genotypes. Testing the timing of QTL expression and its interaction with age based on equations (5.13) and (5.14) can be helpful to our understanding of the way in which QTL trigger an effect on growth and development.

All the above tests (15.34)–(5.14) can be made in terms of the areas under the curves. The area in a time interval $[t_1, t_2]$ for QTL genotype j is the integral of its curve, i.e.,

$$\begin{aligned} U_j[t_1\ t_2] &= \int_{t_1}^{t_2} \frac{a_j}{1 + b_j e^{-r_j t}} dt \\ &= \frac{a_j}{r_j} \left[\ln(b_j + e^{r_j t_2}) - \ln(b_j + e^{r_j t_1}) \right] \end{aligned}$$

The genotypic differences in time (t_I) and growth $[g(t_I)]$ at the inflection point of maximum growth rate can be tested. The test for the genotypic difference is based on the restriction

$$\frac{\log b_j}{r_j} \equiv \frac{\log b}{r} \tag{5.15}$$

for t_I, and

$$\frac{a_j}{2} \equiv \frac{a}{2} \tag{5.16}$$

for $g(t_I)$.

Example 5.1

In a QTL mapping experiment for the mouse, Cheverud et al. (1996) founded an F_2 mapping population of about 550 mice by two strains, Large (LG/J) and Small (SM/J), and constructed a linkage map with microsatellites. Body mass for each mouse was measured at different ages from week 1 to 10. After the data were adjusted for several important covariates like sex, functional mapping implemented with the growth equation (5.4) and AR(1) was used to map QTLs that control growth curves of body mass. Figure 5.2 illustrates the profile of log-likelihood ratios (LR) over mouse chromosome 1. The genomic position corresponding to the peak of a

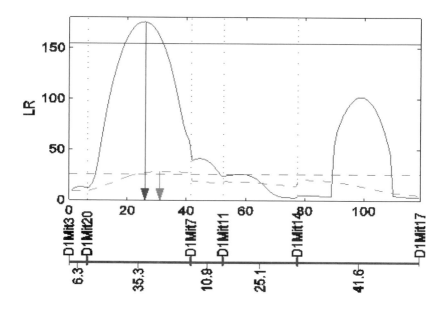

FIGURE 5.2

The profile of the log-likelihood ratios (LR) between the full model (there is a QTL) and reduced (there is no QTL) model for body weight growth across mouse chromosome 1, estimated by functional mapping (solid curve) and interval mapping (broken curve). The vertical broken lines indicate the positions of markers on this chromosome shown beneath. Two horizontal lines indicate the thresholds (the upper one for functional mapping and lower one for interval mapping).

profile is the maximum likelihood estimate of the QTL localization. Permutation tests based on 1000 simulation replicates were performed to determine the critical threshold for declaring the existence of a QTL (Churchill and Doerge, 1994). The 99.9th percentile of the distribution of the maximum LR is used as an empirical critical value at the 0.001 significance level.

Functional mapping detected a significant QTL between markers D1Mit20 and D1Mit7. Curve parameters of the three genotypes at this QTL were estimated and used to draw growth curves (Fig. 5.3). Statistical tests suggest that the three genotypic curves are significantly different before ages 4 weeks, their differences are reduced between ages 4 and 6 weeks, and then increased differentiation occurs after age 6 weeks. This indicates that the QTL detected triggers varying impacts on growth depending on stages of mouse development. The convergence of two homozygotes implies that the additive effect plays a gradually increased role in shaping growth trajectories, whereas the divergence of the heterozygote from the two homozygotes implies that the dominant effect on body weight is increased with age.

☐

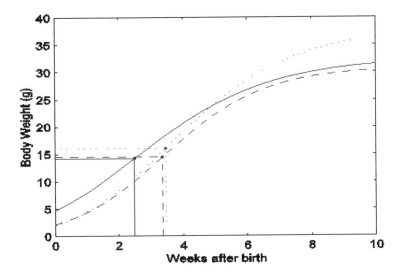

FIGURE 5.3

Three growth curves, each presenting a group of genotypes at the QTL detected on mouse chromosome 1 in the F_2 population. The solid, slash, and dot curves represent the homozygote for the allele from the Large strain, the homozygote for the allele from the Small strain, and the heterozygote, respectively. The timing and growth at the inflection points for the three genotypic curves are indicated by the vertical and horizontal lines, respectively.

As a comparison, the same data were also analyzed by interval mapping. Functional mapping has greater power to detect a growth QTL in terms of the width of the peak of the LR profile than interval mapping. Functional mapping gives a narrower peak and, therefore, can locate the QTL at a smaller region, than interval mapping. More importantly, the effects of the detected QTL on growth rate can be tested by functional mapping. The inflection point, at which growth rate is maximal, occurs about one week earlier for the homozygote for the allele from the Large strain than the homozygote for the allele from the Small strain and about one day earlier for the homozygote for the allele from the Small strain than for the heterozygote (Fig. 5.3). It is possible that this difference in development causes the homozygote for the Large strain to reach the asymptotic growth earlier than the other genotypes. By testing the shape and pattern of the three growth curves, we can investigate possible pleiotropic effects of this growth QTL on many different developmental events, such as the timing of sexual maturity and reproductive fitness, or biomedically important traits, such as metabolic rate and fatness. Functional mapping can, therefore, integrate growth and development, which are historically regarded as two different biological problems, into a comprehensive framework under which their common or unique underlying genetic machineries are identified.

5.5   Transform-Both-Sides Model of Functional Mapping

Precise modeling of the autocorrelation structure of the residual sequences is one of the most important parts for functional mapping. Because of its simple structure and the existence of closed forms, the AR(1) model has been used to model the covariance matrix. But its strong assumptions of variance and correlation stationarity may not be held for most longitudinal data sets. Wu et al. (2004c) adopted the transform-both-sides (TBS) model of Carroll and Ruppert (1984) to remove the time-dependent heterogeneity of covariance (see Section 5.2.2.2). By log-transforming equation (5.3) at both sides, we have

$$\log y_i(t) = \sum_{j=1}^{J} \xi_{ij} \log u_j(t) + \varepsilon_i(t), \tag{5.17}$$

or equivalently, defining $z_i(t) = \log[y_i(t)]$,

$$z_i(t) = \sum_{j=1}^{J} \xi_{ij} h_j(t) + \varepsilon_i(t),$$

where

$$h_j(t) = \log u_j(t) = \log\left[\frac{a_j}{1 + b_j e^{-r_j t}}\right],$$

and $\varepsilon_i(t)$ is the residual error after the log-transformation of phenotypic values, distributed as $N(0, \sigma_\varepsilon^2(t))$.

Now, the AR(1) is used to model the structure of the covariance matrix for the residual errors associated with log-transformed longitudinal measurements. As shown in a poplar example (Wu et al., 2004c), the variance of log-transformed phenotypic values tends to be stable across time points, suggesting that it is close to the variance stationarity assumption of the AR(1) model. Wu et al. (2004c) derived the EM algorithm to estimate the parameter vector $(\Omega_{u_j} = (a_j, b_j, r_j), \Omega_v = (\rho, \sigma^2))$ that are the same as those for the functional mapping model (5.6) on the original scale. In the E step, define the posterior probability of progeny j to bear on a QTL genotype j by

$$P_{j|i} = \frac{\omega_{j|i} f_j(\mathbf{z}_i)}{\sum_{j'=0}^{J} \omega_{j'|i} f_{j'}(\mathbf{z}_i)}. \tag{5.18}$$

In the M step, estimate the parameters in terms of $P_{j|i}$ using

$$a_j' = \frac{\sum_{i=1}^{n} P_{j|i} \left[\sum_{t=1}^{T} (z_i(t) + A_j(t)) + \frac{\rho}{1-\rho}(z_i(1) + A_j(1) + z_i(T) + A_j(T)) \right]}{\left[T + \frac{2\rho}{1-\rho} \right] \sum_{i=1}^{n} P_{j|i}},$$

$$b'_j = \frac{b_j \left[\sum_{t=1}^{T-1} \delta_j^h(t) \left(\frac{e^{-r_j t}}{G_j(t)} - \frac{\rho e^{-r_j(t+1)}}{G_j(t+1)} \right) + \frac{(1-\rho^2)e^{-T r_j} h_j(T)}{G_j(T)} \right] \sum_{i=1}^{n} P_{j|i}}{\sum_{i=1}^{n} P_{j|i} \sum_{t=1}^{T-1} \left[\delta_i^z(t) \left(\frac{e^{-r_j t}}{G_j^2(t)} - \frac{\rho e^{-r_j(t+1)}}{G_j^2(t+1)} \right) + \frac{(1-\rho^2)y_i(T)e^{-r_j T}}{G_j^2(T)} \right]},$$

$$r'_j =$$
$$\frac{r_j \sum_{i=1}^{n} P_{j|i} \left[\sum_{t=1}^{T-1} \delta_i^z(t) \left(\frac{t e^{-r_j t}}{G_j^2(t)} - \frac{(t+1)\rho e^{-r_j(t+1)}}{G_j^2(t+1)} \right) + \frac{(1-\rho^2)T z_i(T)e^{-r_j T}}{G_j^2(T)} \right]}{\left[\sum_{t=1}^{T-1} \delta_j^h(t) \left(\frac{t e^{-r_j t}}{G_j^2(t)} - \frac{(t+1)\rho e^{-r_j(t+1)}}{G_j^2(t+1)} \right) + \frac{(1-\rho^2)T h_j(T)e^{-T r_j}}{G_j^2(T)} \right] \sum_{i=1}^{n} P_{j|i}},$$

$$(\sigma_\varepsilon^2)' = \frac{1}{nT(1-\rho^2)} \sum_{j=0}^{J} \sum_{i=1}^{n} P_{j|i} \left[\sum_{t=1}^{T-1} \Lambda_{ij}^2(t) + (1-\rho^2)(z_i(T) - u_j(T))^2 \right],$$

$$\rho' = \left[1 - \frac{(nT-n-B)\rho^3}{A} - \frac{(B+C-nT+n)\rho}{A} \right]^{1/2},$$

(5.19)

where the symbol $'$ denotes the new estimator of the parameters in the next iterative step, and

$$G_j(t) = \log(1 + b_j e^{-r_j t})$$
$$\delta_i^z = z_i(t) - \rho z_i(t+1)$$
$$\delta_j^h = h_j(t) - \rho h_j(t+1)$$
$$\Lambda_{ij}(t) = (z_i^{(t)} - u_j(t)) - \rho(z_i(t+1) - u_j(t+1))$$
$$A = \sum_{j=1}^{2} \sum_{i=1}^{n} P_{j|i} \left[\sum_{t=1}^{T-1} (z_i(t) - u_j(t))(z_i(t+1) - u_j(t+1)) \right],$$
$$B = \sum_{j=1}^{2} \sum_{i=1}^{n} P_{j|i} \left[\sum_{t=2}^{T} (z_i(t) - u_j(t))^2 - (z_i(T) - u_j(T))^2 \right],$$
$$C = \sum_{j=1}^{2} \sum_{i=1}^{n} \frac{P_{j|i}}{\sigma_\varepsilon^2} \left[\sum_{t=1}^{T-1} \Lambda_{ij}^2(t) + (1-\rho^2)(z_i(T) - u_j(T))^2 \right].$$

The E and M steps are iterated to obtain the MLEs of the parameters.

5.6 Structured Antedependence Model of Functional Mapping

5.6.1 Antedependence Model

The TBS-based model displays the potential to relax the assumption of variance stationarity, but the covariance stationarity issue remains unsolved. Zimmerman and Núñez-Antón (2001) proposed a so-called structured antedependence (SAD) model to model the age-specific change of correlation in the analysis of longitudinal traits. The SAD model has been employed in several studies and displays many favorable properties for genetic mapping of dynamic traits (Zhao et al., 2005).

The antedependence model, originally proposed by Gabriel (1962), states that an observation at a particular time t depends on the previous ones, with the degree of dependence decaying with time lag. If an observation at time t is independent of all observations before $t - r$, this antedependence model is thought to be of order r. Assume that a growth trait (y) is measured for a finite set of times, $1, \cdots, T$. Multivariate normal observations $y(1), \cdots, y(T)$ are rth-order antedependent if the conditional distribution of $y(t)$ given $y(t-1), \cdots, y(1)$ depends on $y(t-1), \cdots, y(t-r)$, for all $t \geq r$ (Gabriel, 1962). This concept is equivalent to $y(1), \cdots, y(T)$ having a Markovian dependence of order r. The order r serves as a memory gauge, where $r = 0$ corresponds to independence and $r = T - 1$ to arbitrary multivariate dependence. Observed growth for progeny i at time i can be expressed as

$$y_i(t) = \xi_i \mu_j(t) + \sum_{s=1}^{r^*} \phi_{t,t-s}[y_i(t-s) - \xi_i \mu_j(t-s)] + e_i(t), \qquad (5.20)$$

where $r^* = \min(r, t-1)$, $\phi_{t,t-s}$'s are unrestricted antedependence parameters, and independent normal random variable $e_i(t)$ may have time-dependent variances, $\sigma^2(t)$, termed *innovation variances*.

5.6.2 Structured Antedependence Model

Like the AR model, the model (5.20) allows for serial correlation within subjects, but unlike AR models it does not assume that the variances are constant nor that correlations between measurements equidistant in time are equal. The antedependence model (5.20) is called the unstructured antedependence model of order r (UAD(r)) because $(r+1)(2T-r)/2$ parameters, $\sigma^2(t)$ and $\sigma(t, t-s)$, are not expressed as a function of a smaller set of parameters.

To make the UAD(r) model more parsimonious, the SAD model is proposed (Núñez-Antón and Woodworth, 1994; Núñez-Antón and Zimmerman, 2000). One useful class models the autoregressive coefficients with the Box-Cox power law and models the innovation variances with a parametric function, i.e.,

$$\begin{aligned}
\phi_{t,t-s} &= \phi_s^{w(T_t; \lambda_s) - w(T_{t-s}; \lambda_s)}, \ t = r+1, \cdots, T; s = 1, ..., r, \ \phi_s > 0, \\
\sigma^2(t) &= \sigma^2 \kappa(T_t; \psi), \ s = r+1, \cdots, T; \ \sigma^2 > 0, \ \{\psi : \kappa(T_t; \psi)\} > 0,
\end{aligned} \qquad (5.21)$$

where T_t and T_{t-s} are measurement times, $w(T; \lambda)$ equals $(T^\lambda - 1)/\lambda$ if $\lambda \neq 0$ and equals $\log T$ if $\lambda = 0$, and $\kappa(\cdot)$ is a function of relatively few parameters (e.g., a low-order polynomial). Thus, different from the original treatment of Gabriel (1962), we only need to estimate three parameters to model the innovation variances if a quadratic polynomial is used, regardless of the number of time points.

As a simplified example with the SAD(1) model in which innovation variances are constant over time points, for $t \geq s$, Jaffrézic et al. (2003) derived the analytical forms for variance and correlation functions, expressed, respectively, as

$$\sigma(t) = \frac{1 - \phi^{2T_t}}{1 - \phi^2} \sigma^2, \tag{5.22}$$

$$\sigma(t,s) = \phi^{T_t - T_s} \sqrt{\frac{1 - \phi^{2T_s}}{1 - \phi^{2T_t}}}, \tag{5.23}$$

It can be seen that even for constant innovation variances, the residual variance can change with time (Jaffrézic et al., 2003). Also, the correlation function is non-stationary even for the simplest SAD model because the correlation does not depend only on the time interval $T_t - T_s$ but also depends on the start and end points of the interval T_t, T_s.

5.6.3 Model Selection

Jaffrézic et al. (2003) proposed an ad hoc approach for model selection. Their strategy is to increase the antedependence order until the additional antedependence coefficient is close to zero. Núñez-Antón and Zimmerman (2000) proposed using the AIC information criterion to select the best model. Hurvich and Tsai (1989) showed that AIC can drastically underestimate the expected Kullback-Leibler information when only few repeated measurements are available. Instead, they derived a corrected AIC, expressed as

$$\text{AIC}_\text{C} = T \left[\log \hat{\sigma}^2 + \frac{T + r}{T - (r + 2)} \right] \tag{5.24}$$

where $\hat{\sigma}^2$ is the white noise variance, T is the number of repeated measurements and r is the order of the model. The number of parameters is heavily penalized so that models selected by AIC_C are typically much more parsimonious than those selected by AIC (Hurvich and Tsai, 1989).

5.7 An Optimal Strategy of Structuring the Covariance

Functional mapping displays strong statistical robustness due to the estimation of fewer parameters that model the mean-covariance structure (Ma et al., 2002). In the previous sections, we introduced two approaches for modeling the structure of the covariance matrix, including the autoregressive (AR) (Diggle et al., 2002) and structured antedependence (SAD) model (Núñez-Antón and Zimmerman, 2000) . Whereas the AR model assumes variance stationarity and correlation stationarity, these two assumptions are relaxed for the SAD model that approximates the variance and correlation as a function of time. Zhao et al. (2005) found that the AR and SAD models cannot be replaced by each other, although the SAD seems to be more general in terms of its non-stationary nature. In statistical modeling of longitudinal variables, many other models have been available to structure the covariance matrix, all of which can be potentially used for functional mapping. Zimmerman and Núñez-Antón (2001) systematically reviewed various approaches for estimating covariance functions of growth curves. Meyer (2001) evaluated the applications of some of these approaches to cattle genetic and breeding studies. Yang et al. (2008) incorporate several most commonly used covariance-structuring approaches into the framework of functional mapping, thus providing a general strategy for the choice of an optimal approach for structuring the covariance matrix for practical data sets.

According to the theory of matrix algebra, we can decompose Σ into the product of standard deviations and correlations expressed as

$$\Sigma = V^{1/2}RV^{1/2} \tag{5.25}$$

where $V^{1/2}$ is the diagonal matrix which is composed of the residual standard deviation $\sigma(t)$ for $t = 1, \ldots, T$, and $\mathbf{R}$ is the correction matrix. Approaches have been available to model the structures of $\mathbf{V}$ and $\mathbf{R}$.

5.7.1   Standard Deviation Function

A number of variance functions have been proposed to model heterogeneous variances across different time points (Foulley et al., 1998). These functions includes a step function or a polynomial function as follows:

$$\sigma^2(t) = \begin{cases} \sigma^2 \left(1 + \sum_{r=1}^{v} b_r t^r\right), \\[2mm] \sigma^2 e^{\left(1 + \sum_{r=1}^{v} b_r t^r\right)}, \\[2mm] e^{\left(\sigma^2 + \sum_{r=1}^{v} b_r t^r\right)}, \text{ or } \ln \sigma^2(t) = \sigma^2 + \sum_{r=1}^{v} b_r t^r \end{cases} \tag{5.26}$$

where σ^2 is the variance at the intercept, b_r is the coefficient of the variance function and v is the order of polynomial fit. The standard deviation function is the square root of $\sigma^2(t)$.

5.7.2   Correlation Function

Correlations in R between different time points can be modeled as a function of time interval. Correlation functions with one or more parameters can be either stationary

TABLE 5.1

Stationary models for correlation functions.

Model	Correlation Function	Range of Parameter
Compound symmetry (CS)	ρ	$-1 < \rho < 1$
First-order AR	ρ^l	$-1 < \rho < 1$
Exponential (EXP)	$e^{-\rho l}$	$\rho > 0$
Gaussian (GAU)	$e^{-\rho l^2}$	$\rho > 0$
Cauchy (CAU)	$(1 + \rho l^2)^{-1}$	$\rho > 0$
Hyperbolic cosine (HYP)	$(\cosh(\pi \rho l / 2))^{-1}$	$\rho > 0$
Uniform distribution (UNI)	$\sin(\rho l)/\rho l$	$\rho > 0$
Triangular distribution (TRI)	$(1 - \cos(\rho l))/\rho l$	$\rho > 0$
"Damped" exponential (DEX)	$e^{-\rho l^\phi}$	$\rho > 0, 0 < \phi < 2$

Note: l denotes the lag in time for a pair of time points.

or non-stationary. The simplest correlation function is those specified by a single parameter. Table 5.1 lists several stationary models for correlation functions. Equation (5.25) can be reduced to the AR(1) covariance structure when $\sigma^2(t)$ is equal to σ^2 and when the correlation function is specified by the AR(1) model. All the parameters that describe the standard deviation and correlation functions are arrayed in Ω_v.

More complicated models, such as autoregressive moving average (ARMA), can be considered to model correlation functions. The ARMA model typically contains two parts, an autoregressive (AR) and a moving average (MA). The model is usually referred to as an ARMA(p, q) model where p is the order of the autoregressive part and q is the order of the moving average part. For some data, non-stationary models perform better than stationary models.

5.7.3 Model Selection

For a practical data set, we should know which model is optimal to explain the structure of the covariance matrix. The selection of an optimal covariance-structuring model is crucial for increasing the power of functional mapping and its estimation precision. In statistics, there are several criteria for model selection although none of them perform the best under all circumstances. Here, we used the commonly used Bayesian information criterion (BIC) (Schwarz, 1978) as the model selection criterion of the optimal covariance-structuring model. The BIC is defined as

$$\text{BIC} = -2 \ln L(\widehat{\Omega}_{u_j}, \widehat{\Omega}_v | \text{M}) + \text{dimension}(\Omega_{u_j}, \Omega_v | \text{M}) \ln(n), \text{ for } j = 2, 1, 0.$$

where $\widehat{\Omega}_{u_j}$ and $\widehat{\Omega}_v$ are the MLEs of parameters under a particular model (M), dimension $(\Omega_{u_j}, \Omega_v | r)$ represents the number of independent parameters under this model, and n is the total number of observations at a particular time point. The optimal model is one that displays the minimum BIC value.

5.8   Functional Mapping Meets Ontology

An important challenge that faces modern developmental biology is to estimate and test the genetic control of the timing of major milestones and their relationships during development (Rice, 1997, 2002; Raff, 1998; Pujar et al., 2006). As a statistical tool, functional mapping can well be used to unravel the genetic and molecular mechanisms controlling temporal patterns of developmental events and examine how genes pleiotropically affect different biological processes. The implication of functional mapping can be strengthened when it is incorporated into an emerging discipline of systems biology–ontology (Jaiswal, 2005; Bard et al., 2005; Pujar et al., 2006; Ilic et al., 2007; Avraham et al., 2008).

Ontology is an explicit specification of a conceptualization. It is a description of the concepts and relationships that can exist for an agent or a community of agents. This concept has been recently used in genomic and bioinformation studies with the increasing availability of data from several genome sequencing projects (including humans, mouse, and rice). In plants, the Plant Ontology Consortium (POC: http://www.plantontology.org) was launched to develop standard vocabularies that can be used to generically describe plant anatomy, growth, and developmental stages, and to further use those vocabularies to annotate various datasets from plant genomics and genetics projects (Bruskiewich et al., 2002). Plant ontology provides a promising direction for investigating genes and their functional and sequence homologs involved in plant development.

Below is shown the usage of functional mapping to better understand the developmental pattern of plants by using rice as an example. Rice growth undergoes complex developmental stages (Fig. 5.4) (Moldenhauer and Slaton, 2004) that include

(1) Vegetative (germination to panicle initiation),

(2) Reproductive (panicle initiation to heading),

(3) Grain filling and ripening or maturation (heading to maturity).

Each of these stages can be further decomposed into a variety of sub-stages that determines grain yield uniquely by influencing the number of panicles per unit land area, the average number of grain produced per panicle, and the average weight of the individual grains. The Plant Ontology Consortium has developed a comprehensive package of vocabularies to describe detailed developmental events at different organizational levels in rice (Jaiswal, 2005). Figure 5.4 is one such illustration for different stages of morphological development in rice.

Functional mapping can be used to detect the genetic control of QTLs over the length of any of these stages and any development events. To show how functional mapping can be used to integrate different physiological and developmental changes during an organism's lifetime, we construct a diagram of statistical tests for the possible pleiotropic effect of a QTL on vegetative growth processes and time-to-event

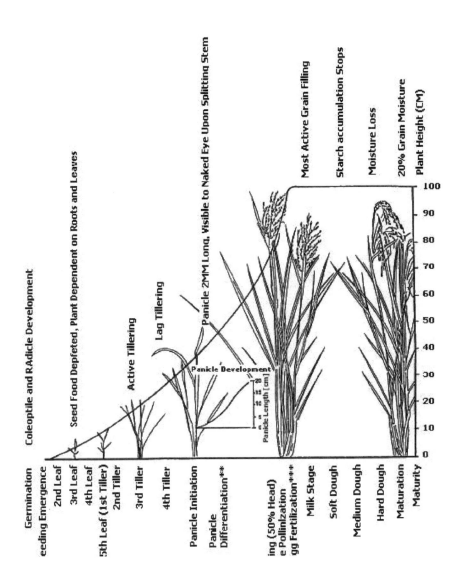

FIGURE 5.4
Diagram for different developmental stages in rice. Adapted from Moldenhauer and Slaton (2004).

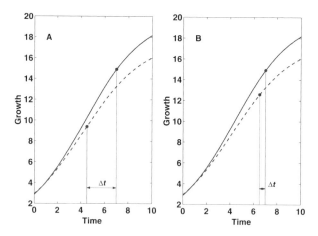

FIGURE 5.5

Model that illustrates how a QTL pleiotropically affects vegetative growth and reproductive behavior. In **A**, the QTL affects both features because there is a significant
difference (denoted by Δt) in the time to first flower between two growth QTL genotypes (each represented by a curve), whereas, in **B**, the growth QTL does not govern
reproductive behavior because such a difference is nonsignificant.

phenotypes, such as survival and age-at-onset (Fig. 5.5). Age-at-onset traits can be
related to reproductive behaviors including the time to first flower or the time to form
mature seeds. A QTL that determines the growth process may or may not trigger an
effect on the time to first flower. In Fig. 5.5**A**, the time to first flower displays a
pronounced difference between two QTL genotypes with discrepant growth curves,
suggesting that this QTL affects both growth and reproductive features. But a hypothesized QTL for growth curves in Fig. 5.5**B** does not affect pleiotropically the
time to first flower. In practice, the two cases illustrated in Figure 5.5 can be tested
by the model (5.1) that is extended to include reproductive information.

6

Functional Mapping of Pharmacokinetics and Pharmacodynamics

Functional mapping was originally conceived from a quantitative trait locus (QTL) mapping study of growth curves in an experimental cross. It has now been extensively extended to map any other dynamic or longitudinal traits, such as drug response (Wang and Wu, 2004; Gong et al., 2004). Also, the type of mapping population used can not be only derived from a controlled cross, but also random samples from a natural population. Functional mapping was further integrated with genetic association studies of single nucleotide polymorphisms (SNPs) to characterize haplotype variants that encode longitudinal responses (Lin et al., 2005). With such integration, functional mapping has been developed from QTL mapping to the identification of quantitative trait nucleotides (QTNs).

To map or haplotype drug response with functional mapping, we need to have an understanding of the mechanisms of drug deposition and drug action, and the relationships between drug concentration and effect. Pharmacological properties of a drug include interactions between chemicals and living tissue, which can be quantified by two different but interactive biochemical processes–pharmacokinetics (PK) and pharmacodynamics (PD) (Derendorf and Meibohm, 1999). The PK is a study of what the body does to the drug, whereas the PD is the study of what the drug does to the body. During the last decade, pharmacokinetic and pharmacodynamic modeling technologies have been increasingly useful in contemporary clinical pharmacology. These technologies have been applied during the process of developing new drug entities as well as for better insights into clinical actions of drugs that are already marketed. The intensity of therapeutic response produced by a drug is related to the concentration of the drug at the site of action which is in turn affected by a variety of factors including genes and environments. Thus, through identifying these factors and their control mechanisms for the pharmacokinetic and pharmacodynamic aspects of drug response, we are able to determine the dosage regimen for an individual that is likely to achieve the desirable therapeutic response with minimum risk of toxic effects.

In this chapter, we will show how functional mapping can be used to identify genes that control PK and PD processes. We will provide introductory information about the mathematical aspects of PK and PD related to pharmacological and clinical principles of drug deposition, drug metabolism, and drug action. By using several representative mathematical equations of PK and PD, we will then demonstrate the utilization of functional mapping to reveal the genetic secrets hidden in biochemical pathways of drug response. A number of hypothesis tests regarding the interplay between genetic, developmental, and pharmacological interactions are described.

6.1 Mathematical Modeling of Pharmacokinetics and Pharmacodynamics

6.1.1 Modeling Pharmacokinetics

Pharmacokinetics is a combination of Greek "pharmacon" (meaning drug) and "kinetikos" (meaning putting in motion). It is the study of the movement of drugs in the body, including the processes of absorption, distribution, localization in tissues, biotransformation, and excretion. In many cases, pharmacological action, as well as toxicological action, is related to plasma concentration of drugs. Consequently, through the study of pharmacokinetics, the pharmacist will be able to individualize therapy for the patient. More specifically, the pharmacokinetic process characterizes the change of drug concentration over time through four phases, i.e.,

(1) Absorption, the movement of drug from the site of adminstration to the blood circulation,

(2) Distribution, the diffusion and transferring of drug from intravascular space to extravascular space,

(3) Metabolism, a biotransformation reaction process by which the chemical structure of a drug is altered,

(4) Excretion, the removal of drugs and biotransformation products from the body.

These four phases (abbreviated as ADME) form a general sequential PK process, but they are not discrete events and may occur simultaneously. The ADME can be sorted into two classes, drug input (absorption) and drug output (distribution, metabolism, and excretion).

Mathematically, the pharmacokinetic process behind these four phases can be described by a set of parameters that are related to physiological, clinical, physical, or chemical properties of drug (Hochhaus and Derendorf, 1995). A typical example of the PK process is illustrated in Fig. 6.1, whose shape and rate can be determined by a mathematical model.

6.1.2 Modeling Pharmacodynamics

Kinetic-dynamic modeling uses mathematical methods to directly evaluate the relationship of drug concentrations (c) to drug effect (E), aimed to determine how much variability in drug effect is attributable to measurable drug concentrations. The mainstay of modeling such pharmacodynamics is the Hill, or sigmoid E_{max}, equation,

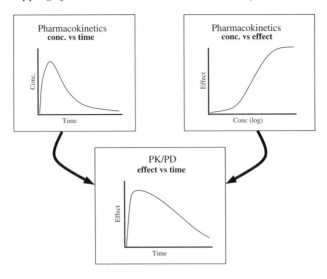

FIGURE 6.1

Profiles that describe the concentration change of a drug following its administration (pharmacokinetics, PK), the change of drug effect with drug concentration (pharmacodynamics, PD), and the time course of drug effect (PK-PD link).

which postulates the following relationship between drug concentration (c) and drug effect (E) (Fig. 6.2) (Giraldo, 2003):

$$E(c) = E_0 + \frac{E_{max} c^H}{EC_{50}^H + c^H},$$
(6.1)

where E_0 is the constant or baseline value for the drug response parameter, E_{max} is the asymptotic (limiting) effect, EC_{50} is the drug concentration that results in 50% of the maximal effect, and H is the slope parameter that determines the slope of the concentration-response curve. The larger H, the steeper the linear phase of the log-concentration-effect curve (Fig. 6.2). When the effect is a continuous variable, estimates of E_{max}, EC_{50}, and H are usually obtained by extended least squares or iteratively re-weighted least squares when there is sufficient data for analysis of individual subjects.

These models were derived using data obtained from patients receiving increasing doses of drug, until they achieved target response or predetermined maximum dose. There is a standard clinical procedure to collect these data (including various physiological parameters, such as heart rate, and systolic/diastolic blood pressure) at each dose level during successive time intervals.

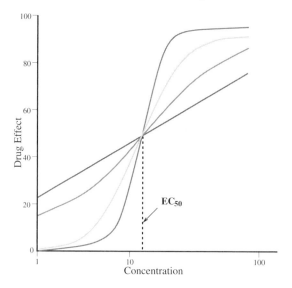

FIGURE 6.2

The shapes of the sigmoid Emax model, as affected by different H coefficients. Larger H values correspond to steeper curves.

6.1.3 Linking Pharmacokinetics and Pharmacodynamics

As the two main principles that determine the relationship between drug dose and response, pharmacokinetics and pharmacodynamics can be linked directly or indirectly (Hochhaus and Derendorf, 1995). The direct link applies to a situation in which the measured concentration (or the plasma concentration c_p) is proportional to drug concentration at the effect site (biophase). The plasma concentration for a single dose in the direct link is calculated for different cases, which include

$$c_p = \begin{cases} \dfrac{d}{V_d} e^{-k_e t} & \text{for intravenous (IV) bolus} \\[2ex] \dfrac{dk_a}{V_d(k_a - k_e)}(e^{-k_a t} - e^{-k_e t}) & \text{for first-order absorption} \\[2ex] \dfrac{k_o}{k_e V_d}(e^{k_e \tau} - 1)e^{-k_e t} & \text{for zero-order absorption} \end{cases} \quad (6.2)$$

where d is the bioavailable dose, V_d is the volume of distribution (which is a constant for a known drug), k_e is the elimination rate constant, k_a is a first-order absorption rate constant, k_o is a zero-order absorption rate constant, τ is the duration of zero-order absorption ($\tau = t$ during absorption and constant in the post-absorption phase) and t is the time after the last dose was administrated.

The indirect approach for linking PK and PD assumes that there is a hypothetical effect compartment and it is based on the concentration (c_e) in the compartment hypothesized to cause the drug effect. The relationship between c_e and time can be described by

$$
c_e = \begin{cases}
\dfrac{dk_{co}}{V_d(k_{co} - k_e)}(e^{-k_c t} - e^{-k_{co} t}) & \text{for intravenous (IV) bolus} \\[2em]
\dfrac{dk_a k_{co}}{V_d}\left[\dfrac{e^{-k_c t}}{(k_a - k_e)(k_{co} - k_e)} + \dfrac{e^{-k_a t}}{(k_e - k_a)(k_{co} - k_a)} \right. \\
\left. + \dfrac{e^{-k_{co} t}}{(k_e - k_{co})(k_a - k_{co})}\right] & \text{for first-order absorption} \\[2em]
\dfrac{k_o}{k_e V_d(k_{co} - k_e)}[k_{co}(e^{k_e \tau} - 1)e^{-k_e t} - k_c(e^{k_{co} \tau} - 1)e^{-k_{co} t}] \\
& \text{for zero-order absorption}
\end{cases}
\tag{6.3}
$$

where k_{co} is the rate constant for distributing drug from the compartment in which drug concentration causes the effect.

The PK and PD processes can be jointly modeled by substituting pharmacokinetic model (6.2) or (6.3) to pharmacodynamic model (6.1). With such a substitution, the effect-time-dosage relationship can be derived. For model (6.2), the measured concentration is used, whereas for model (6.3), the concentration hypothesized to cause the drug effect is used. For example, an integrative PK-PD model in the case of an IV bolus injection without an effect compartment can be expressed as

$$
E = \frac{E_{\max} d e^{-k_e t}}{EC_{50} V_d + d e^{-k_e t}}.
\tag{6.4}
$$

Figure 6.1 shows how PK and PD are linked to describe the time course of drug effect. A landscape of drug effect as function of time and dose is drawn in Fig. 6.3 for three hypothesized genotypes. As seen from this figure, the three landscapes are different among the three genotypes, which suggests a possible genetic component involved in the PK-PD link. Thus, by testing differences in a set of parameters that define the landscape, we can detect whether there is a gene involved and how this gene controls drug response. Functional mapping can be used to estimate genotype-specific parameters that define differences in the landscape of drug response.

6.2 Functional Mapping of Pharmacokinetics

It is statistically straightforward to extend functional mapping described in Chapter 5 to map or haplotype pharmacokinetic processes of drug response. Zhu et al. (2003) and Wang and Wu (2004) used two different types of kinetic models for functional

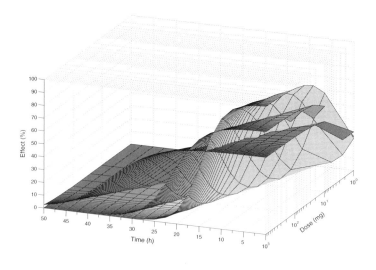

FIGURE 6.3
Landscapes of drug effects varying as a function of dosage and time for three hypothesized genotypes.

mapping of QTLs in a natural population. Although their work purported to detect genes for HIV viral load trajectories in a host, the kinetic models they used are principally relevant to a pharmacokinetic process.

6.2.1   Kinetic Derivation of a Bi-Exponential Model

We show how a kinetic model can be derived to describe the change of the amount of HIV viral loads with time in a human host. The relationship between drug concentration and time can be derived in a similar way. The HIV pathogenesis is the consequence of biological interactions between human cells and viruses in different compartments. These compartments include

(1) Uninfected target cells (T),

(2) "Mysterious" infected cells (T_m), such as tissue langerhans and microglial cells,

(3) Long-lived infected cells (T_s),

(4) Latently infected cells (T_l),

(5) Productively infected cells (T_p),

(6) Infectious virus (V_l),

(7) Noninfectious virus (V_{NI}).

Under the infection by V_I, T may become T_m, T_s, T_l, and T_p with respective proportions, and also T_l may become T_p with a rate of δ_l and T_m, T_s, and T_p are killed at the rates of δ_m, δ_s, and δ_p, respectively. In addition, T_m, T_s, and T_l may naturally die at the rates of μ_m, μ_s, and μ_l, respectively, without producing viruses. Both infectious and noninfectious viruses are eliminated with the same elimination rate of λ_c.

After patients receive the treatment of antiviral drugs, such as protease inhibitor (PI) or reverse transcriptase inhibitor (RTI), the interactive relationships among different compartments will be changed. In general, the T cell count would recover slightly during the first 2 to 4 weeks of antiviral treatment, with the recovery rate of λ_r. Wu and Ding (1999) constructed dynamic models for different compartments after combined treatments with PI and RTI. They further derived a comprehensive model for predicting total virus loads in infected individuals at time t, which is expressed as

$$V(t) = P_0 + P_1 e^{-\delta_p t} + P_2 e^{-\lambda_m t} + P_3 e^{-\lambda_s t} + P_4 e^{-\lambda_l t} +$$
$$(P_5 + P_6 t)e^{-\lambda_c t} + P_7 e^{-\lambda_r t} + P_8 e^{-(\lambda_c + \lambda_r)t}, \tag{6.5}$$

where P_k ($k = 0, ..., 8$) represent the initial viral production rate, with $P_5 = V(0) - P_0 - P_1 - P_2 - P_3 - P_4 - P_7 - P_8$, and the λ's represent the exponential decay rates of virus in the corresponding compartments, with $\lambda_m = \delta_m + \mu_m$, $\lambda_s = \delta_s + \mu_s$, and $\lambda_l = \delta_l + \mu_l$. Figure 6.4 illustrates the results of different phases of viral decline following drug treatment *in vivo* predicted by model (6.5). These phases include:

(1) Several hours of intracellular and pharmacological delay after the first dose of antiviral drugs,

(2) Transition phase reflecting the decay of free viruses and T_p,

(3) Rapid decay phase reflecting the decay of T_p, T_s, and/or T_l,

(4) Slow decay phase reflecting the decay of T_s and/or T_l and other residual infected cells,

(5) The levelling off to steady state reflecting residual virus production and virus replenishment from other reservoirs.

All parameters in model (6.5) can be identified if accurate measurements on virus loads are made over a long time period. However, since this condition may not be met in practice, only a small number of λ's can be realistically fitted to the data. Based on the exponential curve peeling method advocated by Van Liew (1967) and Wu and Ding (1999) proposed a simplified model for estimating clinically interesting parameters by collapsing some decay phases. This new model is expressed as

$$V(t) = P_0 + P_1 e^{-\delta_p t} + P_2 e^{-\lambda t} + (P_3 + P_4 t)e^{-\lambda_c t},$$

where λ is a possibly confounded clearance rate of long-lived and latently infected cells, and $(P_3 + P_4 t)e^{-\lambda_c t}$ is the term due to the clearance rate of free viruses. Because

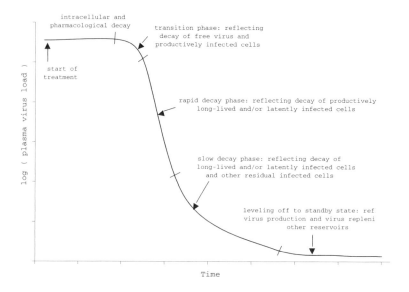

FIGURE 6.4
Diagram for different phases of viral dynamics, described by model (6.5), after antiviral drug treatment. The data for rapid and slow decay phases can be modeled by model (6.6).

the clearance of free viruses is observed to be very rapid (less than 6 hours of half-life), the term $(P_3 + P_4 t)e^{-\lambda_c t}$ can be neglected compared to other terms due to the clearance of infected cells after a few days of treatment. The above model can be further simplified as

$$
\begin{aligned}
V(t) &= P_0 + P_1 e^{-\delta_p t} + P_2 e^{-\lambda t} \quad \text{for } t \geq t_c, \\
&= P_0 + P_1 e^{-\lambda_1 t} + P_2 e^{-\lambda_2 t}
\end{aligned}
\tag{6.6}
$$

where t_c is the time required for the term $(P_3 + P_4 t)e^{-\lambda_c t}$ to be negligible. Let $\delta_p = \lambda_1$ and $\lambda = \lambda_2$. The single exponential models and bi-exponential models constructed from clinical data are special cases of the model (6.6). The two dynamic models (6.5) or (6.6), shown in Fig. 6.4, can be used to characterize the trajectories of HIV concentration.

6.2.2 QTL Mapping with a Bi-Exponential Model

6.2.2.1 Model Structure

Consider a random sample of size n drawn from a natural human population at Hardy–Weinberg equilibrium. Multiple polymorphic markers, e.g., single nucleotide

TABLE 6.1

Frequencies of marker-QTL genotypes and diplotypes in terms of haplotype frequencies.

	AA	Aa $A\|a + a\|A$	aa	Total
MM	p_{11}^2	$2p_{11}p_{10}$	p_{10}^2	n_2
Mm	$2p_{11}p_{01}$	$2p_{11}p_{00} + 2p_{10}p_{01}$	$2p_{10}p_{00}$	n_1
mm	p_{01}^2	$2p_{01}p_{00}$	p_{00}^2	n_0

polymorphisms (SNPs), are typed to detect significantly associated QTLs that affect the kinetic relationship between the concentration of a drug or virus and time. In practice, drug concentrations are measured at a finite set of times, $1, ..., T$, which can be considered as a multivariate trait vector, $y_i(1), \cdots, y_i(T)$. This finite set of data can be modeled by a kinetic model (6.5) or (6.6).

Assume a QTL with alleles A (in a frequency of q) and a (in a frequency of $1 - q$) affecting concentration-time curves. For a particular genotype j ($j = 2$ for AA, 1 for Aa, and 0 for aa) of this QTL, the parameters describing its HIV dynamics are denoted by $\Omega_{u_j} = (\{P_{kj}\}_{k=0}^8, \delta_{pj}, \lambda_{mj}, \lambda_{sj}, \lambda_{lj}, \lambda_{cj}, \lambda_{rj})$ for the comprehensive model (6.5), or $\Omega_{u_j} = (P_{0j}, P_{1j}, P_{2j}, \lambda_{1j}, \lambda_{2j})$ for the simplified model (6.6). We hope to find out this QTL by analyzing a marker that is genetically associated with it. This marker is assumed to have two alleles M (in a frequency of p) and m (in a frequency of $1 - p$), leading to three observable genotypes MM, Mm, and mm. Let D be the coefficient of (gametic) linkage disequilibrium between the marker and QTL. Then, marker-QTL haplotype frequencies are expressed as

$$\begin{aligned}
p_{11} &= pq + D && \text{for } MA, \\
p_{10} &= p(1 - q) - D && \text{for } Ma, \\
p_{01} &= (1 - p)q - D && \text{for } mA, \\
p_{00} &= (1 - p)(1 - q) + D && \text{for } ma.
\end{aligned}$$

A mixture-based likelihood similar to equation (5.6) is formulated as

$$L(\Omega) = \prod_{i=1}^{n} \left[\sum_{j=0}^{2} \omega_{j|i} f_j(\mathbf{y}_i) \right], \tag{6.7}$$

with $\Omega = (\{\omega_{j|i}, \Omega_{u_j}\}_{j=0}^2, \Omega_v)$, where the proportions of mixture proportions ($\omega_{j|i}$), i.e., the conditional probabilities of QTL genotypes given the marker genotype of a subject i, are now expressed in terms of the linkage disequilibrium between the marker and QTL, which are derived from marker-QTL joint genotype frequencies given in Table 6.1, and Ω_v is a set of parameters that model the covariance structure.

Of these genotypes, double heterozygote $MmAa$ includes two different diplotypes (a set of haplotype pairs) each with a different frequency of formation. Let n_2, n_1, and n_0 be the sample sizes of three marker genotypes, respectively.

6.2.2.2 Model Estimation

Each of the three marker genotypes is a mixture of the three latent QTL genotypes. Thus, the complete data set permitting explicit estimators of haplotype frequencies include the observable marker data and missing marker-QTL configurations. The EM algorithm is used to estimate the model parameters. Let $\omega_{j|i}$ denote the probability with which subject i carries QTL genotype j. This (prior) probability takes the values of Table (6.1), depending on the specific marker genotype subject i carries.

In the E step, the posterior probability with which subject i has QTL genotype j is calculated by

$$P_{j|i} = \frac{\omega_{j|i} f_j(\mathbf{y}_i)}{\sum_{j'=0}^{2} \omega_{j'|i} f_{j'}(\mathbf{y}_i)}, \tag{6.8}$$

In the M step, the haplotype frequencies are estimated from the calculated posterior probability by

$$
\begin{aligned}
\hat{p}_{11} &= \frac{1}{2n}\left[\sum_{i=1}^{n_2}(2P_{2|i}+P_{1|i}) + \sum_{i=1}^{n_1}(P_{0|i}+\phi P_{1|i}) \right], \\
\hat{p}_{10} &= \frac{1}{2n}\left\{ \sum_{i=1}^{n_2}(P_{1|i}+2P_{0|i}) + \sum_{i=1}^{n_1}[P_{0|i}+(1-\phi)P_{1|i}] \right\}, \\
\hat{p}_{01} &= \frac{1}{2n}\left\{ \sum_{i=1}^{n_0}(2P_{2|i}+P_{1|i}) + \sum_{i=1}^{n_1}[P_{1|i}+(1-\phi)P_{1|i}] \right\}, \\
\hat{p}_{00} &= \frac{1}{2n}\left[\sum_{i=1}^{n_0}(2P_{0|i}+P_{1|i}) + \sum_{i=1}^{n_1}(P_{0|i}+\phi P_{1|i}) \right],
\end{aligned}
\tag{6.9}
$$

where $\phi = p_{11}p_{00}/(p_{11}p_{00}+p_{10}p_{01})$.

In the M step, the parameters that model QTL genotype-specific kinetic processes and the covariance structure in terms of AR(1) or SAD(1) models are estimated by the EM algorithm as derived in Ma et al. (2002). As shown in Zhao et al. (2004a), a simplex algorithm may be easier for model implementation than the EM algorithm. Thus, an integrative EM-simplex algorithm can be constructed with equations (6.8) and (6.9).

6.2.2.3 Hypothesis Tests

A number of biologically meaningful hypotheses can be tested with functional mapping. These hypothesis tests can be a *global* test for the existence of a significant QTL, a *local* test for the genetic effect on drug or viral concentration at a static point, a *regional* test for the overall effect of the QTL on a particular period of kinetic processes, or *interaction* test for the change of QTL expression across times. We use kinetic model (6.6) to demonstrate these hypothesis tests.

Global Test

Testing whether a specific QTL exists to affect the shape of kinetic curves is a first step toward the understanding of the genetic architecture of kinetic processes. The genetic control over an entire process can be tested by formulating the following hypotheses:

$$\begin{cases} H_0 : (P_{0j}, P_{1j}, P_{2j}, \lambda_{1j}, \lambda_{2j}) \equiv (P_0, P_1, P_2, \lambda_1, \lambda_2), \quad j = 2, 1, 0 \\ H_1 : \text{Not all these eqalities above hold} \end{cases} \tag{6.10}$$

The H_0 states that there are no QTL affecting kinetic processes (the reduced model), whereas the H_1 proposes that such QTL do exist (the full model). The log-likelihood ratio (LR) test statistic is then calculated with the likelihoods from the H_0 and H_1. An empirical approach for determining the critical threshold is based permutation tests (Churchill and Doerge, 1994).

The global effects of different genetic components, additive and dominant, can also be tested on the shapes of entire kinetic curves. The hypothesis for testing the additive effect of the QTL on overall curve shapes can be formulated as

$$\begin{cases} H_0 : V_2(t) = V_0(t) \\ H_1 : V_2(t) \neq V_0(t), \end{cases} \tag{6.11}$$

which is equivalent to testing the difference of the full model with no restriction and the reduced model with a restriction $V_2(t) = V_0(t)$. Thus, under the reduced model (H_0) of (6.11) the data can be fit by five unknown parameters fewer than under the full model (H_1). An empirical approach for determining the critical threshold for the hypothesis test of (6.11) is based on simulation studies.

The test for the dominant effect of the QTL is equivalent to testing the difference of the full model with no restriction and the reduced model with a restriction:

$$2V_1(t) = V_2(t) + V_0(t).$$

Local Test

The local test can test for the significance of the additive or dominant effect of the QTL on kinetic curves measured at a time point (t^*) of interest. The tests of additive and dominance effects can be made on the basis of the corresponding restrictions as given above. For example, the hypothesis for testing the additive effect of the QTL on kinetic curves at a given time t^* can be formulated as

$$\begin{cases} H_0 : V_2(t^*) = V_0(t^*) \\ H_1 : V_2(t^*) \neq V_0(t^*), \end{cases} \tag{6.12}$$

which is equivalent to testing the difference of the full model with no restriction and the reduced model with a restriction $V_2(t^*) = V_0(t^*)$.

Regional Test

Sometimes we are interested in testing the difference of kinetic curves in a time interval rather than simply at a time point. The question of how QTL exerts its effects on a period of kinetic process $[t_1, t_2]$ can be tested using a regional test approach based on the areas,

$$G_j[t_1, t_2] = \int_{t_1}^{t_2} V_j(t)dt,$$

covered by kinetic curves. The hypothesis test for the genetic effect on a period of HIV dynamics is equivalent to testing the difference between the full model without a restriction and the reduced model with a restriction. The types of restriction used can be similarly given, depending on the additive or dominant effect.

Interaction Test

The effects of QTL may change with time, which suggests the occurrence of QTL × time interaction effects on kinetic curves. The differentiation of $V(t)$ with respect to time t represents a slope of kinetic curve (decay rate). If the slopes at a particular time point t^* are different between the curves of different QTL genotypes, this means that significant QTL × time interaction occurs between this time point and next. The test for QTL × time interaction can be formulated with the restriction:

$$\frac{dV_2(t^*)}{dt} = \frac{dV_0(t^*)}{dt}. \tag{6.13}$$

for the additive effect of the QTL. The tests for QTL × time interactions due to the dominant effect can be similarly formulated.

The effect of QTL × time interaction on kinetic curves can be examined during an entire time course. The global test of QTL × time interaction due to the additive effect of QTL $\mathcal{Q}_k$ can be formulated with the restriction:

$$\int_0^m \left[\frac{dV_2(t)}{dt} - \frac{V_0(t)}{dt} \right]^2 dt = 0. \tag{6.14}$$

The restriction (6.14) means that there is the same slope at every time point between the two kinetic curves of QTL genotypes AA and aa, thus suggesting that the additive effect of the QTL does not lead to significant QTL × time interaction on entire kinetic processes.

Test for Clinically Important Parameters

There are a number of parameters that can be used to evaluate the efficacy of a drug. These parameters, called "derived variables," include:

(1) Two viral decay rates (δ_p and λ),

(2) Viral load change (VLC) from baseline to a pre-specified time (say day 7) in a log scale,

(3) The area under the curve (AUC) of viral load trajectory,

(4) The proportion of virus produced by long-lived and/or latently infected cells in the total virus pool, approximated by $P_2/(P_1 + P_2)$,

(5) The eradication times for the two compartments, productively infected cells, and long-lived/or latently infected cells, approximated by $\log P_1/\delta_p$ and $\log P_2/\lambda$, respectively.

Wu and Ding (1999) analyzed the potency of anti-HIV therapies in AIDS clinical trials based on δ_p, λ, VLC, and AUC. All of these derived variables are likely to be controlled by genetic factors. For example, Wu and Ding (1999) observed considerable inter-patient variation in the two decay rates. Whether or not the QTL detected for overall viral load trajectories also affects a derived variable can be tested by the likelihood analysis. The null hypothesis for these tests is that a given derived variable has identical genotypic values among the three QTL genotypes.

Example 6.1

In general, the amount of virions after initiation of treatment with potent antiretroviral agents follows a certain dynamic pattern that can be characterized by mathematical functions. Based on a firm understanding of biological interactions using real virological data from clinical trials, some simplified forms of HIV dynamic models have been derived (Wu and Ding, 1999). Figure 6.5 illustrates the dynamics of viral load for 53 HIV-1-infected patients after the treatment with highly active antiretroviral therapy (HAART) from the AIDS Clinical Trials Group (ACTG) Protocol 315, supported by the National Institute of Allergy and Infectious Diseases. Viral loads of these patients were measured on days 0, 2, 7, 10, 14, 21, 28 and weeks 8 and 12. This ACTG 315 dataset is of typical longitudinal nature and has been subjected to extensive statistical analyses (Wu and Ding, 1999; Wu, 2002). In particular, Wu and Ding (1999) showed that HIV-1 dynamics for the ACTG 315 data can be adequately fitted by a bi-exponential function (6.6).

Whether or not there exists a major genetic determinant to control HIV-1 dynamics or curves in hosts can be tested by using functional mapping although there are no SNP data available for these subjects. The hypothesis tests embedded within the functional mapping context allow the exploration of the control mechanism of a genetic determinant in regulating the time-dependent curve changes of HIV-1 load. Each component in the mixture model of functional mapping is assigned to be a genotype for a major gene assumed, although such a component may be founded by other factors, such as sex, ages, and sexual behavior, without the use of molecular marker data. Suppose there is a major genetic determinant with two alleles A and a that affects HIV-1 dynamics. Let q and $1-q$ denote the allele frequencies for A and a, respectively. The three genotypes at the major gene, AA (2), Aa (1), and aa (0), have genotypic frequencies expressed as $\omega_2 = q^2 + D$, $\omega_1 = 2q(1-q) - 2D$, and $\omega_0 = (1-q)^2 + D$, respectively, where D is the coefficient of Hardy–Weinberg disequilibrium at the major gene. For an arbitrarily chosen patient, he or she must carry one (and only one) of the three genotypes, with a probability represented by ω_2, ω_1, or ω_0. The mixture model for all individuals based on these three latent genotypes establishes a theoretical foundation for characterizing an individual's major genotype.

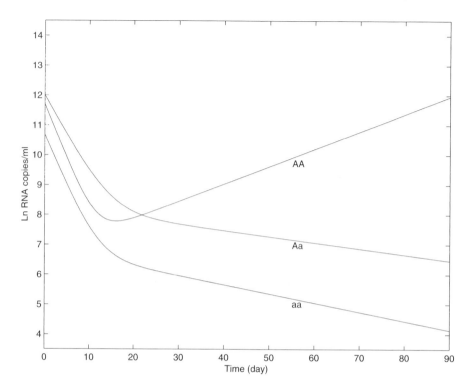

FIGURE 6.5
Load trajectories of HIV-1 virions (measured in viral RNA copies) for 53 patients
(with shadow curves) during the first 12 weeks after the initiation of HAART for the
ACTG 315 data.

The log-likelihood ratio (LR) test statistic was calculated as 31.42, greater than
the critical threshold (24.67, $\alpha = 0.01$) determined from simulated data under the
null hypothesis that there is no major gene. The three genotypes at the detected
major gene displayed marked differentiations in their viral load trajectories (Fig.
6.5). Their curves are described by

$$V_2(t) = e^{11.6959 - 0.3677t} + e^{6.6988 + 0.0584t} \quad \text{for genotype } AA,$$

$$V_1(t) = e^{11.9814 - 0.2726t} + e^{8.2763 - 0.0202t} \quad \text{for genotype } Aa,$$

and

$$V_0(t) = e^{10.6641 - 0.3477t} + e^{6.8662 - 0.0303t} \quad \text{for genotype } aa.$$

The heterozygote (*Aa*) and one homozygote (denoted by *aa*) that together account
for an overwhelming majority of patients ($\sim$90%) were found to decline consistently
with time in viral load after initiation of antiretroviral drugs. For the second homozygote (*AA*) in the frequency of $\sim$10%, viral load turns out to increase from day

20 following a short period of decline after the drug administration. We estimated allele frequencies for the major gene detected, which are 0.32 for the allele causing the increase of viral load after day 20 and 0.68 for the allele maintaining a consistent decline. No significant Hardy–Weinberg disequilibrium (D) was detected according to the log-likelihood ratio test.

The two decay rates of different virus compartments, λ_1 and λ_2, have clinical values in AIDS trials (Wu and Ding, 1999). The half-lives of productively infected cells ($t_{1/2} = \log 2/\lambda_1$) were estimated as (1.9, 2.5, 2.0) days for three major genotypes AA, Aa, and aa, respectively. The corresponding $t_{1/2}$ values of long-lived and/or latently infected cells ($\log 2/\lambda_2$) were estimated as (11.9, 34.4, 22.9) days. While the major gene detected has a nonsignificant effect on the half-life of productively infected cells ($LR_{\lambda_1} = 3.29$, d.f. $= 2$, $P > 0.05$), it triggers a significant effect on the half-life of long-lived and/or latently infected cells ($LR_{\lambda_2} = 10.30$, d.f. $= 2$, $P < 0.01$). The control of the detected major gene over the half-times of long-lived and/or latently infected cells suggests a possibility of altering second-phase viral load trajectories through a gene therapy strategy.

We calculated the eradication times for the two compartments, productively infected cells, and long-lived and/or latently infected cells, approximated by $\tau_1 = \log \lambda_1/\lambda_1$ and $\tau_2 = \log \lambda_2/\lambda_2$, respectively (Wu and Ding, 1999). The estimated eradication times for the three QTL genotypes are (6.7, 9.1, 6.8) days for the first virus compartment and (32.6, 104.8, 63.6) days for the second virus compartment. We found that the major gene detected displayed significant effects on the eradication times for the two compartments ($LR_{\tau_1} = 6.84$, d.f. $= 2$, $P < 0.05$ and $LR_{\tau_2} = 13.19$, d.f. $= 2$, $P < 0.01$). The genetic regulation of eradication times may assist clinicians in optimizing drug therapy on the basis of each patient's genetic constitution.

Additional note: In example 6.1, a single major gene is assumed to affect HIV dynamics. This simplified assumption can be relaxed by modeling any more genes involved. This will need the determination of an optimal number of major genes. When the number of genes, m, (and therefore the number of genotypes, $J = 3^m$) increases, the likelihood value based on the mixture model will increase. Various criteria have been proposed to measure a model's suitability by balancing model fit and model complexity (Lo et al., 2001; Chen et al., 2004). Celeux and Soromenho (1996) developed a method for choosing the number of mixture components based on normalized entropy criterion (NEC). According to NEC, the log-likelihood, $\log L_J(\Omega|\mathbf{y})$ (simplified as L_J), with J components can be decomposed into two parts, i.e.,

$$L_J = C_J + E_J,$$

where

$$C_J = \sum_{i=1}^{n} \sum_{j=1}^{3^m} P_{j|i} \log \left[\omega_j f_j(\mathbf{y}_i) \right]$$

is the classification log-likelihood, with $P_{j|i} = \dfrac{\omega_j f_j(\mathbf{y}_i)}{\sum_{j'=1}^{J} \omega_{j'} f_j(\mathbf{y}_i)}$, and

$$E_J = -\sum_{i=1}^{n}\sum_{j=1}^{3^m} P_{j|i}\log P_{j|i}$$

is the entropy that measures the overlap of the mixture components.

The normalized entropy criterion to be minimized for assessing the number of mixture components is given by

$$\mathrm{NEC}_J = \frac{E_J}{L_J - L_1}.$$

But since NEC_1 is not defined as can be seen from above, we are unable to compare situations $J = 1$ vs. $J > 1$ directly with NEC_J. Here, we use a general procedure for the one component case to decide between $J =$ and $J > 1$ (Biernacki et al., 1999). Let J^* be the value that minimizes NEC_J ($2 \le J \le J_{sup}$) with J_{sup} being an upper bound for the number of mixture components. We choose J^* if $\mathrm{NEC}_{J^*} \le 1$, otherwise we declare one component for the data. $\quad\square$

6.2.3 Functional Mapping Based on Ho et al.'s Kinetic Model

By assuming the steady state for a system, Ho et al. (1995), Wei et al. (1995), and Perelson et al. (1996) derived a different type of kinetic model to describe the time-dependent total concentration ($V(t)$) of plasma virions (including infectious, $V_I(t)$, and noninfectious, $V_{NI}(t)$) after antiviral treatment (reviewed in Perelson and Nelson (1999)). This function is expressed as

$$V(t) = V_0 e^{-ct} + \frac{cV_0}{c-\delta}\left\{\frac{c}{c-\delta}\left[e^{-\delta t} - e^{-ct}\right] - \delta t e^{-ct}\right\}, \qquad (6.15)$$

with

$$V_I(t) = V_0 e^{-ct}, \qquad (6.16)$$

for infectious virions and

$$V_{NI}(t) = \frac{cV_0}{c-\delta}\left\{\frac{c}{c-\delta}\left[e^{-\delta t} - e^{-ct}\right] - \delta t e^{-ct}\right\}, \qquad (6.17)$$

for noninfectious virions, where V_0 are the concentration of viral particles in plasma at time 0 (the time of onset of the drug effect), δ is the rate of loss of virus-producing cells, and c is the rate constant for virion clearance. This model is different from the equation derived by Wei et al. (1995) in which the loss of infected cells and virion clearance due to the use of inhibitors of HIV-1 protease cannot be distinguished.

Using the mathematical model for viral dynamics (6.15) and non-linear least squares fitting of the data, Perelson et al. (1996) obtained five distinct curves, redrawn in Fig.

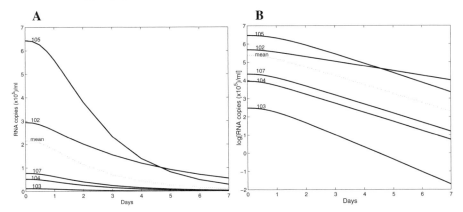

FIGURE 6.6

HIV-1 dynamic curves (plasma concentrations, copies per milliliter) for 5 patients 102, 103, 104, 105, and 107 measured at 16 time points. The dot curve represents the mean curve of the 5 patients. The variation in viral load among the 5 patients after drug treatment decreases substantially with time for the untransformed curves (**A**), but it is roughly constant over time for the log-transformed curves (**B**). The constant variation is a prerequisite for the use of mathematically tractable AR(1) model to approximate the 16-dimensional residual (co)variance matrix (Σ).

6.6, each corresponding to a different patient. These differences in curve shape can be explained by differences in the three estimated curve parameters (V_0, c, δ); for example, the estimated c values ranged from 2.06 to 3.81 day^{-1}, with a coefficient of variation (CV) of 21%, and the estimated δ values ranged from 0.26 to 0.68 day^{-1}, with a CV of 27% (see Table 1 of Perelson et al. (1996)). These authors further estimated several clinically important variables which include the average life spans ($1/c$) and average half-lives ($\ln 2/c$) of plasma virions, the average life spans ($1/\delta$) and average half-lives ($\ln 2/\delta$) of productively infected cells, the average viral generation time (defined as the time from the release of a virion until it infects another cell and causes the release of a new generation of viral particles and calculated by $1/c + 1/\delta$), and the duration of the HIV-1 life cycle (defined as the time from the release of a virion until the release of its first progeny virus). As shown in Table 2 of Perelson et al. (1996), the 5 patients examined display remarkable variation in these clinical variables, with the coefficients of variation ranging from 8% to 36%. These discrepancies suggest that genes exist in humans to affect HIV-1 dynamics *in vivo*.

Kinetic models (6.15), (6.16), and (6.17) have been incorporated in functional mapping to map quantitative trait loci (QTLs) for time-dependent change of infectious virions and infectious virions (Zhu et al., 2003). A simulation study was performed to test statistical behavior of functional mapping. Phenotypic longitudinal and marker data were simulated through a QTL that links these two types of data. When HIV dynamics has a small heritability ($H^2 = 0.1$), the power of functional mapping to detect the genetic effect of an underlying QTL is about 50%, but it can in-

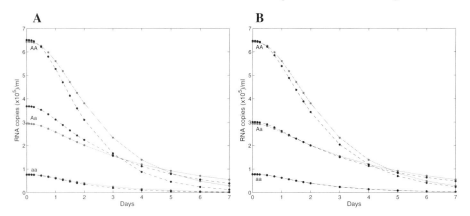

FIGURE 6.7

Estimated HIV dynamics curves (broken, black) for each of the three QTL genotypes, *AA*, *Aa*, and *aa*, in a comparison with the given curves (solid, red) used to
simulate individual curves, under heritabilities of $H^2 = 0.1$ (**A**) and $H^2 = 0.4$ (**B**).
The dots denote the time points at which patients were measured by mimicking in
Perelson et al. (1996).

crease to 100% with a higher heritability ($H^2 = 0.4$). The curve parameters (V_0, c, δ)
for each QTL genotype can be estimated with reasonably high accuracy, having the
estimated values more consistent with the hypothesized values under a heritability of
$H^2 = 0.4$ than 0.1 (Fig. 6.7).

As described by equation (6.15), the total concentration of virions in plasma contains two components due to infectious (equation (6.16)) and noninfectious virions
(equation (6.17)). The QTL detected with equation (6.15) governs the curves of the
total viral load dynamics, which does not necessarily affect the dynamics of infectious and noninfectious virions, respectively. But the effects of the detected QTL
on these two compartments can be tested by formulating the null hypothesis that the
three QTL genotypes have an identical infectious or noninfectious viral curve, i.e.,
$V_{I,AA}(t) = V_{I,Aa}(t) = V_{I,aa}(t)$ or $V_{NI,AA}(t) = V_{NI,Aa}(t) = V_{NI,aa}(t)$.

6.3 Functional Mapping of Pharmacodynamics

6.3.1 Patterns of Genetic Control in Pharmacodynamics

As a biochemical process, pharmacodynamics is controlled by genes. It is possible
that many different types of genes are involved in the determination of this process,
but broadly these genes can be grouped into three types (Fig. 6.8). Each type of
genes controls variation in response–dose curves that represent the change in drug

effect as a function of dose. Assuming that there are two different genotypes at a gene, these three types of genes can be described as follows:

(1) **Faster-slower genes** control variation in the dose at which drug effect is maximal. While one genotype at a gene (faster) has maximal drug effect at higher dose levels, the other (slower) has maximal drug effect at lower doses. The faster-slower gene determines how quickly patients reach their physiological limits in drug response;

(2) **Higher-lower genes** determine variation in the maximum drug effect reached. A higher genotype can tolerate higher physiological limits than a lower genotype;

(3) **Earlier-later genes** are responsible for variation in the dose at which drug effect increases exponentially. One genotype responds to lower dose levels more quickly than to higher dose levels (earlier), whereas the genotype responds to higher dose levels more quickly than to earlier dose levels (later).

It is possible that these three types of genes may exist simultaneously in the same population so that such a population may contain mixtures of genotypes that vary along different axes of variation. Also, a single gene may be characterized by these three features of expression. A central challenge is to quantify how different components of variation controlled by specific genes contribute to the total genetic variation for response-dose curves in a population.

6.4 Sequencing Pharmacodynamics

6.4.1 Basic Model

Lin et al. (2005) incorporated the mathematical equation of pharmacodynamics into functional mapping to characterize DNA sequence variants that regulate the shape of response–dose curves. Consider a group of subjects randomly sampled from a Hardy–Weinberg population. Each subject is genotyped for single nucleotide polymorphisms (SNPs) at specific regions or the entire genome and measured repeatedly for the effect parameters of a drug at different C dosage levels. Let $\mathbf{y} = (y(1), ..., y(C))$ denote a phenotypic vector and $\mathbf{S}$ denote a SNP dataset. Consider two SNPs with the notation of alleles, haplotypes, diplotypes, and genotypes and their frequencies being given in Section 2.1 and Table 2.1. The observations of each SNP genotype are denoted by $n_{r_1 r_1'/r_2 r_2'}$ $(r_1 \geq r_1', r_2 \geq r_2' = 1, 0)$. Similar to equation (2.2) for a static trait, we formulate a joint likelihood for the longitudinal data ($\mathbf{y}$) and marker data ($\mathbf{S}$) as

$$\log L(\boldsymbol{\Theta}_p, \boldsymbol{\Theta}_q | \mathbf{y}, \mathbf{S}) = \log L(\boldsymbol{\Theta}_p | \mathbf{S}) + \log L(\boldsymbol{\Theta}_q | \mathbf{y}, \mathbf{S}, \boldsymbol{\Theta}_p), \qquad (6.18)$$

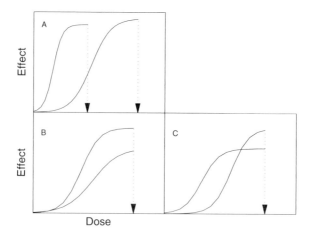

FIGURE 6.8
Hypothetical patterns of variation in response–dose curves. Whether or not there is a specific gene that controls these variation patterns can be tested and estimated by functional mapping.

where

$$\log L(\boldsymbol{\Theta}_p|\mathbf{S}) = \text{constant}$$
$$+2n_{11/11}\log p_{11}$$
$$+n_{11/10}\log(2p_{11}p_{10})$$
$$+2n_{11/00}\log p_{10}$$
$$+n_{10/11}\log(2p_{11}p_{01})$$
$$+n_{10/10}\log(2p_{11}p_{00}+2p_{10}p_{01})$$
$$+n_{10/00}\log(2p_{10}p_{00})$$
$$+2n_{00/11}\log p_{01}$$
$$+n_{00/10}\log(2p_{01}p_{00})$$
$$+2n_{00/00}\log p_{00}$$

$$\log L(\boldsymbol{\Theta}_q|\mathbf{y},\mathbf{S},\boldsymbol{\Theta}_p) =$$
$$\sum_{i=1}^{n_{11/11}}\log f_2(\mathbf{y}_i)$$
$$+\sum_{i=1}^{n_{11/10}}\log f_1(\mathbf{y}_i)$$
$$+\sum_{i=1}^{n_{11/00}}\log f_0(\mathbf{y}_i)$$
$$+\sum_{i=1}^{n_{10/11}}\log f_1(\mathbf{y}_i)$$
$$+\sum_{i=1}^{n_{10/10}}\log[\phi f_1(\mathbf{y}_i)+(1-\phi)f_0(\mathbf{y}_i)]$$
$$+\sum_{i=1}^{n_{10/00}}\log f_0(\mathbf{y}_i)$$
$$+\sum_{i=1}^{n_{00/11}}\log f_0(\mathbf{y}_i)$$
$$+\sum_{i=1}^{n_{00/10}}\log f_0(\mathbf{y}_i)$$
$$+\sum_{i=1}^{n_{00/00}}\log f_0(\mathbf{y}_i)$$

$$(6.19)$$

assuming that [11] is a risk haplotype, where $f_j(\mathbf{y}_i)$ is a multivariate normal distribution density function of composite diplotype j ($j = 2, 1, 0$).

Lin et al. (2005) used the Emax model (6.1) to model the mean vector contained in $f_j(\mathbf{y}_i)$ so that a set of composite diplotype-specific curve parameters $\boldsymbol{\Omega}_{u_j} = (E_0, E_{\max}, EC_{50}, H)$ are estimated. In practice, longitudinal data can be adjusted to remove inter-individual differences in the baseline parameter E_0. Thus, adjusted data are modeled using a new Emax equation,

$$E(c) = \frac{E_{\max} c^H}{EC_{50}^H + c^H},\tag{6.20}$$

so that parameters $\Omega_{u_j} = (E_{\max j}, EC_{50j}, H_j)$ are estimated for composite diplotype j.

The EM algorithm described in Section 2.3 is implemented to estimate haplotype frequencies, whereas the simplex algorithm is implemented to estimate the curve parameters Ω_{u_j} and the parameters that model the covariance structure by the AR(1) or SAD(1). An optimal risk haplotype is determined as one that provides the largest likelihood value among all possible assumed risk haplotypes. Hypothesis tests are performed following to the procedure described in Section 6.2.2.3.

6.4.2   A Pharmacogenetic Study of Heart Rate Responses

6.4.2.1   Introduction

Cardiovascular disease, principally heart disease and stroke, is the leading killer for both men and women among all racial and ethnic groups. Dobutamine is a medication that is used to treat congestive heart failure by increasing heart rate and cardiac contractility, with actions on the heart similar to the effect of exercise. Dobutamine is also commonly used to screen for heart disease in those unable to perform an exercise stress test. It is this latter use for which the study participants received dobutamine in this study. It is a synthetic catecholamine that primarily stimulates β-adrenergic receptors (βAR), which play an important role in cardiovascular function and responses to drugs (Nabel, 2003; Johnson and Terra, 2002).

Both the β1AR and β2AR genes have several polymorphisms that are common in the population (Genet, 2002). Two common polymorphisms are located at codons 49 (Ser49Gly) and 389 (Arg389Gly) for the β1AR gene and at codons 16 (Arg16Gly) and 27 (Gln27Glu) for the β2AR gene (Nabel, 2003). The polymorphisms in each of these two receptor genes are in linkage disequilibrium, which suggests the importance of taking into account haplotypes, rather than a single polymorphism, when defining biologic function. This study attempts to detect haplotype variants within these candidate genes, which determine the response of heart rate to varying concentrations of dobutamine.

6.4.2.2   Study Design

A group of 163 patients ranging in age from 32 to 86 years old participated in this study. Patients had a wide range of testing (untreated) heart rate. Each of these subjects was genotyped for SNP markers at codons 49 and 389 within the β1AR gene and at codons 16 and 27 within the β2AR gene. Dobutamine was injected into these subjects to investigate their response in heart rate to this drug. The subjects received increasing doses of dobutamine, until they achieved target heart rate response or predetermined maximum dose. The dose levels used were 0 (baseline), 5, 10, 20, 30, and 40 mcg/min, at each of which heart rate was measured. The time interval of 3 minutes is allowed between two successive doses for subjects to reach a plateau in response to that dose. Only those (107) in whom there were heart rate data at all the 6 dose levels were included for data analyses.

6.4.2.3   Results

By assuming that one haplotype is different from the rest of the haplotypes, we hope to detect a particular DNA sequence associated with the response of heart rate to dobutamine. The phenotypic data for drug response were normalized as percentages to remove the baseline effect, which is due to between-subject differences in heart rate prior to the test. At the $\beta 1AR$ gene, we did not find any haplotype that contributed to inter-individual difference in heart rate response. A significant effect was observed for haplotype Gly16(G)–Glu27(G) within the $\beta 2AR$ gene (Table 6.2). The log-likelihood ratio (LR) test statistics for the difference of GG from the other three haplotypes was 30.03, which is significant at $p = 0.021$ based on the critical threshold determined from 1000 permutation tests. The LR values when selecting any haplotype rather than GG as a reference gave no significant results ($p = 0.16$–0.40). We used a second testing criterion based on the area under curve (AUC) to test the haplotype effect. This test supports the result from the first test.

The maximum likelihood estimates (MLEs) of the population genetic parameters, such as haplotype frequencies, allele frequencies, and linkage disequilibrium between the two SNPs within the $\beta 2AR$ gene were obtained from our model. As indicated by the sampling errors of the MLEs based on the approach of Louis (1982), these estimates display reasonable precision. The allele frequencies within this gene are estimated as 0.62 for Gly16 at codon 16 and 0.40 for Glu27 at codon 27. The MLE of the linkage disequilibrium between the two SNPs is 0.1303. These suggest that the two SNPs identified within the $\beta 2AR$ gene display a pretty high heterozygosity and linkage disequilibrium.

The MLEs of the quantitative genetic parameters were obtained, also with reasonable estimation precision (Table 6.2). Using the estimated response parameters, the profiles of heart rate response to increasing dose levels of dobutamine were drawn for three composite genotypes comprising of haplotypes GG and non-GG (symbolized by $\overline{GG}$) (Fig. 6.9). The composite homozygote [GG][GG] displayed consistently higher heart rate across all dose levels, especially at higher dose levels than the composite homozygote [$\overline{GG}$][$\overline{GG}$]. But the composite heterozygote had consistently the lowest curve at all dose levels tested. We used AUC to test in which gene action mode (additive or dominant) haplotypes affect drug response curves for heart rate. The testing results suggest that both additive and dominant effects are important in determining the shape of the response curve (Table 6.3), together accounting for about 14% of the observed variation in drug response. We did not detect evidence for haplotypes to have an effect on curve parameters, H and EC_{50}, for the heart rate response.

TABLE 6.2

Log-likelihood ratio (LR) test statistics of different haplotype models and the corresponding maximum likelihood estimates (MLEs) of population genetic (SNP allele frequencies and linkage disequilibria) and quantitative genetic parameters (drug response and covariance-structuring parameters) in a sample of 107 subjects within the $\beta 2AR$ gene. The asymptotic sampling errors of the MLEs are given in parentheses.

Composite genotype	Parameters	Risk haplotype $[A_{k_1}^1 A_{k_2}^2]$			
		[AC]	[AG]	[GC]	**[GG]**
	LR	12.14	7.41	3.97	**30.03**
	p-value	0.34	0.50	0.76	**0.02**
	Population genetic parameters				
	$\hat{p}_{A_{k_1}^1}$	0.38	0.38	0.62	**0.62(0.037)**
	$\hat{p}_{A_{k_2}^2}$	0.60	0.40	0.60	**0.40(0.039)**
	$\hat{D}$	0.13	-0.13	-0.13	**0.13(0.014)**
	Quantitative genetic parameters				
$[A_{k_1}^1 A_{k_2}^2][A_{k_1}^1 A_{k_2}^2]$	$\hat{E}_0$	0.11	0.00	0.10	**0.11(0.021)**
	$\hat{E}_{\max}$	0.37	0.80	0.37	**0.75(0.263)**
	$\widehat{EC}_{50}$	23.72	13.14	21.35	**42.10(15.903)**
	$\hat{H}$	1.93	0.01	2.34	**1.73(0.288)**
$[A_{k_1}^1 A_{k_2}^2][\overline{A_{k_1}^1 A_{k_2}^2}]$	$\hat{E}_0$	0.10	0.11	0.10	**0.10(0.012)**
	$\hat{E}_{\max}$	0.37	0.88	0.40	**0.39(0.062)**
	$\widehat{EC}_{50}$	25.87	54.98	25.25	**29.27(4.902)**
	$\hat{H}$	1.95	2.35	2.12	**2.01(0.249)**
$[\overline{A_{k_1}^1 A_{k_2}^2}][\overline{A_{k_1}^1 A_{k_2}^2}]$	$\hat{E}_0$	0.11	0.10	0.11	**0.10(0.013)**
	$\widehat{E}_{\max}$	0.50	0.41	0.47	**0.39(0.042)**
	$\hat{EC}_{50}$	31.09	26.68	32.27	**23.57(2.683)**
	$\hat{H}$	1.99	1.99	1.82	**2.04(0.203)**
	$\hat{\rho}$	0.89	0.89	0.89	**0.88(0.014)**
	$\hat{\sigma}^2$	0.01	0.01	0.01	**0.01(0.001)**

The LR tests for the significance of an overall genetic effect triggered by the two SNPs. The optimal haplotype model detected on the basis of the LR test is indicated in boldface. There are two alleles Arg16 (A) and Gly16 (G) at codon 16 and 2 alleles Gln27 (C) and Glu27 (G) at codon 27.

TABLE 6.3
Testing results for two drug response parameters, H and EC_{50}, and total
genetic, additive and dominant effects based on AUC in 107 subjects under
the optimal haplotype model [GG].

Test	H	EC_{50}	Genetic	Additive	Dominance
LR	0.7566	4.1837	17.6391	7.2456	19.8279
p-value	>0.05	>0.05	<0.01	<0.01	<0.01

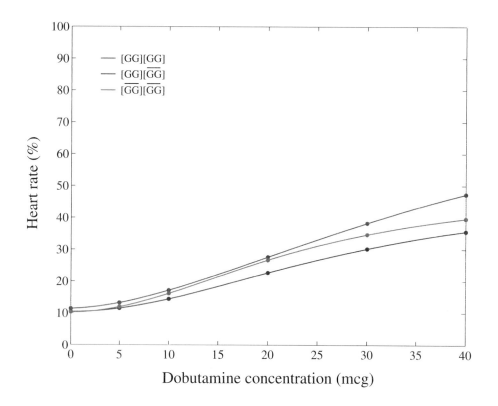

FIGURE 6.9
Profiles of heart rate in response to different dosages of dobutamine (indicated by
dots) for three composite genotypes (foreground) identified at two SNPs within the
β2AR gene. The profiles of 107 studied subjects from which the 3 different composite genotypes were detected are also shown (background).

6.4.2.4 Prospect

Results from the pharmacogenetic study of heart rate responses suggest that people carrying different diplotypes for two SNPs at codons 16 and 27 typed within the β2AR candidate gene may respond differently to dobutamine in heart rate. People with the diplotype consisting of two copies of haplotype GG are more sensitive in heart rate to increasing dosages of dobutamine than those with other haplotypes. This model provides a powerful tool for elucidating the genetic variants of drug response and, ultimately, designing personalized medications based on each patient's genetic constitution.

7

Haplotyping Drug Response by Linking Pharmacokinetics and Pharmacodynamics

Functional mapping described in Chapter 6 models drug response separately through pharmacokinetic (PK) and pharmacodynamic (PD) equations. This is a reasonable treatment because PK and PD data are generally established in separate, parallel studies to assist in the design of dosage schedules for subsequent evaluation in clinical trials (Toutain and Lees, 2004). An increasing trend is to introduce the concept of PK/PD modeling, an approach in which PK and PD data are generated in the same study and then modeled jointly to better quantifying the relationship between pharmacological effects of the drug and its dose as well as the time after the drug is administrated (Hochhaus and Derendorf, 1995). The PK-PD integration can provide a basis for selecting clinically rational dosage regimens (both dose and dosing interval) for subsequent evaluation in disease models and clinical trials. Functional mapping as an approach for studying dynamic genetic control has power to integrate these two pharmacological aspects as a whole through a genetic network.

In this chapter, we will describe a statistical model and computational algorithm for detecting and identifying the genetic variants that control an integrative PK and PD process. This model capitalizes on the information of DNA sequence generated by single nucleotide polymorphisms (SNPs) and embeds mathematical functions of linking PK and PD processes into a mixture model framework. Also, we will introduce a covariance-structuring model based on autoregressive moving average (ARMA) models. In particular, the ARMA-based mapping model provides a general platform for vigorous modeling of the covariance matrix and can largely enhance computational efficiency.

7.1 A Unifying Model for Functional Mapping

7.1.1 Clinical Design

A clinical trial is designed to determine the genetic control of inter-individual variability in drug response and its PK and PD mechanisms. The design of PK studies involves the administration of drug, and the assay of the concentration of the drug and its metabolites sampled from biological fluids such as blood, plasma, urine or

saliva at pre-specified times. PD studies collect the data that characterize the pharmacological effect of the drug described by a biomarker, surrogate marker or clinical
end point. For a PK-PD linking study, drug effects are measured at a series of time
points after the drug was administrated with each of many dosage levels.

Suppose there are n subjects randomly sampled from a human population at Hardy-
Weinberg equilibrium. These are genotyped for a battery of single nucleotide polymorphisms (SNPs) from candidate regions of chromosomes or the entire genome,
and measured for the effect of a drug at T time points and D doses after it is administrated. Let $y_i(d,t)$ be the observed drug effect for subject i at a particular time t
($t = 1,\ldots,T$) and dosage d ($d = 1,\ldots,D$).

Genes that control PK and PD aspects of drug response may be complicated. Yet,
for the sake of simplicity, the description of the haplotyping model will focus on
a quantitative trait nucleotide (QTN) composed of two different SNPs. Multiple
QTNs, each of which may be composed of more than two SNPs, as well as their
genetic interactions can be modeled in a similar way. Each of the two SNPs studied
has two alleles, labeled as 1 and 0, leading to four diplotypes [11], [10], [01], and
[00]. Use the same notation for allele, haplotype, diplotype and genotype frequencies
as given in Section 2.1 and Table 2.1. Let $n_{r_1 r_1'/r_2 r_2'}$ ($r_1 \geq r_1', r_2 \geq r_2' = 1,0$) be the
observations of each SNP genotype. To detect haplotype effects on drug response
to different dosage during a time course, we need to incorporate a PK-PD linking
model into functional mapping.

We assume that [11] is the risk haplotype (denoted as A) and [10], [01] and [00]
are the non-risk haplotype (denoted as $\bar{A}$) for the PK process, while [10] is the risk
haplotype (denoted as B) and [11], [01] and [00] are the non-risk haplotype (denoted
as $\bar{B}$) for the PD process. Under above assumptions, three possible composite diplo-
types are AA, $A\bar{A}$, and $\bar{A}\bar{A}$ for PK and BB, $B\bar{B}$, and $\bar{B}\bar{B}$ for PD. Thus, the expected
drug response of composite diplotypes in terms of PK and PD can be denoted as $\mathbf{u}_{j_1 j_2}$
($j_1, j_2 = 2$ for AA or BB, 1 for $A\bar{A}$ or $B\bar{B}$, and 0 for $\bar{A}\bar{A}$ or $\bar{B}\bar{B}$, respectively (Table
7.1).

7.1.2 Likelihood

At a particular time t and dose d, the relationship between the observed and expected
drug effects for subject i can be described by a linear regression model,

$$y_i(d,t) = \sum_{j_1=0}^{2} \sum_{j_2=0}^{2} \xi_{ij_1 j_2} u_{j_1 j_2}(d,t) + e_i(d,t),   \tag{7.1}$$

where $\xi_{ij_1 j_2}$ is the indicator variable denoted as 1 if a particular composite diplotype combination $j_1 j_2$ is considered for subject i and 0 otherwise and $e_i(d,t)$ is the
residual errors at time i and dose d that is iid normal with mean zero and variance
$\sigma^2(d,t)$. The covariance matrix Σ among different times and doses is factorized into
different blocks, expressed as

TABLE 7.1
Possible diplotype configurations of nine genotypes at two SNPs that affect pharmacokinetics (PK) and pharmacodynamics (PD).

Genotype	Diplotype	Diplotype freq.	Relative diplotype freq. within Genotypes	Haplotype Composition 11	12	21	22	Observation	Genotypic Mean Vector	Parameters for Genotypic Mean Vector PK	PD
11/11	[11][11]	p_{11}^2	1	1	0	0	0	$n_{11/11}$	$\mathbf{u}_{20}$	k_{e2}	(E_{max0}, EC_{500})
11/12	[11][12]	$2p_{11}p_{12}$	1	$\frac{1}{2}$	$\frac{1}{2}$	0	0	$n_{11/12}$	$\mathbf{u}_{11}$	k_{e1}	(E_{max1}, EC_{501})
11/22	[12][12]	p_{12}^2	1	0	1	0	0	$n_{11/22}$	$\mathbf{u}_{12}$	k_{e1}	(E_{max2}, EC_{502})
12/11	[11][21]	$2p_{11}p_{21}$	1	$\frac{1}{2}$	0	$\frac{1}{2}$	0	$n_{12/11}$	$\mathbf{u}_{10}$	k_{e1}	(E_{max0}, EC_{500})
12/12	$\begin{cases} [11][22] \\ [12][21] \end{cases}$	$\begin{cases} 2p_{11}p_{22} \\ 2p_{12}p_{21} \end{cases}$	$\begin{cases} \omega \\ 1-\omega \end{cases}$	$\frac{1}{2}\omega$	$\frac{1}{2}(1-\omega)$	$\frac{1}{2}(1-\omega)$	$\frac{1}{2}\omega$	$n_{12/12}$	$\begin{cases} \mathbf{u}_{10} \\ \mathbf{u}_{01} \end{cases}$	$\begin{cases} k_{e1} \\ k_{e0} \end{cases}$	$\begin{cases} (E_{max0}, EC_{500}) \\ (E_{max1}, EC_{501}) \end{cases}$
12/22	[12][22]	$2p_{12}p_{22}$	1	0	$\frac{1}{2}$	0	$\frac{1}{2}$	$n_{12/22}$	$\mathbf{u}_{01}$	k_{e0}	(E_{max1}, EC_{501})
22/11	[21][21]	p_{21}^2	1	0	0	1	0	$n_{22/11}$	$\mathbf{u}_{00}$	k_{e0}	(E_{max0}, EC_{500})
22/12	[21][22]	$2p_{21}p_{22}$	1	0	0	$\frac{1}{2}$	$\frac{1}{2}$	$n_{22/12}$	$\mathbf{u}_{00}$	k_{e0}	(E_{max0}, EC_{500})
22/22	[22][22]	p_{22}^2	1	0	0	0	1	$n_{22/22}$	$\mathbf{u}_{00}$	k_{e0}	(E_{max0}, EC_{500})

$\omega = \frac{p_{11}p_{22}}{p_{11}p_{22}+p_{12}p_{21}}$ where p_{11}, p_{12}, p_{21} and p_{22} are the haplotype frequencies of [11], [12], [21], and [22], respectively. The PK and PD are assumed to be affected by different risk haplotypes, [11] for the PK and [12] for the PD.

$$\Sigma = \begin{pmatrix} \Sigma_1 & \cdots & \Sigma_{1D} \\ \vdots & \ddots & \vdots \\ \Sigma_{D1} & \cdots & \Sigma_D \end{pmatrix}, \tag{7.2}$$

where $\Sigma_1, \ldots, \Sigma_D$ are each the $(T \times T)$ covariance matrices among different time points under D different drug dose levels, respectively, and $\Sigma_{d_1 d_2}$ is a across-time covariance matrix between dosage d_1 and d_2 $(d_1, d_2 = 1, ..., D)$.

As like in Section 2.2, we construct a joint likelihood for longitudinal drug response data ($\mathbf{y}$) and SNP data ($\mathbf{S}$) to estimate population (Ω_p) and quantitative genetic parameters (Ω_q), which is expressed as

$$\log L(\Omega_p, \Omega_q | y, \mathbf{S}) = \log L(\Omega_p | \mathbf{S}) + \log L(\Omega_q | y, \mathbf{S}, \Omega_p). \tag{7.3}$$

The likelihood $(\log L(\Omega_p | \mathbf{S}))$ based on a multinomial function formulated to estimate population genetic parameters has the same form as equation (13.5) for a simple two-SNP model. The likelihood $(\log L(\Omega_q | \mathbf{y}, \mathbf{S}, \Omega_p))$ based on a Gaussian mixture model for estimating quantitative genetic parameters includes modeling longitudinal drug effects expressed as a function of dose and time. For the assumed risk haplotypes for PK and PD (Table 7.1), $\log L(\Omega_q | y, \mathbf{S}, \Omega_p)$ is given by

$$\begin{aligned} \log L(\Omega_p, \Omega_q | \mathbf{y}, \mathbf{S}) =\ & \sum_{i=1}^{n_{11/11}} \log f_{20}(\mathbf{y}_i) + \sum_{i=1}^{n_{11/12}} \log f_{11}(\mathbf{y}_i) \\ & + \sum_{i=1}^{n_{11/22}} \log f_{12}(\mathbf{y}_i) + \sum_{i=1}^{n_{12/11}} \log f_{10}(\mathbf{y}_i) \\ & + \sum_{i=1}^{n_{12/12}} \log[\omega f_{10}(\mathbf{y}_i) + (1-\omega) f_{01}(\mathbf{y}_i)] \\ & + \sum_{i=1}^{n_{12/22}} \log f_{01}(\mathbf{y}_i) + \sum_{i=1}^{n_{22/11}+n_{22/12}+n_{22/22}} \log f_{00}(\mathbf{y}_i), \end{aligned} \tag{7.4}$$

where

$$\mathbf{y}_i = \left[\underbrace{y_i(1,1), \ldots, y_i(1,T)}_{\text{dose 1}}, \ldots, \underbrace{y_i(D,1), \ldots, y_i(D,T)}_{\text{dose } D} \right]$$

is the vector for longitudinal drug effects measured at all times and under all dosages for patient i.

Because the phenotypic measurements of drug effect are continuously variable, it is reasonable to assume that $\mathbf{y}_i$ follows a multivariate normal distribution, as used in general quantitative genetic studies (Lynch and Walsh, 1998). The multivariate normal distribution for patient i who carries composite diplotype combination $j_1 j_2$ is expressed as

$$f_{j_1 j_2}(\mathbf{y}_i; \mathbf{u}_{j_1 j_2}, \Sigma) = \frac{1}{(2\pi)^{DT} |\Sigma|^{1/2}} \exp\left[-\frac{1}{2}(\mathbf{y}_i - \mathbf{u}_{j_1 j_2}) \Sigma^{-1}(\mathbf{y}_i - \mathbf{u}_{j_1 j_2})' \right],$$

with

$$\mathbf{u}_{j_1 j_2} = \left[\underbrace{u_{j_1 j_2}(1,1),\ldots,u_{j_1 j_2}(1,T)}_{\text{dose } 1},\ldots,\underbrace{u_{j_1 j_2}(D,1),\ldots,u_{j_1 j_2}(D,T)}_{\text{dose } D} \right]$$

being a vector of expected values for composite diplotype combination $j_1 j_2$ at different doses and times.

For the likelihood function (7.3), population genetic parameter vector Ω_p includes SNP haplotype frequencies (and therefore allele frequencies and linkage disequilibrium between two SNPs), and quantitative genetic parameter vector Ω_q contains the mean vectors of drug effects for different composite genotypes and the residual covariance matrix among different time points and dosages.

Although it is not perfect, the assumption of independence among D different dosages can facilitate our modeling and analysis. With this assumption, the likelihood (7.4) can be re-written as

$$
\begin{aligned}
&\log L(\Omega_p,\Omega_q|\mathbf{y},\mathbf{S}) \\
&= \sum_{i=1}^{n_{11/11}} \left\{ \log f_{20}[\mathbf{y}_i(1)]+\ldots+\log f_{20}[\mathbf{y}_i(D)] \right\} \\
&+ \sum_{i=1}^{n_{11/12}} \left\{ \log f_{11}[\mathbf{y}_i(1)]+\ldots+\log f_{11}[\mathbf{y}_i(D)] \right\} \\
&+ \sum_{i=1}^{n_{11/22}} \left\{ \log f_{12}[\mathbf{y}_i(1)]+\ldots+\log f_{12}[\mathbf{y}_i(D)] \right\} \\
&+ \sum_{i=1}^{n_{12/11}} \left\{ \log f_{10}[\mathbf{y}_i(1)]+\ldots+\log f_{10}[\mathbf{y}_i(D)] \right\} \\
&+ \sum_{i=1}^{n_{12/12}} \Big(\log\{\omega f_{10}[\mathbf{y}_i(1)]+(1-\omega)f_{01}[\mathbf{y}_i(1)]\}+\ldots \\
&\quad + \log\{\omega f_{10}[\mathbf{y}_i(D)]+(1-\omega)f_{01}[\mathbf{y}_i(D)]\} \Big) \\
&+ \sum_{i=1}^{n_{11/22}} \left\{ \log f_{01}[\mathbf{y}_i(1)]+\ldots+\log f_{01}[\mathbf{y}_i(D)] \right\} \\
&+ \sum_{i=1}^{n_{22/11}+n_{22/12}+n_{22/22}} \left\{ \log f_{00}[\mathbf{y}_i(1)]+\ldots+\log f_{00}[\mathbf{y}_i(D)] \right\}, \quad (7.5)
\end{aligned}
$$

where

$$
\begin{cases}
\mathbf{y}_i(1) = [y_i(1,1),\ldots,y_i(1,T)], \\
\quad\vdots \\
\mathbf{y}_i(D) = [y_i(D,1),\ldots,y_i(D,T)],
\end{cases}
$$

with dosage-specific mean vectors for composite genotype j specified by

$$
\begin{cases}
\mathbf{u}_{j_1 j_2}(1) = [u_{j_1 j_2}(1,1),\ldots,u_{j_1 j_2}(1,T)] \\
\vdots \\
\mathbf{u}_{j_1 j_2}(d) = [u_{j_1 j_2}(d,1),\ldots,u_{j_1 j_2}(d,T)] \\
\vdots \\
\mathbf{u}_{j_1 j_2}(D) = [u_{j_1 j_2}(D,1),\ldots,u_{j_1 j_2}(D,T)],
\end{cases}
\tag{7.6}
$$

and dosage-specific covariance matrices denoted by

$$
\Sigma_1,\ldots,\Sigma_d,\ldots,\Sigma_D,
\tag{7.7}
$$

respectively.

We will not directly estimate the elements in the mean vector and covariance matrix, rather than estimate the parameters that model the mean-covariance structures. This will depend on two factors. First, the parameters that model dynamic drug effects should be biologically meaningful. Second, statistical modeling of the mean-covariance structures will be informative and can be easily implemented in computing programs.

7.1.3 Modeling the Mean Vector

The integrative PK and PD process can be described by equation (6.4). This can be incorporated into the mean vector of drug effects at different time points for different dosages in equation (7.6). For a composite diplotype combination, we will have the unknown vector for three PK and PD parameters $\Omega_{u_{j_1 j_2}} = (k_{e j_1}, E_{\max j_2}, EC_{50 j_2})$ to describe time-dependent drug effects for a particular dosage d, expressed as

$$
\mathbf{u}_{j_1 j_2}(d) = \left(\frac{E_{\max j_2} d e^{-k_{e j_1} t_1}}{EC_{50 j_2} V_d + d e^{-k_{e j_1} t_1}}, \ldots, \frac{E_{\max j_2} d e^{-k_{e j_1} t_T}}{EC_{50 j_2} V_d + d e^{-k_{e j_1} t_T}} \right).
$$

Note that the PK and PD are controlled by different risk haplotypes so that their parameters are fit by different composite diplotypes (Table 7.1).

If different composite genotypes have different combinations of these parameters, this implies that this nucleotide sequence plays a role in governing the differentiation of the PK-PD link. Thus, by testing for the difference of $\Omega_{u_{j_1 j_2}}$ among different composite genotypes, we can determine whether there exists a specific sequence variant that confers an effect on the PK-PD link.

7.1.4 Modeling the Covariance Structure

The AR and structured antedependence (SAD) models have been proposed to model the structure of the covariance matrix for longitudinal traits measured at multiple time points (Diggle et al., 2002; Zimmerman and Núñez-Antón, 2001). For the joint modeling of PK and PD, the autoregressive moving average (ARMA) model will be incorporated to structure the covariance structure. There are three major advantages for ARMA model in approximating the covariance function:

(1) The closed-form of the autocovariance matrix for the ARMA process has been derived (Haddad, 2004), which allows the expression of autocovariance function as a function of the AR $(\phi_1, \ldots, \phi_r)$ and MA coefficients $(\theta_1, \ldots, \theta_s)$,

(2) The derivation of the inverse and determinant of the covariance matrix for the ARMA process by Haddad (2004) has largely facilitated the computation of the likelihood function (7.5),

(3) As shown above, the ARMA model is a general form of the commonly used AR model, which makes the ARMA-based structuring approach more useful in practice.

At a given dosage d, the residual error at time t in equation (7.1) depends on its previous errors (deterministic) and on a random disturbance (opportunistic). For all the following notation with dosage, such as $e_i(d,t)$, we will omit the symbol d unless it is specified. If the dependence of $e_i(t)$ on the previous r errors is further assumed to be linear, we can write

$$e_i(t) = \phi_1 e_i(t-1) + \phi_2 e_i(t-2) + \ldots + \phi_r e_i(t-r) + \widetilde{\varepsilon}_i(t), \qquad (7.8)$$

where constants $(\phi_1, \ldots, \phi_r)$ are called autoregressive (AR) coefficients, and $\widetilde{\varepsilon}_i(t)$ is the disturbance at time t and is usually modelled as a linear combination of zero-mean, uncorrelated random variables or a zero-mean white noise process, $\varepsilon_i(t)$, i.e.,

$$\widetilde{\varepsilon}_i(t) = \varepsilon_i(t) + \theta_1 \varepsilon_i(t-1) + \theta_2 \varepsilon_i(t-2) + \ldots + \theta_s \varepsilon_i(t-s), \qquad (7.9)$$

in which $\varepsilon(t)$ is a white noise process with mean 0 and variance σ^2 if and only if $E[\varepsilon(t)] = 0$, $E[\varepsilon^2(t)] = 0$ for all t, and $E[\varepsilon(t_1), \varepsilon(t_2)] = 0$ if $t_1 \neq t_2$, where E denotes the expectation. The constants $(\theta_1, \ldots, \theta_s)$ are called the moving average (MA) coefficients. Combining equations (7.8) and (7.9) we have

$$\begin{aligned} & e_i(t) - \phi_1 e_i(t-1) - \ldots - \phi_r e_i(t-r) \\ &= \varepsilon_i(t) + \theta_1 \varepsilon_i(t-1) + \ldots + \theta_s \varepsilon_i(t-s), \end{aligned} \qquad (7.10)$$

which is defined as a zero-mean ARMA process of order r and s, or ARMA(t,s). We will omit the symbol d for the following notation unless it is specified. For a nonzero stationary ARMA process, two sequences at different time points can be connected through the backward shift operator $\mathcal{B}$, i.e.,

$$\mathcal{B}^k e_i(t) = e_i(t-k).$$

Define the autoregressive polynomial $\phi(x)$ and moving average polynomial $\theta(x)$ as

$$\begin{aligned} \phi(x) &= 1 - \phi_1 x - \phi_2 x^2 - \ldots - \phi_r x^r, \\ \theta(x) &= 1 + \theta_1 x + \theta_2 x^2 + \ldots + \theta_s x^s, \end{aligned}$$

with the assumption that $\phi(x)$ and $\theta(x)$ have no common factors. Equation (7.10) can be written in form

$$\phi(\mathcal{B})e(t) = \theta(\mathcal{B})\varepsilon(t). \tag{7.11}$$

Two special cases: When $s = 0$ only the AR part remain and equation (7.10) reduces to a pure autoregressive process of order r denoted by AR(r). Similarly, if $r = 0$, we obtain a pure moving average process of order s, MA(s). For these two cases, we have

$$e_i(t) - \phi_1 e_i(t-1) - \ldots - \phi_r e_i(t-r) = \phi(\mathcal{B})e(t) = \varepsilon(t) \text{ for AR}(r)$$
$$e_i(t) = \varepsilon_i(t) + \theta_1 \varepsilon_i(t-1) + \ldots + \theta_s \varepsilon_i(t-s) = \theta(\mathcal{B})\varepsilon(t) \text{ for MA}(s).$$

When neither r nor s is zero, an ARMA(r,s) model is sometimes referred to as a "mixed model".

Several efficient methods can be used to compute the exact ARMA covariance, Σ_d, for a particular dosage d. Here, we use the general matrix representation of Van der Leeuw (1994) in closed form to describe the ARMA covariance. Following Van der Leeuw (1994), we define two special types of Toeplitz matrices. The first one is a square $(T \times T)$ Toeplitz matrix which is given for the AR and MA parameters is, respectively, by

$$\mathbf{R} = \begin{bmatrix} \mathbf{R}_1 & \mathbf{O} \\ \mathbf{R}_2 & \mathbf{R}_3 \end{bmatrix}, \quad \mathbf{S} = \begin{bmatrix} \mathbf{S}_1 & \mathbf{O} \\ \mathbf{S}_2 & \mathbf{S}_3 \end{bmatrix} \tag{7.12}$$

where

$$\mathbf{R}_1 = \begin{bmatrix} 1 & 0 & 0 & \ldots & 0 \\ \phi_1 & 1 & 0 & \ldots & 0 \\ \phi_2 & \phi_1 & 0 & \ldots & 0 \\ \vdots & \vdots & \vdots & \ddots & \vdots \\ \phi_{r-1} & \phi_{r-2} & \phi_{r-3} & \ldots & 1 \end{bmatrix}_{r \times r} \quad \mathbf{S}_1 = \begin{bmatrix} 1 & 0 & 0 & \ldots & 0 \\ \theta_1 & 1 & 0 & \ldots & 0 \\ \theta_2 & \theta_1 & 0 & \ldots & 0 \\ \vdots & \vdots & \vdots & \ddots & \vdots \\ \theta_{s-1} & \theta_{s-2} & \theta_{s-3} & \ldots & 1 \end{bmatrix}_{s \times s},$$

$$\mathbf{R}_2 = \begin{bmatrix} \phi_r & \phi_{r-1} & \phi_{r-2} & \ldots & \phi_1 \\ 0 & \phi_r & \phi_{r-1} & \ldots & \phi_2 \\ 0 & 0 & \phi_r & \ldots & \phi_3 \\ \vdots & \vdots & \vdots & \ddots & \vdots \\ 0 & 0 & 0 & \ldots & \phi_r \\ 0 & 0 & 0 & \ldots & 0 \\ \vdots & \vdots & \vdots & \ddots & \vdots \\ 0 & 0 & 0 & \ldots & 0 \end{bmatrix}_{(T-r) \times r} \quad \mathbf{S}_2 = \begin{bmatrix} \theta_r & \theta_{r-1} & \theta_{r-2} & \ldots & \theta_1 \\ 0 & \theta_r & \theta_{r-1} & \ldots & \theta_2 \\ 0 & 0 & \theta_r & \ldots & \theta_3 \\ \vdots & \vdots & \vdots & \ddots & \vdots \\ 0 & 0 & 0 & \ldots & \theta_r \\ 0 & 0 & 0 & \ldots & 0 \\ \vdots & \vdots & \vdots & \ddots & \vdots \\ 0 & 0 & 0 & \ldots & 0 \end{bmatrix}_{(T-s) \times s}$$

$$\mathbf{R}_3 = \begin{bmatrix} 1 & 0 & 0 & \cdots & 0 & 0 & \cdots & 0 \\ \phi_1 & 1 & 0 & \cdots & 0 & 0 & \cdots & 0 \\ \phi_2 & \phi_1 & 1 & \cdots & 0 & 0 & \cdots & 0 \\ \vdots & \vdots & \vdots & \ddots & \vdots & \vdots & \ddots & \vdots \\ \phi_r & \phi_{r-1} & \phi_{r-2} & \cdots & 1 & 0 & \cdots & 0 \\ \phi_r & \phi_{r-1} & \phi_{r-2} & \cdots & \phi_1 & 1 & \cdots & 0 \\ \vdots & \vdots & \vdots & \ddots & \vdots & \vdots & \ddots & \vdots \\ 0 & 0 & 0 & & \phi_r & \phi_{r'} & \phi_{r'-1} & \cdots & 1 \end{bmatrix}_{(T-r) \times (T-r)}$$

$$\mathbf{S}_3 = \begin{bmatrix} 1 & 0 & 0 & \cdots & 0 & 0 & \cdots & 0 \\ \theta_1 & 1 & 0 & \cdots & 0 & 0 & \cdots & 0 \\ \theta_2 & \theta_1 & 1 & \cdots & 0 & 0 & \cdots & 0 \\ \vdots & \vdots & \vdots & \ddots & \vdots & \vdots & \ddots & \vdots \\ \theta_s & \theta_{s-1} & \theta_{s-2} & \cdots & 1 & 0 & \cdots & 0 \\ \theta_s & \theta_{s-1} & \theta_{s-2} & \cdots & \theta_1 & 1 & \cdots & 0 \\ \vdots & \vdots & \vdots & \ddots & \vdots & \vdots & \ddots & \vdots \\ 0 & 0 & 0 & & \theta_s & \theta_{s'} & \theta_{s'-1} & \cdots & 1 \end{bmatrix}_{(T-s) \times (T-s)}.$$

The second defined matrix has dimension $T \times r$ for the AR parameters and $T \times s$ for the MA parameters, whose upper parts are a square ($r \times r$) or ($s \times s$) Toeplitz matrix. For these two types of parameters, we define, respectively,

$$\mathbf{U} = \begin{bmatrix} \mathbf{U}_1 \\ \mathbf{O} \end{bmatrix}, \quad \mathbf{V} = \begin{bmatrix} \mathbf{V}_1 \\ \mathbf{O} \end{bmatrix}, \tag{7.13}$$

where

$$\mathbf{U}_1 = \begin{bmatrix} \phi_r & \phi_{r-1} & \cdots & \phi_1 \\ 0 & \phi_r & \cdots & \phi_2 \\ \vdots & \vdots & \ddots & \vdots \\ 0 & 0 & \cdots & \phi_r \end{bmatrix}_{r \times r}, \quad \mathbf{V}_1 = \begin{bmatrix} \theta_s & \theta_{s-1} & \cdots & \theta_1 \\ 0 & \theta_s & \cdots & \theta_2 \\ \vdots & \vdots & \ddots & \vdots \\ 0 & 0 & \cdots & \theta_s \end{bmatrix}_{s \times s}$$

$$\mathbf{O} = \begin{bmatrix} 0 & 0 & \cdots & 0 \\ \vdots & \vdots & \ddots & \vdots \\ 0 & 0 & \cdots & 0 \end{bmatrix}_{(T-r) \times r}, \quad \mathbf{V}_1 = \begin{bmatrix} 0 & 0 & \cdots & 0 \\ \vdots & \vdots & \ddots & \vdots \\ 0 & 0 & \cdots & 0 \end{bmatrix}_{(T-s) \times s}$$

Given the error vectors,

$$\mathbf{e} = [e(t-1), e(t-2), \ldots, e(t-r+1), e(t-r)]^{\mathrm{T}},$$
$$\varepsilon = [\varepsilon(s-1), \varepsilon(s-2), \ldots, \varepsilon(t-s+1), \varepsilon(t-s)]^{\mathrm{T}},$$

we define the corresponding auxiliary vectors by

$$\bar{\mathbf{e}} = [e(-r+1), e(-r+2), \ldots, e(-1), e(0)]^{\mathrm{T}},$$
$$\bar{\varepsilon} = [\varepsilon(-s+1), \varepsilon(-s+2), \ldots, \varepsilon(-1), \varepsilon(0)]^{\mathrm{T}}.$$

We then write equation (7.10) in matrix form

$$[\mathbf{U} \ \mathbf{R}] \begin{bmatrix} \bar{\mathbf{e}} \\ \mathbf{e} \end{bmatrix} = [\mathbf{V} \ \mathbf{S}] \begin{bmatrix} \bar{\varepsilon} \\ \varepsilon \end{bmatrix} \text{ or } \mathbf{Re} = \mathbf{V}\bar{\varepsilon} - \mathbf{S}\varepsilon - \mathbf{U}\bar{\mathbf{e}},$$

from which a series of matrix operations can be performed to prove that the covariance matrix Σ_d corresponding to the ARMA(r, s) error specification is solution to the equation (Van der Leeuw, 1994),

$$\mathbf{R}\Sigma_d\mathbf{R}^{\mathrm{T}} = \mathbf{S}\mathbf{S}^{\mathrm{T}} + \mathbf{V}\mathbf{V}^{\mathrm{T}} + [\mathbf{U} \ \mathbf{O}]\Sigma_d[\mathbf{U} \ \mathbf{O}]^{\mathrm{T}} - [\mathbf{V} \ \mathbf{O}]\mathbf{S}^{\mathrm{T}}\Sigma_d^{-\mathrm{T}}[\mathbf{U} \ \mathbf{O}]^{\mathrm{T}}$$
$$- [\mathbf{U} \ \mathbf{O}]\mathbf{R}^{-1}\mathbf{S}[\mathbf{V} \ \mathbf{O}]^{\mathrm{T}}. \tag{7.14}$$

As shown in Van der Leeuw (1994), if the invertibility condition for the AR part holds, the covariance equation (7.14) has an unique solution, which is

$$\Sigma_d = \mathbf{R}^{-1}$$
$$\times [\mathbf{S}\mathbf{S}^{\mathrm{T}} + (\mathbf{R}\mathbf{V} - \mathbf{S}\mathbf{U})(\mathbf{P}_1^{\mathrm{T}}\mathbf{P}_1 - \mathbf{U}_1^{\mathrm{T}}\mathbf{U}_1)^{-1}(\mathbf{R}\mathbf{V} - \mathbf{S}\mathbf{U})^{\mathrm{T}}]\mathbf{R}^{-\mathrm{T}}. \tag{7.15}$$

By substituting $\mathbf{R} = \mathbf{I}$ and $\mathbf{Q} = \mathbf{0}$ in equation (7.15), the covariance matrix for the MA(s) model is obtained as

$$\Sigma_d = \mathbf{S}\mathbf{S}^{\mathrm{T}} + \mathbf{V}\mathbf{V}^{\mathrm{T}}. \tag{7.16}$$

By substituting $\mathbf{S} = \mathbf{O}$ and $\mathbf{V} = \mathbf{O}$, next pre-multiplying both sides of equation (7.14) and post-multiplying by its transpose, the covariance matrix for the AR(r) model is derived as

$$\Sigma_d = (\mathbf{R}^{\mathrm{T}}\mathbf{R} - \mathbf{U}\mathbf{U}^{\mathrm{T}})^{-1}. \tag{7.17}$$

Based on the expression for the inverse of the sum of two matrices (Rao 1973), we obtain the inverse of Σ_d as

$$\Sigma_d^{-1} = \mathbf{R}^{\mathrm{T}}\mathbf{S}^{-\mathrm{T}}[\mathbf{I}_T - \mathbf{M}(\mathbf{M}^{\mathrm{T}}\mathbf{M} + \mathbf{R}_1^{\mathrm{T}}\mathbf{R}_1 - \mathbf{U}_1\mathbf{U}_1^{\mathrm{T}})^{-1}\mathbf{M}^{\mathrm{T}}]\mathbf{S}^{-\mathrm{T}}\mathbf{R}, \tag{7.18}$$

with $\mathbf{M} = \mathbf{S}^{-1}\mathbf{R}\mathbf{V} - \mathbf{U}$. Given that the value of the determinant of $\mathbf{S}^{-1}\mathbf{R}$ is equal to one, we have

$$|\Sigma_d|$$
$$= \left| \mathbf{I}_T - \mathbf{S}^{-1} \begin{bmatrix} \mathbf{R}_1\mathbf{V}_1 - \mathbf{S}_1\mathbf{U}_1 \\ \mathbf{O} \end{bmatrix} (\mathbf{R}_1^{\mathrm{T}}\mathbf{R}_1 - \mathbf{U}_1\mathbf{U}_1^{\mathrm{T}})^{-1} \begin{bmatrix} \mathbf{R}_1\mathbf{V}_1 - \mathbf{S}_1\mathbf{U}_1 \\ \mathbf{O} \end{bmatrix}^{\mathrm{T}} \mathbf{S}^{-1} \right|$$
$$= \left| \mathbf{I}_r + (\mathbf{R}_1^{\mathrm{T}}\mathbf{R}_1 - \mathbf{U}_1\mathbf{U}_1^{\mathrm{T}})^{-1}(\mathbf{R}_1\mathbf{V}_1 - \mathbf{S}_1\mathbf{U}_1)^{\mathrm{T}}\widetilde{\mathbf{S}}_1^{\mathrm{T}}\widetilde{\mathbf{S}}_1(\mathbf{R}_1\mathbf{V}_1 - \mathbf{S}_1\mathbf{U}_1) \right|, \tag{7.19}$$

where $\widetilde{\mathbf{S}}_1$ is the $(T \times r)$ matrix, consisting of the first r columns of $\mathbf{S}^{-1}$. For the AR(r) model, the determinant is reduced to

$$\left|(\mathbf{R}_1^T \mathbf{R}_1 - \mathbf{U}_1 \mathbf{U}_1^T)^{-1}\right|,$$

which is independent of T.

With the above derivations, it can be seen that the parameters that model the covariance matrix are arrayed by $\boldsymbol{\Omega}_v = \{\phi_1(d), \ldots, \phi_r(d), \theta_1(d), \ldots, \theta_s(d), \sigma^2\}_{d=1}^{D}$ for D dose levels when the ARMA(r, s) model is implemented. The unknown vector can reduce to the AR(r) or MA(s) model, depending on practical problems.

7.2 Algorithms and Determination of Risk Haplotypes

Three groups of unknown parameters in the integrated sequence and functional mapping model need to be estimated. These parameters are haplotype frequencies, arrayed in $(\boldsymbol{\Omega}_p)$, that characterize population structure and diversity with multilocus markers, the curve parameters, arrayed in $(\boldsymbol{\Omega}_{u_j})$, that model the mean vector with biologically meaningful mathematical equations, and the parameters, arrayed in $(\boldsymbol{\Omega}_v)$, that model the structure of a large longitudinal covariance. The EM algorithm, originally proposed by Dempster et al. (1977), has been derived to obtain the maximum likelihood estimates (MLEs) of these parameters (Ma et al., 2002; Wu et al., 2004c). In some case, the EM algorithm derived to estimate $(\boldsymbol{\Omega}_p)$, and the simplex algorithm implemented to estimate $(\boldsymbol{\Omega}_{u_j})$ and $(\boldsymbol{\Omega}_v)$ are integrated for the solution of functional mapping (Zhao et al., 2004a).

The EM algorithm contains a loop of two steps (see Section 2.3). In the E step, the prior (ϕ) and posterior probabilities (Π_i) with which the double heterozygote 10/10 carries diplotype [11][00] are estimated, respectively. In the M step, these calculated probabilities are then used to estimate the haplotype frequencies and mean vector- and covariance-structuring parameters.

For either PK or PD, there are four possible choices for their risk haplotypes. In total, the PK-PD link needs 4^2 combinations of risk haplotypes. For each combination, a likelihood under the full model is calculated. The optimal combination of risk haplotypes for PK and PD is one in which there is the largest likelihood.

7.3 Hypothesis Tests

Functional mapping for linking the PK-PD process also allows the tests of a number of biologically or clinically meaningful hypotheses at the interface between gene actions and the pattern of drug response. The existence of specific genetic variants that affect the integrated PK-PD process can be tested by formulating the hypotheses as follows:

$$
\begin{cases}
H_0 : \Omega_{u_{j_1 j_2}} \equiv \Omega_u, \quad j_1, j_2 = 2, 1, 0 \\
H_1 : \text{at least one of the equalities above does not hold,}
\end{cases}
\tag{7.20}
$$

where H_0 corresponds to the reduced model, in which the data can be fit by a single drug response curve, and H_1 corresponds to the full model, in which there exist different dynamic curves to fit the data. The test statistic for testing the hypotheses (7.20) is calculated as the log-likelihood ratio (LR) of the reduced to the full model:

$$
\text{LR} = -2[\log L(\tilde{\Omega}_u, \tilde{\Omega}_v | \mathbf{y}) - \log L(\hat{\Omega}_p, \hat{\Omega}_{u_j}, \hat{\Omega}_v | \mathbf{y}, \mathbf{S})],
\tag{7.21}
$$

where the tildes and hats denote the MLEs of the unknown parameters under the H_0 and H_1, respectively. The LR is asymptotically χ^2-distributed with the number of degrees of freedom that is equivalent to the difference in unknown parameters between the H_1 and H_0. In practice, an empirical approach based on permutation tests can be used to determine the critical threshold. By repeatedly shuffling the relationships between marker genotypes and phenotypes, a series of the maximum log-likelihood ratios are calculated, from which the critical threshold is determined.

An alternative approach for testing the existence of DNA sequence variants that are responsible for the integrated PK-PD process can be based on the volume under the landscape (VUL). The VUL for a landscape for a given composite diplotype combination $j_1 j_2$ can be calculated by

$$
\begin{aligned}
\text{VUL}_{j_1 j_2} &= \int_{d_1}^{d_D} \int_{t_1}^{t_T} \frac{\text{E}_{\max j_2} d e^{-k_{e j_1} t}}{\text{EC}_{50 j_2} V_d + d e^{-k_{e j_1} t}} \, dt \cdot dd \\
&= \int_{d_1}^{d_D} \left[\frac{\text{E}_{\max j_2}}{k_{e j_1}} \ln \left(1 + \frac{d}{\text{EC}_{50 j_2} V_d} \right) \right] dd \quad \text{as } t_1 \to 0 \text{ and } t_T \to \infty
\end{aligned}
$$

The null hypothesis for the existence of a significant DNA sequence variant based on the VUL is expressed as $\text{VUL}_j \equiv \text{VUL}$. A similar LR test statistic can be calculated using equation (7.21) for statistical testing.

Although the PK (k_e) and PD parameters ($\text{E}_{\max}$ and EC_{50}) are assumed to be the same among different dose levels, functional mapping can be generalized to allow these parameters to vary over dosage. It is possible that different risk haplotypes are operational for pharmacodynamic effects under different dosages, which is the cause of haplotype $\times$ dosage interaction effects. For any pair of dosages, d_1 and d_2, this haplotype $\times$ dosage interaction can be tested by formulating the hypothesis tests of drug effect as follows:

$$
\begin{cases}
H_0 : \text{E}_{j_1 j_2}(d_1, t) - \text{E}_{j_1 j_2}(d_2, t) = \text{E}_{j'_1 j'_2}(d_1, t) - \text{E}_{j'_1 j'_2}(d_2, t), \quad j_1 \neq j'_1 \text{ or } j_2 \neq j'_2 \\
H_1 : \text{at least one of the equalities above does not hold,}
\end{cases}
$$

$$
\tag{7.22}
$$

where $E_{j_1 j_2}(d_p, t)$ or $E_{j'_1 j'_2}(d_p, t)$ ($p = 1, 2$) is the effect–time profile for combined composite genotype $j_1 j_2$ or $j'_1 j'_2$ under dosage d_1 or d_2, respectively, which is described by equation (6.4). The parameters under the H_0 of (15.34) can be estimated by posing appropriate constraints as shown in Lin and Wu (2005). If the null hypothesis is rejected, this means that the same haplotype may trigger different impacts on drug effect under different dosages.

Similarly, we can make a hypothesis test about haplotype × time interactions. The null hypothesis for this test is based on the effect–dosage profile with the relationship like $E_{j_1 j_2}(d, t_1) - E_{j_1 j_2}(d, t_2) = E_{j'_1 j'_2}(d, t_1) - E_{j'_1 j'_2}(d, t_2)$ for two given time points t_1 and t_2. In addition, individual curve parameters, such as $E_{\max}$, EC_{50}, and k_e, can be tested. The tests of these parameters are important for the design of personalized drugs to control particular diseases.

7.4 Computer Simulation

Computer simulation was used to examine the statistical properties of functional mapping used to haplotype the integrated PK-PD process. We simulate three different composite diplotypes with the PK and PD parameters that generate three landscapes of drug effect as a function of time and time (shown in Fig. 6.3) with equation (6.4). The three landscapes are over-crossed, suggesting that there are interactions among haplotypes, doses, and times.

The simulation will be based on the ARMA(r, s) model (7.10). For computational simplicity, we only consider the ARMA(1,1) parameters so that only ϕ_1, θ_1, and σ^2 are used to model the structure of the residual covariance matrix. A total of 100 unrelated individuals are assumed to randomly sample from a human population at Hardy–Weinberg equilibrium with respect to haplotypes. Table 7.1 tabulates the distribution of the frequencies of haplotypes constructed by two SNPs in terms of allele frequencies and linkage disequilibrium. The diplotypes derived from four haplotypes at these two SNPs affect the integrated PK and PD process. As two different processes, it is possible that PK and PD are controlled by different DNA sequence variants. By assuming arbitrarily different risk haplotypes for the PK (say [11]) and PD (say [10]), we can specify the mean vectors for each composite genotype (Table 7.1).

Longitudinal PK and PD data are simulated at six time points for four dose levels under the multivariate normal distribution using the PK (k_e) and PD curve parameters ($E_{\max}$ and EC_{50}) for three composite diplotypes, as given in Table 7.2. These curve parameters are determined in the ranges of empirical estimates of these parameters from pharmacological studies. Using the genetic variance due to the composite diplotypic difference in the volume under landscape, we calculate the residual variances under different heritability levels ($H^2 = 0.1$ and 0.4). These residual variances, plus given AR and MA coefficients, form a structured residual covariance matrix Σ.

TABLE 7.2

Maximum likelihood estimates of SNP population genetic parameters (allele frequencies and linkage disequilibrium), the curve parameters and matrix-structuring parameters for pharmacokinetics (PK) and pharmacodynamics (PD) responses. The numbers in parentheses are the square root mean square errors for the estimates.

Parameters	Composite genotype	True value	Heritability 0.1	0.4
	Population genetic parameters			
$p_1^{(1)}$		0.60	0.606(0.034)	0.606(0.035)
$p_1^{(2)}$		0.60	0.601(0.035)	0.602(0.039)
Δ		0.08	0.078(0.020)	0.080(0.020)
	Curve parameters: Pharmacokinetics (IV Bolus Model)			
k_e (h^{-1})	$[11][11]$	0.20	0.200(0.005)	0.200(0.002)
	$[11]\overline{[11]}$	0.30	0.300(0.006)	0.300(0.002)
	$\overline{[11]}\,\overline{[11]}$	0.40	0.402(0.015)	0.400(0.005)
	Curve parameters: Pharmacodynamics (Emax Model)			
E_{max}	$[10][10]$	0.60	0.610(0.074)	0.602(0.025)
	$[10]\overline{[10]}$	0.80	0.802(0.020)	0.800(0.009)
	$\overline{[10]}\,\overline{[12]}$	1.00	1.002(0.013)	0.100(0.005)
EC_{50} (mg/l)	$[10][10]$	0.01	0.011(0.006)	0.010(0.001)
	$[10]\overline{[10]}$	0.02	0.020(0.001)	0.020(4e-4)
	$\overline{[10]}\,\overline{[10]}$	0.03	0.030(0.001)	0.030(2e-4)
	Matrix-structuring parameters (ARMA(1,1) Model)			
ϕ_1		0.60	0.576(0.079)	0.571(0.076)
θ_1		0.90	0.897(0.111)	0.877(0.100)
Dose $d = 1$mg				
σ^2		2e-4	2e-4(2e-5)	
		3e-5		3e-5(3e-6)
Dose $d = 10$mg				
σ^2		0.006	0.006(5e-4)	
		0.001		0.001(8e-5)
Dose $d = 100$mg				
σ^2		0.039	0.039(0.003)	
		0.007		0.007(4e-4)
Dose $d = 1000$mg				
σ^2		0.111	0.111(0.009)	
		0.018		0.019(0.002)

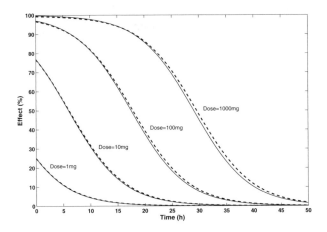

FIGURE 7.1
Estimated drug effects (dash) for a drug with a pharmacokinetic IV bolus model and a pharmacodynamic Emax model under four different doses in subjects with genotype 11/11. The consistency between the estimated and hypothesized (solid) curves suggests that this model can provide the precise estimation of the genetic control over response curves in patients.

The population genetic parameters of the SNPs can be estimated with reasonably high precision using the closed-form solution approach (Table 7.2). The estimates of these parameters are only dependent on sample size and are not related to the size of heritability. This model allows for the identification of correct risk haplotypes for the PK and PD processes. By considering all possible combinations of risk haplotypes for these two processes, we calculated the LR test statistics using equation (15.34) for each combination. The maximum LR value corresponds to the case in which the risk haplotypes for these two processes are consistent with the given ones (see Table 7.2).

Figure 7.1 visualizes the effect–time profile for the subjects with genotype 11/11 under four different doses in a comparison between the given and estimated curves. The estimated curves are consistent with the given curves, suggesting that this model can provide reasonable estimates of PK and PD curves. The parameters for the PK and PD models of each composite genotype can be estimated accurately and precisely (Table 7.2). As expected, the estimation precision (assessed by square roots of the MSEs) increases remarkably when the heritability increases from 0.1 to 0.4. The estimates of the ARMA(1,1) parameters that model the structure of the covariance matrix Σ also display reasonably high precision (Table 7.2). But it seems that their estimation precision is independent of the size of heritability.

In each of 100 simulations, we calculated the log-likelihood ratios (LR) for the hypothesis test of the presence of a genetic variant affecting both PK and PD processes. The LR values in each simulation under both heritability levels are strikingly higher than the critical threshold estimated from 100 replicates of simulations under the null hypothesis that there is no PK/PD-associated genetic variant. This suggests that this model has sufficient power to detect the genetic variant under given SNPs, curve and matrix-structuring parameters for the simulation.

7.5   Genetic and Statistical Considerations

The genetic control of pharmacokinetics (PK) and pharmacodynamics (PD) is one of the most important aspects in current pharmacogenomic or pharmacogenetic research (Watters and McLeod, 2003). Historically, this kind of study has been complicated by two factors. First, there was no adequate information available about the DNA structure and organization of the human genome. The recent release of the haplotype map (HapMap) constructed by SNPs (Gibbs et al., 2003, 2005; Frazer et al., 2007) has made it possible to identify polymorphic sites at the DNA sequence level and further associate the DNA sequence variants with complex diseases or drug response. Second, drug response is a dynamic process, affected by time and drug concentration. It is insufficient to study the genetic basis for drug effects at a single time point and dosage because this is likely to provide biased results about the genetic regulation of pharmacodynamic actions of a drug.

This chapter describes a statistical model for characterizing genetic control mechanisms that underlie PK and PD processes. By considering the influences of both time and dosage, we have pushed a curve-based longitudinal problem to its landscape extension (see Fig. 6.3). This model is constructed with a finite mixture model framework founded on the tenets of our sequence (Lin and Wu, 2005) and functional mapping (Wu et al., 2002a, 2004a; Ma et al., 2002). This model has particular power to discern the discrepancy between the genetic mechanisms for PK and PD. With the aid of HapMap, this model allows for the genome-wide scan of genetic variants that are responsible for these two different but physiologically related processes.

This model can directly detect DNA sequences that code PK and PD processes. Thus, genetic information extracted from observed data by our model is more informative and precise than that from traditional approaches. In particular, clinically meaningful mathematical functions (Hochhaus and Derendorf, 1995; Giraldo, 2003; McClish and Roberts, 2003) have been integrated to a statistical framework, thus displaying significant advantages. First, it facilitates statistical analysis and strengthens power to detect significant genetic variants because fewer parameters are needed to be estimated, as opposed to traditional multivariate analysis. Second, the results it produces are close to the biological realm given that the mathematical functions used are founded on a firm understanding of pharmacology (Derendorf and Meibohm, 1999). Based on differences in the PK-PD landscape, as shown in Figure 6.3, we can test for the genetic control regarding how pharmacokinetic actions change over time (the time surface) and how pharmacodynamic effects vary over dosage (the dosage surface) by estimating the slopes of each surface. The statistical power of our model is further increased by the attempt to structure the covariance matrix with autoregressive models (Diggle et al., 2002).

This model has relied upon the modeling and analysis of the residual covariance matrix. A time series model–autoregressive moving average (ARMA) model–was incorporated into the mixture model context. The ARMA model has been commonly used in econometric research aimed to approach the time-dependent influence of errors (Van der Leeuw, 1994). This model is very flexible in that any order of a polynomial that model dynamic errors can be determined. Furthermore, there is a surprisingly simple form for its description of the covariance matrix. The closed forms of the inverse and determinant of the matrix can greatly enhance computational efficiency (Van der Leeuw, 1994; Haddad, 2004). With the ARMA model, the AR(1) model that has been used to model the covariance structure (Ma et al., 2002; Lin and Wu, 2005) can be viewed as a special case.

This model can also be extended to consider various measurement schedules, in particular with uneven concentration lags varying from patient to patient. For these irregular measurement schedules, it is needed to formulate individualized likelihood functions when the mean vector and covariance matrix are modeled (Núñez-Antón and Woodworth, 1994; Núñez-Antón and Zimmerman, 2000; Zimmerman and Núñez-Antón, 2001). With these extensions, our model will assist in the discovery and characterization of a network of genes that influence drug response in terms of pharmacokinetics and pharmacodynamics.

8

Functional Mapping of Biological Clocks

Rhythmic phenomena have been thought to be an ubiquitous mechanism for every organism populating the earth to respond to daily and seasonal changes resulting from the planet's rotation and orbit around the sun (Goldbeter, 2002). There is a widely accepted view that the normal function of biological rhythms is strongly correlated with the genes that control them. For example, several studies have identified several so-called clock genes and clock-controlled transcription factors through gene mutants in animal models (Takahashi, 1993; Reppert and Weaver, 2002). The implications of these detected genes in clinical trials will hold a great promise for the determination of an individualized optimal body time for drug administration based on a patient's genetic makeup. It has been suggested that drug administration at the appropriate body time can improve the outcome of pharmacotherapy by maximizing potency and minimizing the toxicity of the drug (Levi et al., 1997), whereas drug administration at an inappropriate body time can induce severe side effects (Ohdo et al., 2001). In practice, body time-dependent therapy, termed "chronotherapy" (Labrecque and Belanger, 1991), can be optimized by implementing the patient's genes that control expression levels of his/her physiological variables during the course of a day.

With the completion of the Human Genome Project, it has been possible to draw a comprehensive picture of genetic control for the functions of the biological clock and, in the ultimate, integrate genetic information into routine clinical therapies for disease treatment and prevention. To achieve this goal, Liu et al. (2007) extend functional mapping to detect genes or quantitative trait loci that determine circadian rhythms as a dynamic complex trait. They incorporated mathematical models of rhythmic oscillations generated by complex cellular feedback processes (Scheper et al., 1999) into a mapping framework, aimed to better characterize the molecular mechanisms and functions of biological rhythms. These mathematical models have proven to be useful for investigating the dynamical bases of physiological disorders related to perturbations of biological behavior.

In this chapter, we will describe a statistical model for haplotyping patterns of rhythmic responses with random samples drawn from a natural population. The model is implemented with functional mapping and a system of differential equations for biological clocks. We will focus our description on the principle and procedure of constructing this model rather than the estimation of parameters contained in the model. We hope that the introduction of this chapter leads to a promising area in which mathematical algorithms for solving differential equations are integrated with functional mapping to address fundamental questions about the genetic control of a periodic biological system.

8.1   Mathematical Modeling of Circadian Rhythms

In all organisms studied so far, circadian rhythms that allow the organisms to adapt to their periodically changing environment originate from the negative autoregulation of gene expression. Scheper et al. (1999) illustrated and analyzed the generation process of the circadian rhythm that involves a cascade of the reaction loop as depicted in Figure 8.1A. The reaction loop consists of the production of the effective protein from its mRNA and a negative feedback from the effective protein on the production of its mRNA. The protein production process involves the translation and subsequent processing steps, such as phosphorylation, dimerization, transport and nuclear entry. It is assumed that the protein production cascade and the negative feedback are nonlinear processes in the reaction loop (Fig. 8.1**B**), with a delay time between the protein production and subsequent processing. It is these nonlinearities and the delay time that critically determine the free-running periodicity in the feedback loop.

Scheper et al. (1999) proposed a coupled differential equation to analyze the circadian behavior of the intracellular oscillator, expressed as

$$\frac{dM}{dt} = \frac{r_M}{1 + \left(\frac{P}{k}\right)^h} - q_M M$$
$$\frac{dP}{dt} = r_P M(t - \tau)^m - q_P P \tag{8.1}$$

where M and P are the relative concentration of mRNA and the effective protein measured at a particular time, respectively, r_M is the scaled mRNA production rate constant, r_P is the protein production rate constant, q_M and q_P are the mRNA and protein degradation rate constants, respectively, h is the Hill coefficient, m is the exponent of the nonlinearity in the protein production cascade, τ is the total duration of protein production from mRNA, and k is a scaling constant.

Equation (8.1) constructs an unperturbed (free-running) system of the intracellular circadian rhythm generator that is defined by seven parameters $\{h, m, \tau, r_M, r_P, q_M, q_P\}$. The behavior of this system can be determined and predicted by the change of these parameter combinations. For a given quantitative trait nucleotide (QTN), if the parameter combination is different among genotypes, this implies that this QTN is involved in the regulation of circadian rhythms in organisms. Statistical models will be developed to infer such genes from observed molecular markers such as single nucleotide polymorphisms (SNPs).

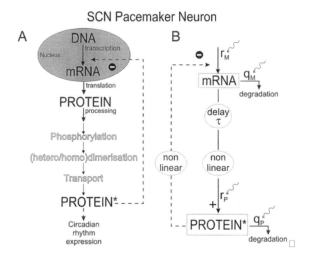

FIGURE 8.1

(**A**) Diagram of the biological elements of the protein synthesis cascade for a circadian rhythm generator. (**B**) Model interpretation of **A** by showing the delay (τ) and nonlinearity in the protein production cascade, the nonlinear negative feedback, as well as the mRNA and protein production (r_M, r_P) and degradation (q_M, q_P).

8.2 Haplotyping Circadian Rhythms

8.2.1 Study Design

To study the genetic control of circadian rhythms and haplotype DNA sequences that are associated with these processes, we sample a group of people from a natural human population at Hardy–Weinberg equilibrium. SNPs are genotyped for these subjects. As usual, our description of a haplotyping model will be based on two associated SNPs (**S**). The extension of the model to include more SNPs can be made with no technical difficulty. Let 1 and 0 denote the alleles at each SNP, and thus four diplotypes formed by these two SNPs are [11], [10], [01], and [00]. Let $n_{r_1 r_1'/r_2 r_2'}$ ($r_1 \geq r_1', r_2 \geq r_2' = 1, 0$) be the observations of each of the nine resulting SNP genotypes. Following the same notation for allele, haplotype, diplotype, and genotype frequencies as given in Section 2.1 and Table 2.1, a multinomial likelihood can be constructed to estimate haplotype frequencies, from which the linkage disequilibrium between the two SNPs can be estimated.

Each sampled subject is measured for the relative concentrations of mRNA (M) and the effective protein (P) at a series of time points, $(1, \ldots, T)$, during a daily light-dark cycle. Thus, two sets of serial measurements are expressed as $[M(1), \ldots, M(T)]$ and $[P(1), \ldots, P(T)]$, respectively. According to differentiate equations (8.1), these two variables, modeled in terms of their change rates, are expressed as differences between two adjacent times, symbolized by

$$\begin{aligned} \mathbf{y} &= [P(2) - P(1), \ldots, P(T) - P(T-1)] \\ &= [y(1), \ldots, y(T-1)] \end{aligned}$$

for the mRNA change and

$$\begin{aligned} \mathbf{z} &= [M(2) - M(1), \ldots, M(T) - M(T-1)] \\ &= [z(1), \ldots, z(T-1)] \end{aligned}$$

for the protein change.

Equation (8.1) models two different but related biological variables, mRNA and protein, using three different groups of physiological parameters. Parameters r_M and q_M are those that specifically determine the mRNA process, whereas parameters r_P and q_P are specific to the protein process. Parameters h, m, and τ are those that determine both processes in an interactive manner. It is reasonable to generally assume that these three groups of physiological parameters are regulated by different risk haplotypes. A procedure can then be derived to test whether risk haplotypes are really parameter-specific in their control of circadian rhythms. Let [11] be the risk haplotype (denoted as A), and [10], [01], and [00] be the non-risk haplotype (denoted as $\bar{A}$) for parameters r_M and q_M. Let [11] be the risk haplotype (denoted as B), and [10], [01], and [00] be the non-risk haplotype (denoted as $\bar{B}$) for parameters r_P and q_P. Let [01] be the risk haplotype (denoted as C), and [11], [10], and [00] be the non-risk haplotype (denoted as $\bar{C}$) for parameters h, m, and τ.

Thus, for each of these three different groups of parameters, we have three composite diplotypes AA, $A\bar{A}$, and $\bar{A}\bar{A}$, BB, $B\bar{B}$, and $\bar{B}\bar{B}$, and CC, $C\bar{C}$, and $\bar{C}\bar{C}$, respectively. The expected mean vectors of composite diplotypes for these two rhythmic processes can be denoted as $\mathbf{u}_{j_1 j_2 j_3}$ ($j_1, j_2, j_3 = 2$ for AA, BB, or CC, 1 for $A\bar{A}$, $B\bar{B}$, or $C\bar{C}$, and 0 for $\bar{A}\bar{A}$, $\bar{B}\bar{B}$, or $\bar{C}\bar{C}$, respectively), which contain the unknown parameters arrayed by $\mathbf{\Omega}_{u_{j_1 j_2 j_3}} = (r_{Mj_1}, q_{Mj_1}, r_{Pj_2}, q_{Pj_2}, h_{j_3}, m_{j_3}, \tau_{j_3})$. Like Table 7.1, we make a Table 8.1 in which different haplotype structures of controlling mRNA (M) and protein (P) properties through different groups of physiological parameters are shown.

8.2.2 Antedependence Model

The time-dependent phenotypic changes of mRNA (**y**) and protein traits (**z**) for subject i measured at time t due to diplotype diversity can be expressed by a bivariate linear statistical model as

$$y_i(t) = \sum_{j_1=0}^{2} \sum_{j_2=0}^{2} \sum_{j_3=0}^{2} \xi_{ij_1 j_2 j_3} u_{M_{j_1 j_2 j_3}}(t) + e_i^y(t)$$

$$z_i(t) = \sum_{j_1=0}^{2} \sum_{j_2=0}^{2} \sum_{j_3=0}^{2} \xi_{ij_1 j_2 j_3} u_{P_{j_1 j_2 j_3}}(t) + e_i^z(t)$$

(8.2)

TABLE 8.1
Possible diplotype configurations of nine genotypes at two SNPs that affect mRNA (M) and protein (P) concentrations.

Genotype	Diplotype Configuration	Frequency	Haplotype Composition 11	10	01	00	Observation	Genotypic Mean Vector	Parameters for Genotypic Mean Vector M	P	Joint
11/11	[11][11]	p_{11}^2	1	0	0	0	$n_{11/11}$	$\mathbf{u}_{200}$	(r_{M2}, q_{M2})	(r_{P0}, q_{P0})	(h_0, m_0, τ_0)
11/10	[11][10]	$2p_{11}p_{10}$	$\frac{1}{2}$	$\frac{1}{2}$	0	0	$n_{11/10}$	$\mathbf{u}_{110}$	(r_{M1}, q_{M1})	(r_{P1}, q_{P1})	(h_0, m_0, τ_0)
11/00	[10][10]	p_{10}^2	0	1	0	0	$n_{11/00}$	$\mathbf{u}_{120}$	(r_{M1}, q_{M1})	(r_{P2}, q_{P2})	(h_0, m_0, τ_0)
10/11	[11][01]	$2p_{11}p_{01}$	$\frac{1}{2}$	0	$\frac{1}{2}$	0	$n_{10/11}$	$\mathbf{u}_{101}$	(r_{M1}, q_{M1})	(r_{P0}, q_{P0})	(h_1, m_1, τ_1)
10/10	$\left\{\begin{array}{l}[11][00]\\ [10][01]\end{array}\right.$	$\left\{\begin{array}{l}2p_{11}p_{00}\\ 2p_{10}p_{01}\end{array}\right.$	$\frac{1}{2}\phi$	$\frac{1}{2}(1-\phi)$	$\frac{1}{2}(1-\phi)$	$\frac{1}{2}\phi$	$n_{10/10}$	$\left\{\begin{array}{l}\mathbf{u}_{100}\\ \mathbf{u}_{011}\end{array}\right.$	$\left\{\begin{array}{l}(r_{M1}, q_{M1})\\ (r_{M0}, q_{M0})\end{array}\right.$	$\left\{\begin{array}{l}(r_{P0}, q_{P0})\\ (r_{P1}, q_{P1})\end{array}\right.$	$\left\{\begin{array}{l}(h_0, m_0, \tau_0)\\ (h_1, m_1, \tau_1)\end{array}\right.$
10/00	[10][00]	$2p_{10}p_{00}$	0	$\frac{1}{2}$	0	$\frac{1}{2}$	$n_{10/00}$	$\mathbf{u}_{010}$	(r_{M0}, q_{M0})	(r_{P1}, q_{P1})	(h_0, m_0, τ_0)
00/11	[01][01]	p_{01}^2	0	0	1	0	$n_{00/11}$	$\mathbf{u}_{002}$	(r_{M0}, q_{M0})	(r_{P0}, q_{P0})	(h_2, m_2, τ_2)
00/10	[01][00]	$2p_{01}p_{00}$	0	0	$\frac{1}{2}$	$\frac{1}{2}$	$n_{00/10}$	$\mathbf{u}_{001}$	(r_{M0}, q_{M0})	(r_{P0}, q_{P0})	(h_1, m_1, τ_1)
00/00	[00][00]	p_{00}^2	0	0	0	1	$n_{00/00}$	$\mathbf{u}_{000}$	(r_{M0}, q_{M0})	(r_{P0}, q_{P0})	(h_0, m_0, τ_0)

$\phi = \frac{p_{11}p_{00}}{p_{11}p_{00}+p_{10}p_{01}}$ where p_{11}, p_{10}, p_{01}, and p_{00} are the haplotype frequencies of [11], [10], [01], and [00], respectively. Different risk haplotypes are assumed for three different groups of physiological parameters for circadian rhythms, [11] for (r_M, q_M), [10] for (r_P, q_P), and [01] for h, m, τ.

where $\xi_{i j_1 j_2 j_3}$ is an indicator variable for the possible composite diplotype combination of subject i and defined as 1 if a diplotype combination $j_1 j_2 j_3$ is indicated and 0 otherwise, $u_{M_{j_1 j_2 j_3}}(t)$ and $u_{P_{j_1 j_2 j_3}}(t)$ are the genotypic values of diplotype combination $j_1 j_2 j_3$ for mRNA and protein changes at time t, respectively, which can be fit using the differentiate equation (8.1), and $e_i^y(t)$ and $e_i^z(t)$ are the residual effects of subject i at time t, including the aggregate effect of polygenes and error effects.

The dynamic features of the residual errors of these two traits can be described by the antedependence model (Gabriel, 1962; Núñez-Antón and Zimmerman, 2000; Zimmerman and Núñez-Antón, 2001). This model states that an observation at a particular time t depends on the previous ones, with the degree of dependence decaying with time lag. Assuming the first-order structured antedependence (SAD(1)) model, the relationship between the residual errors of the two traits y and z at time t for subject i can be modeled by

$$
\begin{aligned}
e_i^y(t) &= \phi_y e_i^y(t-1) + \psi_y e_i^z(t-1) + \varepsilon_i^y(t) \\
e_i^z(t) &= \phi_z e_i^z(t-1) + \psi_z e_i^y(t-1) + \varepsilon_i^z(t)
\end{aligned}
\tag{8.3}
$$

where ϕ_k and ψ_k are the antedependence parameters caused by trait k itself or by the other trait, respectively, and $\varepsilon_i^y(t)$ and $\varepsilon_i^z(t)$ are the time-dependent innovation error terms assumed to be bivariate normally distributed with mean zero and variance matrix,

$$
\Sigma_\epsilon(t) = \begin{pmatrix} \delta_x^2(t) & \delta_x(t)\delta_y(t)\rho(t) \\ \delta_x(t)\delta_y(t)\rho(t) & \delta_y^2(t) \end{pmatrix},
$$

where $\delta_x^2(t)$ and $\delta_y^2(t)$ are termed time-dependent "innovation variances" that can be described by a parametric function like a polynomial of time (Pourahmadi, 1999), but assumed to be constant in this study, and $\rho(t)$ is the correlation between the error terms of the two traits, specified by an exponential function of time t (Lin and Wu, 2005), but is assumed to be time-invariant for this study. It is reasonable to say that there is no correlation between the error terms of two traits at different time points, i.e., $\mathrm{Corr}(\varepsilon_i^y(t_y), \varepsilon_i^z(t_z)) = 0$ $(t_y \neq t_z)$.

Based on these above conditions, the covariance matrix (Σ) of phenotypic values for traits $\mathbf{y}$ and $\mathbf{z}$ can be structured in terms of ϕ_y, ϕ_z, ψ_y, ψ_z, and $\Sigma_\epsilon(t)$ by a bivariate SAD(1) model (Lin and Wu, 2005; Zhao et al., 2005). Also, the closed forms for the determinant and inverse of Σ can be derived and given in the above two publications. We use a vector of parameters arrayed in $\Omega_v = (\phi_y, \phi_z, \psi_y, \psi_z, \delta_y, \delta_z, \rho)$ to model the structure of the covariance matrix involved in the function mapping model.

8.2.3   Likelihood

Population genetic parameters can be estimated by formulating a likelihood ($\log L$ $(\Omega_p | \mathbf{S})$, where $\Omega_p = (p_{11}, p_{10}, p_{01}, p_{00})$ based on a multinomial function, which has the same form as equation (13.5) for a simple two-SNP model. Quantitative genetic parameters are estimated on the basis of the Gaussian mixture likelihood ($\log L$

$(\Omega_q|\mathbf{y},\mathbf{z},\mathbf{S},\Omega_p)$,where $\Omega_q = (\{\Omega_{u_{j_1 j_2 j_3}}\}^2_{j_1,j_2,j_3=0}, \Omega_v)$. For the assumed risk haplotypes for PK and PD (Table 7.1), $\log L(\Omega_q|y,\mathbf{S},\Omega_p)$ is given by

$$\log L(\Omega_q|\mathbf{y},\mathbf{z},\mathbf{S},\hat{\Omega}_p)$$

$$= \sum_{i=1}^{n_{11/11}} \log f_{200}(\mathbf{y}_i,\mathbf{z}_i) + \sum_{i=1}^{n_{11/10}} \log f_{110}(\mathbf{y}_i,\mathbf{z}_i)$$

$$+ \sum_{i=1}^{n_{11/00}} \log f_{120}(\mathbf{y}_i,\mathbf{z}_i) + \sum_{i=1}^{n_{10/11}} \log f_{101}(\mathbf{y}_i,\mathbf{z}_i)$$

$$+ \sum_{i=1}^{n_{10/10}} \log[\phi f_{100}(\mathbf{y}_i,\mathbf{z}_i) + (1-\phi)f_{011}(\mathbf{y}_i,\mathbf{z}_i)]$$

$$+ \sum_{i=1}^{n_{10/00}} \log f_{010}(\mathbf{y}_i,\mathbf{z}_i) + \sum_{i=1}^{n_{00/11}} \log f_{002}(\mathbf{y}_i,\mathbf{z})$$

$$+ \sum_{i=1}^{n_{00/10}} \log f_{001}(\mathbf{y}_i,\mathbf{z}) + \sum_{i=1}^{n_{00/00}} \log f_{000}(\mathbf{y}_i,\mathbf{z}), \qquad (8.4)$$

where

$$(\mathbf{y}_i,\mathbf{z}_i) = (y_i(1),z_i(1),...,y_i(T),z_i(T))$$

is the vector for two longitudinal physiological traits measured at all times for subject *i*.

In the likelihood (8.4), the multivariate normal density of $(\mathbf{y}_i,\mathbf{z}_i)$ contains diplotye-specific mean vectors, expressed as

$$\mathbf{u}_{j_1 j_2 j_3} = (\mathbf{u}_{M_{j_1 j_2 j_3}};\mathbf{u}_{P_{j_1 j_2 j_3}})$$

$$= \left\{ u_{M_{j_1 j_2 j_3}}(t);u_{P_{j_1 j_2 j_3}}(t) \right\}^{T-1}_{t=1}$$

$$= \left(\frac{r_{M_{j_1}}}{1 + \left(\frac{P(t+1)}{k_{j_3}}\right)^{h_{j_3}}} - q_{M_{j_1}} M(t+1); \right.$$

$$\left. r_{P_{j_2}}M(t+1-\tau_{j_3})^{m_{j_3}} - q_{P_{j_2}}P(t+1) \right)^{T-1}_{t=1} \qquad (8.5)$$

and the covariance matrix Σ modeled by the SAD(1).

8.2.4 Algorithm and Determination of Risk Haplotypes

A closed form for the EM algorithm is used to obtain the maximum likelihood estimates (MLEs) of haplotype frequencies p_{11}, p_{10}, p_{01}, and p_{00} (and therefore allele frequencies of the markers and their linkage disequilibrium). Diplotye-specific mathematical parameters in $\mathbf{u}_{j_1 j_2 j_3}$ for two differentiate equations of circadian rhythms and the parameters that specify the structure of the covariance matrix Σ can be theoretically estimated by implementing with the EM algorithm. But it would be difficult to derive the log-likelihood equations for these parameters because they are related in a complicated nonlinear way. The simplex algorithm that only relies upon a target function has proven to be powerful for estimating the MLEs of these parameters and can be used to estimate these parameters. Closed forms for the determinant and inverse of Σ structured by the SAD(1) model should be incorporated into the estimation process to increase computational efficiency.

A search for optimal risk haplotypes, each for a different group of circadian parameters, is important, although it is a tedious task. For each parameter group, there are four possibilities to choose a risk haplotype from two SNPs. Thus, 3 parameter groups will need to choose 4^3 combinations of risk haplotypes. The largest likelihood among these 64 combinations corresponds to the optimal combination of risk haplotypes for the 3 groups of circadian parameters.

8.3   Hypothesis Testing

One of the most significant advantages of functional mapping is that it can ask and address biologically meaningful questions at the interplay between gene actions and trait dynamics by formulating a series of hypothesis tests. Wu et al. (2004b) described several general hypothesis tests for different purposes. Although all these general tests can be directly used in this study, we here propose the most important and unique tests for the existence of a QTN that affect pleiotropically or separately the mRNA and protein changes and for the effects of the QTL on the shape of partial differentiate functions.

8.3.1   Existence of Risk Haplotypes

Testing whether there are risk haplotypes that are associated with differentiate equations (8.1) is a first step toward the understanding of the genetic architecture of circadian rhythms. The genetic control over the entire rhythmic process can be tested by formulating the following hypotheses:

$$
\begin{cases}
H_0 : (r_{M_{j_1}}, q_{M_{j_1}}) \equiv (r_M, q_M); (r_{P_{j_2}}, q_{P_{j_2}}) \\
\qquad \equiv (r_P, r_P); (h_{j_3}, m_{j_3}, \tau_{j_3}) \equiv (h, m, \tau), \\
H_1 : \text{at least one of the equalities above does not hold,}
\end{cases}
\tag{8.6}
$$

for $j_1, j_2, j_3 = 2, 1, 0$. The H_0 states that there are no risk haplotypes that affect circadian rhythms (the reduced model), whereas the H_1 proposes that such risk haplotypes do exist (the full model). The test statistic for testing the hypotheses (8.6) is calculated as the log-likelihood ratio of the reduced to the full model:

$$
\text{LR} = -2[\ln L(\tilde{\Omega}_q | \mathbf{y}, \mathbf{z}) - \ln L(\hat{\Omega}_q | \mathbf{y}, \mathbf{z}, \mathbf{S}, \hat{\Omega}_p,)],
\tag{8.7}
$$

where the tildes and hats denote the MLEs of the unknown parameters under H_0 and H_1, respectively, i.e.,

$$\tilde{\Omega}_q = (\tilde{r}_M, \tilde{q}_M, \tilde{r}_P, \tilde{q}_P, \tilde{h}, \tilde{m}, \tilde{\tau}, \tilde{\phi}_y, \tilde{\phi}_z, \tilde{\psi}_y, \tilde{\psi}_z, \tilde{\delta}_y, \tilde{\delta}_z, \tilde{\rho}),$$

and

$$\hat{\Omega}_q = (\{\hat{r}_{M_{j_1}}, \hat{q}_{M_{j_1}}\}_{j_1=0}^2, \{\hat{r}_{P_{j_2}}, \hat{q}_{P_{j_2}}\}_{j_2=0}^2, \{\hat{h}_{j_3}, \hat{m}_{j_3}, \hat{\tau}_{j_3}\}_{j_3=0}^2, \hat{\phi}_y, \hat{\phi}_z, \hat{\psi}_y, \hat{\psi}_z, \hat{\delta}_y, \hat{\delta}_z, \hat{\rho}).$$

The LR is asymptotically χ^2-distributed with the number of degrees of freedom that is the difference in unknown parameters between the two hypotheses.

8.3.2 Pleiotropic Effect on mRNA or Protein Rhythms

After the existence of significant risk haplotypes that affects circadian rhythms is confirmed, we need to test whether this detected QTN (composed of two SNPs) affects pleiotropically the rhythmic responses of mRNA and protein. The hypothesis for testing the effect of the risk haplotypes on the mRNA response is formulated as

$$\begin{cases} H_0 : \mathbf{u}_{M_{j_1 j_2 j_3}} \equiv \mathbf{u}_M \\ H_1 : \text{At least one of the equalities above does not hold.} \end{cases} \tag{8.8}$$

Similarly, a test is formulated for detecting the effect of the risk haplotypes on the protein rhythm expressed as

$$\begin{cases} H_0 : \mathbf{u}_{P_{j_1 j_2 j_3}} \equiv \mathbf{u}_P \\ H_1 : \text{At least one of the equalities above does not hold.} \end{cases} \tag{8.9}$$

For both hypotheses (8.8) and (8.9), the log-likelihood values under the H_0 and H_1 are calculated, and then the corresponding LR values are estimated in each case. The LR values asymptotically follow a χ^2-distribution. Yet, in practice, an empirical approach based on simulation studies is often used to determine the critical threshold. If the null hypotheses of (8.8) and (8.9) are both rejected, this means that these two SNPs pleiotropically exert an effect on the circadian rhythms of mRNA and protein in a haplotype form.

8.3.3 Risk Haplotypes for the Behavior and Shape of Circadian Rhythms

Two different subspaces of parameters are used to define the features of circadian rhythms, i.e., (h, m, τ) determining the nonlinearity and delay of the system, and (r_M, r_P, q_M, q_P) determining phase-response curves. The null hypotheses regarding the genetic control of the oscillatory behavior of the system and the shape of rhythmic responses are formulated as

$$\begin{cases} H_0 : (h_{j_3}, m_{j_3}, \tau_{j_3}) \equiv (h, m, \tau) \\ H_0 : (r_{M_{j_1}}, r_{P_{j_2}}, q_{M_{j_1}}, q_{P_{j_2}}) \equiv (r_M, r_P, q_M, q_P). \end{cases} \tag{8.10}$$

The oscillatory behavior of circadian rhythms can also be determined by the amplitude of the rhythm defined as the difference in the level between peak and trough values, the phase defined as the timing of a reference point in the cycle (e.g., the peak) relative to a fixed event (e.g., beginning of the night phase) and the period defined as the time interval between phase reference points (e.g., two peaks) is called the period. The genetic determination of all these criteria can be tested.

8.4   Simulation

Results from simulation experiments suggest that a haplotyping model can be used to detect genes for circadian rhythms through incorporating the mathematical functions that describe these biological processes. Simulation studies were based on 200 subjects genotyped for two associated SNPs from a human population at Hardy–Weinberg equilibrium. The phenotypic data of the subjects, the relative concentration of mRNA and protein, are simulated with biological parameters that are determined in the ranges of empirical estimates of these parameters (Scheper et al., 1999). The impact of curve heritability on the estimation was also examined. The covariance structure is modeled by antedependence parameters $(\phi_x, \phi_y, \varphi_x, \varphi_y)$ and innovation variances (δ_x^2, δ_y^2). The main discoveries from simulation studies are summarized below: (1) Risk haplotypes responsible for circadian rhythms can be detected with the haplotyping model. The parameters for rhythmic responses of each composite diplotyep can be estimated accurately with a sample size of 200 and under a modest heritability (0.1). (2) The model is powerful to discern the differences of genetic control triggered by different risk haplotypes, allowing the characterization of a detailed genetic network for circadian rhythms.

8.5   Fourier Series Approximation of Circadian Rhythms

8.5.1   Introduction

Almost all life phenomena populating the earth can be described in terms of periodic rhythms that result from the planet's rotation and orbit around the sun (Goldbeter, 2002). Cell division (Mitchison, 2003), circadian rhythms (Crosthwaite, 2004; Prolo et al., 2005), morphogenesis of periodic structures, such as somites in vertebrates (Dale et al., 2003), and complex life cycles of some microorganisms (Lakin-Thomas and Brody, 2004; Rovery et al., 2005) are all excellent representatives of biological rhythms. From a mechanistic perspective, rhythmic behavior arises in genetic and metabolic networks as a result of nonlinearities associated with various modes of cellular regulation. There have been considerable efforts to describe these nonlinearities by mathematical functions.

Frank (1926) was one of the first researchers who used the so-called Fourier series to approximate periodic and quasi-periodic biological phenomena. Attinger et al. (1966) provided a detailed mathematical formulation of this approximation to a biological rhythmic system through both theoretical and experimental approaches. More recently, Fourier series approximation as an analytical tool has been widely applied to study the mechanisms and patterns of biological rhythmicity, including the cyclic organization of preterm and term neonates during the neonatal period (Fig. 8.2) (Begum et al., 2006) and pharmacodynamics (Mager and Abernethy, 2007).

8.5.2   Fourier Model

An arbitrary time function $g(t)$, which satisfies Dirichlet's conditions can be expanded in the trigonometric series as

$$g(t) = a_0 + \sum_{r=1}^{\infty} \left[a_r \cos\left(\frac{2\pi r}{\tau}t\right) + b_r \sin\left(\frac{2\pi r}{\tau}t\right) \right], \qquad (8.11)$$

where

$$a_0 = \frac{1}{\tau} \int_{-\tau/2}^{\tau/2} g(t)dt,$$

$$a_r = \frac{2}{\tau} \int_{-\tau/2}^{\tau/2} \cos\left(\frac{2\pi r}{\tau}t\right) dt,$$

$$b_r = \frac{2}{\tau} \int_{-\tau/2}^{\tau/2} \sin\left(\frac{2\pi r}{\tau}t\right) dt,$$

and τ is the length of the interval (period) over which $g(t)$ is considered. Although the Dirichlet conditions are always satisfied for a biological system, the applicability of Fourier analysis is limited in a steady system in which there are no effects of transient responses (Attinger et al., 1966).

In equation (8.11), the coefficient a_0 represents the mean value of the function. The terms associated with the fundamental angular frequency $(a_1 \cos(\frac{2\pi}{\tau}t) + b_1 \sin(\frac{2\pi}{\tau}t))$ are commonly called the fundamental or first harmonic of the periodic function $g(t)$ and the remaining terms, whose frequencies are integer multiples of the fundamental, are referred to as the higher harmonics. The magnitude of the coefficients of the higher harmonics usually decreases with frequency so that a given function can be well approximated by only the first few terms of the series. The Fourier series representation for a periodic function $g(t)$ can be written as a finite number of terms, i.e.,

$$g(t) = a_0 + \sum_{r=1}^{R} \left[a_r \cos\left(\frac{2\pi r}{\tau}t\right) + b_r \sin\left(\frac{2\pi r}{\tau}t\right) \right], \qquad (8.12)$$

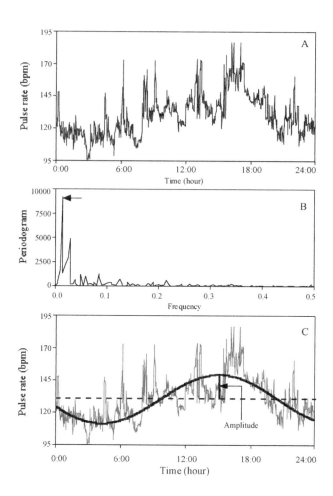

FIGURE 8.2

Fourier approximation of pulse rate. (A) Plot of original data measured once every 10 seconds and averaged into 1 minute time block for 1440 minutes. (B) Periodogram intensities for pulse rate (plotted on linear scale), with the largest peak of the periodogram indicated by the arrow. (C) The corresponding cycle of the largest peak in the periodogram intensities was reconstructed from the Fast Fourier Transform coefficient to fit the sinusoidal function. The detected cycle (period: 1440 minutes = 24 hours) is shown by the bold curve superimposed on the original data. Adapted from Begum et al. (2006).

from which we define a few important parameters that characterize the feature of the Fourier approximation. These parameters are:

$$c_r = \sqrt{a_r^2 + b_r^2}, \tag{8.13}$$

which is the magnitude of the rth harmonic, with c_0 being the constant term or the mean value of the function, and

$$\psi_r = \tan^{-1}\left(\frac{b_r}{a_r}\right), \tag{8.14}$$

which is the phase angle of the rth harmonic.

The exponential form of equation (8.12) can be derived, using the Euler identities for sine and cosine, as

$$g(t) = a_0 + \frac{1}{2}\sum_{r=-R}^{R} c_r \exp\left(i\left(\frac{2\pi r}{\tau}t - \psi_r\right)\right). \tag{8.15}$$

As compared with equation (8.12), equation (8.15) is much more easily manipulated (Goldman, 1963).

8.5.3 Genetic Haplotyping

8.5.3.1 Study Design and Model

We focus on an association study design in which genotype data are available for a dense set of SNPs on a panel of subjects randomly sampled from a natural population at Hardy–Weinberg equilibrium. The sampled subjects have been phenotyped for a periodic quantitative trait, such as pulse rate or respiratory rate, at a series of time points from 1 to T. Let $\mathbf{y}_i = (y_i(1), ..., y_i(T))$ be the serial measurements for subject i. Our strategy is to associate the pattern of haplotype diversity in these SNPs with the dynamic pattern of the longitudinal trait during a period.

For convenience, we consider a pair of SNPs that are linked closely on the same region and use 1 and 0 to denote alternative alleles at each SNP. If one haplotype constructed by these two SNPs is assumed as a risk haplotype (A), we will have three different composite diplotypes, AA (2), $A\bar{A}$ (1), and $\bar{A}\bar{A}$ (0). The haplotyping model being described will be to detect the differences in the dynamic pattern of the periodic trait among these three diplotypes. But because these diplotypes cannot always be distinguished from each other, a mixture-based likelihood should be constructed to estimate the diplotype-specific differences. Different from general mixture model problems, the proportions of components (denoted as Ω_p) within the mixture model constructed here have been estimated from a multinomial likelihood formulated with SNP observations (see Section 2.2).

The relationship between the phenotypic value of the trait ($\mathbf{y}_i$) for subject i and the composite diplotype of this subject at time t is described, by a linear model, as

$$y_i(t) = \sum_{j=0}^{2} \xi_{j|i} u_j(t) + e_i(t), \quad i = 1, ..., n \tag{8.16}$$

where $\xi_{j|i}$ is the indicator variable defined as 1 if a diplotype considered is compatible with subject i and as 0 otherwise, $u_j(t)$ is the genotypic value for composite diplotype j ($j = 2, 1, 0$) at time t, and $e_i(t)$ is the time-dependent residual error distributed as $N(0, \sigma^2(t))$. The errors at different time points can be correlated.

8.5.3.2 Modeling the Mean-Covariance Structure

Based on the idea of functional mapping, we can use the Fourier series approximation to model the periodic changes of genotypic values of different composite diplotypes, if the trait studied is periodically regulated by the underlying haplotypes. According to equation (8.12), the mean vector for composite diplotype j, $\mathbf{u}_j = (u_j(1), ..., u_j(\tau))$, can be modeled as

$$u_j(t) \approx \begin{cases} a_{0j} + a_{1j}\cos\left(\frac{2\pi}{T_j}t\right) + b_{1j}\sin\left(\frac{2\pi}{T_j}t\right), & \text{for the first order;} \\[2ex] a_{0j} + a_{1j}\cos\left(\frac{2\pi}{T_j}t\right) + b_{1j}\sin\left(\frac{2\pi}{T_j}t\right) \\ \quad + a_{2j}\cos\left(\frac{4\pi}{T_j}t\right) + b_{2j}\sin\left(\frac{4\pi}{T_j}t\right), & \text{for the second order;} \\[2ex] a_{0j} + a_{1j}\cos\left(\frac{2\pi}{T_j}t\right) + b_{1j}\sin\left(\frac{2\pi}{T_j}t\right) \\ \quad + a_{2j}\cos\left(\frac{4\pi}{T_j}t\right) + b_{2j}\sin\left(\frac{4\pi}{T_j}t\right) \\ \quad + a_{3j}\cos\left(\frac{6\pi}{T_j}t\right) + b_{3j}\sin\left(\frac{6\pi}{T_j}t\right), & \text{for the third order;} \\[2ex] \cdots, & \text{for any higher order.} \end{cases} \tag{8.17}$$

where a_{0j} is the fundamental frequency of composite diplotype j, a_{1j}, a_{2j}, a_{3j} and b_{1j}, b_{2j}, b_{3j} are the amplitude coefficients of composite diplotype j, which determine the times at which the diplotype achieves peak and trough expression levels of the trait, respectively, and T_j is the period of the trait expression of composite diplotype j.

The covariance matrix of $e_i(t)$ can be modeled by an appropriate approach, such as AR(1) or SAD(1), because it has a certain pattern. The Fourier parameters ($a_{0j}, a_{1j}, b_{1j}, ..., a_{Rj}, b_{Rj}$) that define the periodic changes of different composite diplotypes and the parameters that model the covariance structure can be estimated by the EM algorithm and Nelder-Mead simplex algorithm. All these unknown parameters are arrayed in Ω_q.

8.5.3.3 Model Selection

We need to select the best risk haplotype that explains diplotype variation. This can be done by assigning each of the four haplotypes, constituted by the two SNPs, as a risk haplotype and calculating the corresponding likelihoods. An optimal risk haplotype is one that gives the largest likelihood value.

For each composite diplotype, it is possible that there exists an optimal order of Fourier approximation to model its periodic pattern. Thus, we need an approach to select such an optimal order. One of the leading selection methods is the AIC. This is designed to be an approximately unbiased estimator of the expected Kullback-Leibler information of a fitted model. The minimum AIC produces a selected model which is close to the best possible choice. Within the context of the mixture model, Fourier order r is chosen, which minimizes

$$\text{AIC}(r) = -2\ln L(\hat{\boldsymbol{\Omega}}_q) + 2N(r),$$

where $N(r)$ is the number of independent parameters within the model. In the similar token, the Bayesian Information Criterion (BIC) (Schwarz, 1978) is also available, expressed as

$$\text{BIC}(r) = -2\ln L(\hat{\boldsymbol{\Omega}}_q) + N(r)\ln(n),$$

where $N(r)$ is the number of independent parameters in $\hat{\boldsymbol{\Omega}}_q$. The selected model is the one with the smallest BIC. There is no clear consensus on which criterion is best to use, although the empirical work of Fraley and Raftery (1998) seems to favor BIC. Since information criteria penalize models with additional parameters, the AIC and BIC model order selection criteria are based on parsimony. Notice that since BIC imposes a greater penalty for additional parameters than does AIC, BIC always provides a model with a number of parameters no greater than that chosen by AIC.

8.5.3.4 Hypothesis Testing

Functional mapping generates a number of biologically meaningful hypotheses at the interplay between diplotype variation and developmental patterns. The first important hypothesis concerns overall differences in the periodic trait among different composite diplotypes, which is expressed as

$$\begin{cases} H_0 : (a_{0j}, a_{1j}, b_{1j}, ..., a_{Rj}, b_{Rj}) \equiv (a_0, a_1, b_1, ..., a_R, b_R) \quad \text{for } j = 2, 1, 0 \\ H_1 : \text{At least one of the equalities above does not hold.} \end{cases} \quad (8.18)$$

The log-likelihood ratio (LR) test statistic is then calculated by

$$\text{LR} = -2[\ln L(\tilde{\boldsymbol{\Omega}}_q|\mathbf{y}) - \ln L(\hat{\boldsymbol{\Omega}}_q|\mathbf{y}, \hat{\boldsymbol{\Omega}}_p)], \quad (8.19)$$

where the tildes and hats stand for the MLEs of the unknown parameters under the null and alternative hypotheses, respectively. The critical threshold for claiming distinguishable periodic patterns among different composite diplotypes can be determined on the basis of permutation tests, although the LR value calculated with equation (8.19) is asymptotically χ^2-distributed.

Because each of the Fourier coefficients describes a different aspect of periodic curves, hypotheses can be made for these parameters individually. In addition, the peak to trough ratio, a_r/b_r, reflects the amplitude of periodic profile and can be tested for its differences among the composite diplotypes. For example, if the mean curve is modeled based on the Fourier series of order one, i.e.,

$$u(t) = \frac{1}{2}a_0 + a_1 \cos\left(\frac{2\pi t}{T}\right) + b_1 \sin\left(\frac{2\pi t}{T}\right), \quad (8.20)$$

the hypothesis can be expressed as

$$\begin{cases} H_0 : a_{1j}/b_{1j} \equiv a_1/b_1 \text{ for } j = 2,1,0 \\ H_1 : \text{At least one of the equalities above does not hold.} \end{cases}$$

Simulation studies can be conducted to determine the threshold values for each hypothesis test.

The slope of gene expression profile may change with time, which suggests the occurrence of diplotype $\times$ time interaction effects during a time course. The differentiation of $u_j(t)$ with respect to time t represents a slope of periodic expression for composite diplotype j. If the slopes at a particular time point t^* are different between the curves of different diplotypes, this means that significant diplotype $\times$ time interaction occurs between this time point and next. The test for diplotype $\times$ time interaction can be formulated with the hypotheses:

$$H_0 : \frac{d}{dt}u_j(t^*) \equiv \frac{d}{dt}u(t^*) \text{ vs. } H_1 : \frac{d}{dt}u_j(t^*) \neq \frac{d}{dt}u(t^*), \ j = 2,1,0. \quad (8.21)$$

The effect of diplotype $\times$ time interaction can be examined during a given time course.

It would be interesting to test the effects of haplotypes on two important features of Fourier series approximation, the magnitude (8.13) and phase angle of the rth harmonic (8.14). These can be done by formulating the following null hypotheses:

$$H_0 : a_{rj}^2 + b_{rj}^2 \equiv a_r^2 + b_r^2,$$

and

$$H_0 : \tan^{-1}\left(\frac{b_{rj}}{a_{rj}}\right) \equiv \tan^{-1}\left(\frac{b_r}{a_r}\right).$$

8.6 Further Considerations

Biological rhythms perform key physiological functions and bear on the search for optimal patterns of drug delivery (chronopharmacology). Most of these rhythms occur already at the cellular level. Theoretical models based on biochemical and genetic advances throw light on the conditions in which oscillations occur in regulated metabolic and genetic networks. An important example is provided by circadian rhythms. These rhythms, which have a free-running period close to 24h, are conspicuous by their ubiquity and by the role they play in allowing organisms to adapt to their periodically changing environment. In *Neurospora, Drosophila*, and mammals, circadian rhythms originate from negative autoregulatory feedback exerted by clock proteins on the expression of their genes. Models for such genetic regulatory networks predict the occurrence of sustained circadian oscillations of the clock proteins and their mRNAs. Numerical simulations of the models are used to probe the molecular mechanism of circadian oscillations and to account for the effects of various mutations of the clock genes. An example related to a sleep disorder in human physiology will briefly be discussed.

Oscillations abound in biological systems. Rhythmic behavior arises in genetic and metabolic networks as a result of nonlinearities associated with various modes of cellular regulation. In view of the large number of variables involved and of the complexity of the intertwined feedback processes that generate oscillations, computational models and numerical simulations are needed to clarify the molecular mechanism of cellular rhythms. Computational approaches to two major examples of biological rhythms are examined.

(1) Dictyostelium amoebae aggregate in a wavelike manner after starvation, in response to pulses of cyclic AMP (cAMP) emitted with a periodicity of several minutes by cells behaving as aggregation centers. A model shows that sustained oscillations in cAMP originate from the coupling between a negative feedback loop involving cAMP-induced receptor desensitization and a positive feedback loop due to the activation of cAMP synthesis by extracellular cAMP. The model provides an explanation for the frequency encoding of pulsatile signals of cAMP and for the origin of cAMP oscillations in the course of development. When incorporating diffusion of extracellular cAMP the model shows how concentric and spiral waves of cAMP occur in the course of aggregation.

(2) Among biological rhythms those with a circadian (close to 24h) period are conspicuous by their ubiquity and by the key role they play in allowing organisms to adapt to their periodically varying environment. In all organisms studied so far circadian rhythms originate from the negative autoregulation of gene expression. Computational models of increasing complexity will be presented for circadian oscillations in the expression of clock genes in Drosophila and mammals. When incorporating the effect of light, the models account for phase shifting of circadian rhythms by light pulses and for their entrainment by light-dark cycles. Stochastic simulations permit to test the robustness of circadian oscillations with respect to molecular noise. The model for the mammalian circadian clock will be used to address the dynamical bases of physiological disorders of the human sleep-wake cycle. The example of circadian rhythms shows how theoretical models of genetic regulatory networks can be used to address issues ranging from the molecular mechanism to physiological disorders.

Mathematical models of increasing complexity for the genetic regulatory network producing circadian rhythms in the fly Drosophila predict the occurrence of sustained circadian oscillations of the limit cycle type. When incorporating the effect of light, the models account for phase shifting of the rhythm by light pulses and for entrainment by light-dark cycles. The models also provide an explanation for the long-term suppression of circadian rhythms by a single pulse of light. Stochastic simulations permit to test the robustness of circadian oscillations with respect to molecular noise.

One of the most important aspects of life is the rhythmic behavior that is rooted in the many regulatory mechanisms that control the dynamics of living systems. The most common biological rhythms are circadian rhythms that occur with a period close to 24 h, allowing organisms to adapt to periodic changes in the terrestrial environment (Goldbeter, 2002). With the rapid accumulation of new data on gene, protein, and cellular networks, is is increasingly clear that genes are heavily involved in cellular regulatory interactions of circadian rhythms (Ueda et al., 2001; Tafti et al., 2003). However, a detailed picture of the genetic architecture of circadian rhythms has not been drawn, although ongoing projects like the Human Genome Project will assist in the characterization of circadian genetics.

Alternative to traditional strategies for clock gene identification by the analysis of single gene mutations and characterization of genes identified through cross-species homology (Reppert and Weaver, 2002; Tafti et al., 2003), functional mapping can be used to perform a genome-wide genetic analysis of rhythmic phenotypes, allowing a more thorough examination of the breadth and complexity of genetic influences on circadian behavior throughout the entire genome. With functional mapping, we can better confirm the findings of genetic control by Shimomura et al. (2001) on different discrete aspects of circadian behavior–free-running circadian period, phase angle of entrainment, amplitude of the circadian rhythm, circadian activity level, and dissociation of rhythmicity. As a first attempt of its kind, the haplotyping model described in this chapter has only considered a simple system of circadian rhythms characterized by two variables. This may may be too simple to reflect the complexity of rhythmic behavior. A number of more sophisticated models, governed by a system of 5 (Goldbeter, 1995), 10 (Leloup and Goldbeter, 1998) or 16 kinetic equations (Ueda et al., 2001; Smolen et al., 2001; Leloup and Goldbeter, 2003), have been constructed to describe the detailed feature of a rhythmic system with regard to responses to various interval and external environmental factors.

9

Genetic Mapping of Allometric Scaling

Among a vast range of species from microorganisms to the largest mammals, many biological variables seem to bear a specific quarter-power scaling relationship to overall body size (Fig. 9.1). For example, various biological times (e.g., lifespan and the time between heartbeats) scale with body mass to the 1/4 power, and resting metabolic rate scales with body mass to the 3/4 power (McMahon, 1973; Calder, 1996; Schmidt-Nielsen, 1984; Niklas, 1994; Enquist et al., 1998, 1999; Brown and West, 2000; Niklas and Enquist, 2001). Several attempts have been recently made to derive such general allometric scaling laws based on maximum efficiency (West et al., 1997, 1999b; Banavar et al., 1999) that has been regarded as the fundamental design principle for biological systems. Andresen et al. (2002) argued that maximum efficiency built on evolutionary as well as thermodynamic grounds (Dewey and Delle Donne, 1998) may suffer from internal inconsistencies when it is used to explain scaling laws.

Allometric relationships result from the regulation of scale and proportion in living organisms and are thought to have a genetic component. Wu et al. (2002c) and Ma et al. (2003) presented a statistical model for mapping quantitative trait loci (QTLs) responsible for universal quarter-power scaling laws of structure and function with the entire body size. Long et al. (2006) extended the model to map allometric QTLs based on a more general allometry equation.

In this chapter, we will describe these statistical models and approaches for QTL mapping of allometry and the change of allometry over development. Genetic mapping of allometry in relation to pharmacokinetic and pharmacodynamic processes will be valuable for enhancing biological values of pharmacogenetic studies.

9.1 Allometric Models

Allometric scaling laws can be mathematically described by a power function

$$B = \alpha M^\beta, \tag{9.1}$$

where B is a biological variable, M is the body weight, α is the constant, and β is the scaling exponent.

Although the simple allometry equation (9.1) has been used to model allometric scaling relationships, it is limited for the precise description of many important bio-

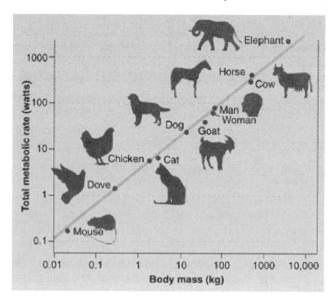

FIGURE 9.1

Allometric scaling laws across different species.

logical phenomena. For example, this equation, which forces two variables to pass through the origin, i.e., when $x = 0$, $y = 0$, cannot describe the relationship between two developmentally asynchronous features, such as reproductive timing and body weight. Many improved allometry equations have been proposed to accommodate general biological phenomena by earlier biologists. Among them, three representative allometry equations are

$$y = \begin{cases} \alpha x^{\beta} + \gamma & \text{(Robb, 1929)} \\ \alpha(x - \gamma)^{\beta} & \text{(Reeve and Huxley, 1945)} \\ \alpha(x - \gamma)^{\beta} + \delta & \text{(Lumer, 1937),} \end{cases} \tag{9.2}$$

where Lumer's four-parameter equation can be viewed as the most general. Ebert and Russell (1994) introduced a model II non-linear regression analysis to estimate the parameters contained in the allometry equations (9.1). Unlike Model I regression that minimizes the distance squared between the coordinate of a data pair on the y-axis and the function, Model II regression minimizes the area connecting the coordinates of the data pair and the function and, thus, it deals with variation in both traits. Model II regression analysis assumes an equal error variance for both traits and is a special case of more general error-in-variables models (Seber and Wild, 1989).

9.2 Allometric Mapping

9.2.1 Genetic Design

Consider a natural population at Hardy–Weinberg equilibrium, from which a random sample of n subjects are drawn. To study the genetic control of allometric scaling, we type these subjects for a number of single nucleotide polymorphisms (SNPs) and associate SNP-constructed haplotypes with allometric traits. Let 1 and 0 denote two alleles at each SNP, respectively. For a pair of SNPs, there are four different haplotypes, [11], [10], [01], and [00], whose frequencies are denoted as p_{11}, p_{10}, p_{01}, and p_{00}, respectively. Section 2.1 gave the formulation of diplotypes and their relationships with genotypes. Nine genotypes are observed for these two SNPs, with observations for each genotype expressed as $n_{r_1 r_1'/r_2 r_2'}$ $(r_1 \geq r_1', r_2 \geq r_2' = 1, 0)$. From Table 2.1, we construct a multinomial likelihood to estimate haplotype frequencies.

All subjects are measured for body mass (M) and a physiological parameter (B) related to drug response, such as metabolic rate. These two traits are related in a power equation (9.1), in which the estimation of the constant parameter α and exponential power β is obtained by a linear regression following the log-transformation, i.e.,

$$\ln B = \alpha + \beta \ln M,$$

or,

$$y = \alpha + \beta x, \tag{9.3}$$

where $y = \ln B$ and $x = \ln M$. A model for haplotyping allometric scaling laws will be based on this log-transformed linear regression function, as done in Wu et al. (2002b).

Figure 9.2 is the plot of a physiological trait B against body mass M for 100 subjects. In general, trait B increases with increasing body mass in a power equation. But it seems that the 100 subjects can be fitted by two distinct power curves, suggesting that a gene may control the scaling relationship between traits B and M. A mapping model to be described is motivated from the existence of multiple curves.

Suppose there is a QTN that controls allometric scaling between B and M. This QTN is composed of two SNPs at which the risk haplotype (A) is combined with the non-risk haplotype ($\bar{A}$) to form three composite diplotypes, AA (**2**), $A\bar{A}$ (**1**), and $\bar{A}\bar{A}$ (**0**). After log-transformation, the two traits B and M for subject i can be related in terms of QTN effects by equation (9.3), i.e.,

$$y_i = \sum_{j=0}^{2} \xi_{ij}(\alpha_j + \beta_j x_i) + e_i, \tag{9.4}$$

where ξ_{ij} is the indicator variable for subject i to carry composite diplotype j that is defined as 1 if subject i and composite diplotype j ($j = \mathbf{2}, \mathbf{1}, \mathbf{0}$) are compatible and 0

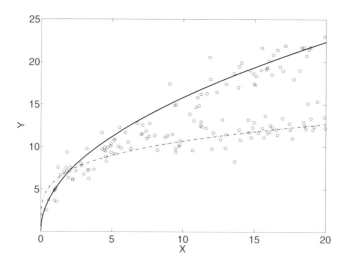

FIGURE 9.2

Diagram for genetic control of allometric scaling between two traits Y and X. Raw data (small dots) can be fit by two distinct power curves, each of which may correspond to a different composite diplotype.

otherwise, α_j and β_j are diplotype-specific coefficients of power equation, and e_i is the residual error of subject i for the transformed B trait, which is assumed to follow a normal distribution with mean 0 and variance σ^2.

9.2.2   Likelihood and Estimation

We follow the notation and definitions of alleles, haplotypes, diplotypes, and genotypes given in Section 2.1. Assuming that [11] is a risk haplotype, we used the information in Table 2.1 to formulate a multinomial likelihood for SNP data $(\mathbf{S})$, aimed to estimate $\Theta_p = (p_{11}, p_{10}, p_{01}, p_{00})$, and a mixture-based likelihood for log-transformed physiological trait (y), log-transformed body weight (x), SNP data, and estimated Θ_p, aimed to estimate haplotype effects on allometric scaling. These two likelihoods are expressed as

$$\log L(\Theta_p|\mathbf{S}) = \text{constant}$$
$$+2n_{11/11}\log p_{11}$$
$$+n_{11/10}\log(2p_{11}p_{10})$$
$$+2n_{11/00}\log p_{10}$$
$$+n_{10/11}\log(2p_{11}p_{01})$$
$$+n_{10/10}\log(2p_{11}p_{00}+2p_{10}p_{01})$$
$$+n_{10/00}\log(2p_{10}p_{00})$$

$$\log L(\Theta_q|x,y,\mathbf{S},\hat{\Theta}_p) =$$
$$\sum_{i=1}^{n_{11/11}}\log f_2(y_i)$$
$$+\sum_{i=1}^{n_{11/10}}\log f_1(y_i)$$
$$+\sum_{i=1}^{n_{11/00}}\log f_0(y_i)$$
$$+\sum_{i=1}^{n_{10/11}}\log f_1(y_i)$$
$$+\sum_{i=1}^{n_{10/10}}\log[\phi f_1(y_i)+(1-\phi)f_0(y_i)]$$
$$+\sum_{i=1}^{n_{10/00}}\log f_0(y_i)$$

$$+2n_{00/11}\log p_{01} \qquad +\Sigma_{i=1}^{n_{00/11}}\log f_0(y_i)$$
$$+n_{00/10}\log(2p_{01}p_{00}) \qquad +\Sigma_{i=1}^{n_{00/10}}\log f_0(y_i)$$
$$+2n_{00/00}\log p_{00} \qquad +\Sigma_{i=1}^{n_{00/00}}\log f_0(y_i),$$

$$(9.5)$$

where $\phi = p_{11}p_{00}/(p_{11}p_{00}+p_{10}p_{01})$, and $f_j(y_i)$ is a normal distribution function with the mean modeled by

$$\mu_{j|i} = \alpha_j + \beta_j x_i, \qquad (9.6)$$

and variance σ^2.

Unlike a traditional treatment of the genotypic mean, $\mu_{j|i}$ here is subscripted by i. This difference can be explained as follows: While traditional mapping estimates the genotypic mean of variable y at a single point, equation (9.6) is to model the genotypic mean as a curve, i.e., a function of genotypic values as body mass. Different genotypes or diplotypes respond differently to body mass, as shown by two power curves in Fig. 9.2. Thus, by comparing the differences in α and β between different diplotypes, we can see how this QTN hypothesized regulates allometric scaling between the two traits.

The EM algorithm can be derived to estimate diplotype-specific power coefficients. In the E step, calculate the posterior probability with which a double heterozygote (i) bears on diplotype [11][00] or [10][01], respectively, by using

$$P_{1|i} = \frac{\phi f_1(y_i)}{\phi f_1(y_i) + (1-\phi)f_0(y_i)}$$
$$P_{0|i} = \frac{\phi f_0(y_i)}{\phi f_1(y_i) + (1-\phi)f_0(y_i)}.$$

$$(9.7)$$

In the M step, estimate the power coefficients and variance with the estimated posterior probability by using

$$\hat{\alpha}_2 = \frac{\Sigma_{i=1}^{n_{11/11}} y_i}{n_{11/11}},$$

$$\hat{\alpha}_1 = \frac{\Sigma_{i=1}^{\dot{n}} y_i + \Sigma_{i=1}^{n_{10/10}} P_{1|i} y_i}{\dot{n} + \Sigma_{i=1}^{n_{10/10}} P_{1|i}},$$

$$\hat{\alpha}_0 = \frac{\Sigma_{i=1}^{\ddot{n}} y_i + \Sigma_{i=1}^{n_{10/10}} P_{0|i} y_i}{\ddot{n} + \Sigma_{i=1}^{n_{10/10}} P_{0||i}},$$

$$\hat{\beta}_2 = \frac{\sum_{i=1}^n P_{1|i} y_i}{\sum_{i=1}^n P_{1|i}}$$

$$\hat{\beta}_1 = \frac{\sum_{i=1}^n P_{1|i} y_i}{\sum_{i=1}^n P_{1|i}}$$

$$\hat{\beta}_0 = \frac{\sum_{i=1}^n P_{1|i} y_i}{\sum_{i=1}^n P_{1|i}}$$

$$\hat{\sigma}^2 = \frac{1}{n} \Bigg\{ \sum_{i=1}^{n_{11/11}} (y_i - \hat{\mu}_2)^2 + \sum_{i=1}^{\dot{n}} (y_i - \hat{\mu}_1)^2 + \sum_{i=1}^{\ddot{n}} (y_i - \hat{\mu}_0)^2$$

$$+ \sum_{i=1}^{n_{10/10}} \left[P_{1|i}(y_i - \hat{\mu}_1)^2 + (1 - P_{0|i})(y_i - \hat{\mu}_0)^2 \right] \Bigg\}.$$

$$(9.8)$$

where

$$\dot{n} = n_{11/10} + n_{10/11},$$

$$\ddot{n} = n_{11/00} + n_{10/00} + n_{01/01} + n_{01/00} + n_{00/00}.$$

Iterations including the E and M steps are repeated between equations (9.7) and (9.8) until the estimates of the parameters converge to stable values. The estimates at convergence are the maximum likelihood estimates (MLEs). By assuming different risk haplotypes (with four possibilities), four possible likelihood values are calculated. An optimal risk haplotype corresponds to the largest likelihood.

9.3   Hypothesis Testing

The hypothesis that a QTN governs allometric scaling can be tested using the following,

$$\begin{cases} H_0 : \ \alpha_2 = \alpha_1 = \alpha_0 = \alpha, \beta_2 = \beta_1 = \beta_0 = \beta, \\ H_1 : \ \text{at least one of the equalities above does not hold.} \end{cases} \quad (9.9)$$

The likelihoods are calculated under the H_0 (with $(\tilde{\alpha}, \tilde{\beta})$) and H_1 (with $(\hat{\alpha}_2, \hat{\beta}_2, \hat{\alpha}_1, \hat{\beta}_1, \hat{\alpha}_0, \hat{\beta}_0)$), respectively, and then a log-likelihood ratio (LR) is estimated. The LR value is compared with the critical threshold from a χ^2-distribution table with four degrees of freedom because the LR is thought to be asymptotically χ^2-distributed. If the sample size used is not adequate, the threshold should be determined from permutation tests.

The additive (a) and dominance genetic effects (d) of the QTN on the allometry detected can be further tested for their significance. These effects are estimated by

$$a_\alpha = \tfrac{1}{2}(\alpha_2 - \alpha_0), \tag{9.10}$$

$$d_\alpha = \alpha_1 - \tfrac{1}{2}(\alpha_2 + \alpha_0), \tag{9.11}$$

$$a_\beta = \tfrac{1}{2}(\beta_2 - \beta_0), \tag{9.12}$$

$$d_\beta = \beta_1 - \tfrac{1}{2}(\beta_2 + \beta_0). \tag{9.13}$$

The null hypotheses for testing the additive and dominance effect are formulated as $H_0 : a_\alpha = a_\beta = 0$ and $H_0 : d_\alpha = d_\beta = 0$, respectively. The critical thresholds for these tests can be obtained from a χ^2-distribution table.

9.4 Allometric Mapping with a Pleiotropic Model

9.4.1 Design

In the above section, we focus on the genetic control of a physiological trait B through a power link with body mass. But a QTN modeled for the physiological trait may also affect body mass. Below, we introduce a haplotyping model that can characterize the genetic effects of a QTN on both physiological trait and body mass.

The pleiotropic model proposes that a common QTN affects variation in both log-transformed traits x and y. At a putative QTN, a natural population can be sorted into three composite diplotypes, AA, $A\bar{A}$, and $\bar{A}\bar{A}$, coded by 2, 1, and 0, respectively. These three diplotypes form three clusters in the coordinate of x and y. If the allometric change of trait y with respect to trait x (described by equation (9.3)) results from the pleiotropic effect of this QTN, then the three points, each representing a pair of the expected mean values of the two traits in one genotype group, should be significantly different from each other but should be on the same line described by the log-transformed allometry equation (2) (Fig. 9.4). With such a linear relationship, the expected mean value of the transformed trait y can be exactly predicted from the mean value of the transformed trait x.

9.4.2 Genetic Model

The differences in x or y among the three diplotypic means reflect the magnitude of the genetic effects of the QTN on the corresponding trait. Denote a and d as the additive and dominant effects of the QTN on x. Thus, the phenotypic values of the two log-transformed traits for individual i are expressed by linear statistical models,

$$x_i = \mu + \xi_i a + \zeta_i d + e_i^x,$$
$$y_i = \alpha + \beta x_i + e_i^y = \alpha + \beta(\mu + \xi_i a + \zeta_i d) + \beta e_i^x + e_i^y,$$

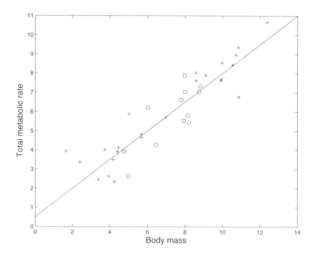

FIGURE 9.3

Allometric scaling laws across different composite diplotypes. Three different diplotypes are assumed, coded by $\times$, o, and +. The log-transformed genotypic values for
the three diplotypes are distributed along a straight line. If the coordinates of the
three diplotypic means do not overlap, this means that there exists a QTN to affect
the allometric law.

where μ is the overall mean for x; ξ_i and ζ_i are the dummy variables indicating the
QTL genotype of individual i, with ξ_i denoted as 1 for AA, 0 for $A\bar{A}$, and -1 for $\bar{A}\bar{A}$ and
ζ_i denoted as 1 for $A\bar{A}$ and 0 for AA or $\bar{A}\bar{A}$; $e_i^x \sim N(0, \sigma_x^2)$ and $e_i^y \sim N(0, \sigma_y^2)$ are the
error terms for traits x and y, respectively, which are correlated among individuals
with correlation coefficients R. Because the genetic effects are fixed effects, the
variances for traits x and y and their correlation can be expressed, respectively, as

$$v_x^2 = \sigma_x^2,$$
$$v_y^2 = \beta\sigma_x^2 + \sigma_y^2,$$
$$\rho = \frac{\sigma_x(\beta\sigma_x + R\sigma_y)}{\sqrt{\beta^2\sigma_x^2 + \sigma_y^2}}.$$

Let μ_j^x and μ_j^y be the mean values of x and y for QTN diplotype j, respectively. According to equation (9.3), the relationship between the means of the two transformed
traits can be modeled by

$$\mu_j^y = \alpha + \beta\mu_j^x, \tag{9.14}$$

for QTN diplotype j.

9.4.3 Statistical Estimation

The likelihoods like (9.5) are constructed to estimate population and quantitative genetic parameters. Assuming that [11] is a risk haplotype, the bi-variate mixture-based likelihood for estimating quantitative genetic parameters ($\Omega_p = (\mu, a, d, \alpha, \beta, \sigma_x^2, \sigma_y^2, R)$) is formulated as

$$\log L(\Omega_q | x, y, \mathbf{S}, \hat{\Omega}_p) = \sum_{i=1}^{n_{11/11}} \log f_2(x_i, y_i) + \sum_{i=1}^{n_{11/10}+n_{10/11}} \log f_1(x_i, y_i)$$

$$+ \sum_{i=1}^{n_{10/10}} \log[\omega f_1(x_i, y_i) + (1-\omega) f_0(x_i, y_i)]$$

$$+ \sum_{i=1}^{n_{11/00}+n_{10/00}+n_{00/11}+n_{00/10}+n_{00/00}} \log f_0(x_i, y_i),$$

where the two-dimensional normal density, $f_j(x_i, y_i)$, is expressed as

$$f_j(x_i, y_i) = \frac{1}{2\pi v_x v_y \sqrt{1-\rho^2}}$$

$$\times \exp \left\{ -\frac{1}{2(1-\rho^2)} \left[\left(\frac{x_i - \mu_j^x}{v_x} \right)^2 - 2\rho \frac{(x_i - \mu_j^x)(y_i - \mu_j^y)}{v_x v_y} + \left(\frac{y_i - \mu_j^y}{v_y} \right)^2 \right] \right\}.$$

$$(9.15)$$

The MLEs of the unknown vector Ω_p can be obtained by implementing the EM algorithm. For detailed iterative EM steps, refer to Wu et al. (2002c).

In addition, models can also be extended to test whether different risk diplotypes are responsible for two allometrically related traits. This needs to incorporate a model selection procedure for choosing an optimal combination of risk haplotypes.

9.4.4 Hypothesis Tests

Several hypotheses about the QTN affecting quarter-power scaling of organisms can be formulated. These hypotheses include:

(1) A QTN affects two allometrically related traits;

(2) Under a best-fitting model, this QTN detected affects the two traits in quarter-power scaling;

(3) The normalization constant α is a characteristic of composite diplotype, as detected at the species or population level (Niklas, 1994).

The overall existence of a QTN is tested by formulating the null hypothesis $H_0 : a = d = \alpha = \beta = 0$ from which a log-likelihood ratio is then calculated. To test whether this QTN is pleiotropic, we need to formulate two null hypotheses, i.e., $H_0 : a = d = 0$

and $H_0 : \alpha = \beta = 0$. Only when each of these two null hypotheses is rejected, we can conclude that this QTN pleiotropically affects physiological trait B and body mass M. After this QTN is found to be pleiotropic, a null hypothesis, $H_0 : \beta = 1/4$ or $H_0 : \beta = 3/4$, is made to test whether this QTN affects the two traits in a quarter-power manner. The formulation of the null hypothesis $H_0 : \alpha = 0$ is to test whether this QTN is associated with the intercept of linearized allometric scaling.

9.5   Allometric Mapping with General Power Equations

Sections 9.2 and 9.2 take advantage of a linear relationship between B and M after the power function is log-transformed and, therefore, is statistically straightforward to derive. When more complicated power functions, for which the log-transformation cannot lead to a linear relationship, are needed to describe allometric scaling laws, a different model based on Taylor's approximation theory is developed (Ma et al., 2003). But Taylor's application to allometric mapping presents a significant problem in determining the order with which Taylor's series are expanded. Although the expansions of higher-orders can theoretically result in a more precise description of the B-M relationship than those of lower-orders, the former tend to be computationally more expensive than the latter.

Long et al. (2006) presented an approximate approach for allometric mapping with a general complicated power equation as those given in equation (9.2). This approach includes two steps: (1) Use a non-linear method to power coefficients in these equations for the mapping population, and then predict dependent variables with the equations, and (2) Map or haplotype allometric scaling based on these predicted values.

9.5.1   Estimating Power Coefficients Using Model II Non-Linear Regression

The observations of subject i for traits x and y present a point (A) with the co-ordinate (x_i, y_i) as shown in Fig. 9.4. To detect an allometry equation that best fits this individual, we define four more points B, C, D, and E, whose co-ordinates are expressed as

$$B = \left[x_i, \alpha(x_i - \delta)^\beta + \gamma \right]$$
$$C = (z_i, y_i)$$
$$D = (x_i, 0)$$
$$E = (z_i, 0),$$

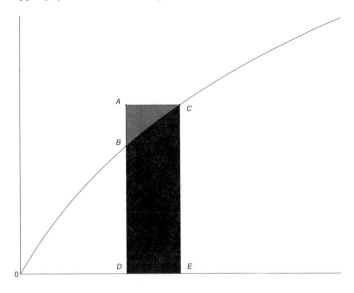

FIGURE 9.4
The diagram of calculating the area under curve using the loss function approach.
Adapted from Ebert and Russell (1994).

where

$$z_i = \begin{cases} \left(\frac{y_i - \gamma}{\alpha}\right)^{\frac{1}{\beta}} & \text{Robb equation} \\[2mm] \left(\frac{y_i}{\alpha}\right)^{\frac{1}{\beta}} + \gamma & \text{Reeve-Huxley equation} \\[2mm] \left(\frac{y_i - \gamma}{\alpha}\right)^{\frac{1}{\beta}} + \delta & \text{Lumer equation,} \end{cases}$$

The area of the rectangle *ADEC* is

$$\mathbf{A}_i(ADEC) = y_i \left| z_i - x_i \right|.$$

The area under the curve, *BDEC*, is

$$\mathbf{A}_i(BDEC)$$

$$= \begin{cases} \int_{x_i}^{z_i} (\alpha x_i^{\beta} + \delta) dx_i & \text{Robb equation} \\[3mm] \int_{x_i}^{z_i} \alpha (x_i - \gamma)^{\beta} dx_i & \text{Reeve-Huxley equation} \\[3mm] \int_{x_i}^{z_i} [\alpha (x_i - \gamma)^{\beta} + \delta] dx_i & \text{Lumer equation} \end{cases}$$

$$
= \begin{cases}
\left| \dfrac{\alpha}{\beta+1} \left| x_i^{\delta+1} - \left(\dfrac{y_i - \delta}{\alpha} \right)^{\frac{\beta+1}{\beta}} \right| + \delta \left| x_i - \left(\dfrac{y_i - \delta}{\alpha} \right)^{\frac{1}{\beta}} \right| \right. \\
\qquad\qquad\qquad\qquad \text{Robb equation} \\[2ex]
\dfrac{\alpha}{\beta+1} \left| (x_i - \gamma)^{\beta+1} - (\dfrac{y_i}{\alpha})^{\frac{\beta+1}{\beta}} \right| \\
\qquad\qquad\qquad\qquad \text{Reeve-Huxley equation} \\[2ex]
\dfrac{\alpha}{\beta+1} \left| (x_i - \gamma)^{\beta+1} - \left(\dfrac{y_i - \delta}{\alpha} \right)^{\frac{\beta+1}{\beta}} - \gamma \right| + \delta \left| x_i - \left(\dfrac{y_i - \delta}{\alpha} \right)^{\frac{\beta+1}{\beta}} - \gamma \right| \\
\qquad\qquad\qquad\qquad \text{Lumer equation,}
\end{cases}
$$

$$(9.16)$$

We define the absolute value of the difference between $\mathbf{A}_i(ADEC)$ and $\mathbf{A}_i(BDEC)$, i.e., area

$$\mathbf{A}_i(ABC) = \mathbf{A}_i(ADEC) - \mathbf{A}_i(BDEC),$$

as the loss function for individual i. The loss function for all subjects is expressed as

$$\mathbf{A}(ABC) = \sum_{i=1}^{n} \mathbf{A}_i(ABC). \qquad (9.17)$$

The estimates of α, β, γ, and δ fitting the allometry equation can be obtained by minimizing the loss function defined by areas (9.16). It is impossible to derive their analytical solutions for this nonlinear function. However, their numerical solutions can be obtained by using the simplex algorithm (Nelder and Mead, 1965). The advantage of the simplex algorithm is that it is derivative-free and easy to implement with current software, such as Matlab. One caution for this algorithm is the possibility to obtain the local optimal solutions for the loss function (9.17). By carefully selecting the initial values, however, this problem can be the minimum if there exists global optimal solutions. If no global optimal solutions exist, we can take the minimum in the space of these parameters.

For a practical data set, it is essential to determine the best allometry equation that can describe the allometric scaling relationship. The criterion for the determination can be based on the values of loss function summed over all individuals. For the same value of loss function, the allometry equation with fewer parameters is better than those with more parameters.

9.5.2 Predicting Dependent Variables

After the parameters (α, β, γ) are estimated from model II non-linear regression by minimizing the loss function (9.17), we substitute these estimates, indexed by $\breve{\alpha}$, $\breve{\beta}$,

$\check{\gamma}$, and $\check{\delta}$, to the likelihood function (9.5). Before doing so, we take a simple change of more allometry equations for subject i:

$$\begin{cases} \ln(y_i - \check{\gamma}) = \check{\alpha} + \check{\beta}\ln(x_i) & \text{Robb equation} \\ \ln(y_i) = \check{\alpha} + \check{\beta}\ln(x_i - \check{\gamma}) & \text{Reeve-Huxley equation} \\ \ln(y_i - \check{\delta}) = \check{\alpha} + \check{\beta}\ln(x_i - \check{\gamma}) & \text{Lumer equation,} \end{cases}$$

or

$$\begin{cases} \ln(y_i') = \check{\alpha} + \check{\beta}\ln(x_i) & \text{Robb equation} \\ \ln(y_i) = \check{\alpha} + \check{\beta}\ln(x_i') & \text{Reeve-Huxley equation} \\ \ln(y_i') = \check{\alpha} + \check{\beta}\ln(x_i') & \text{Lumer equation} \end{cases} \qquad (9.18)$$

where "*'*" denotes the transformation of the two traits. Because $\check{\alpha}$, $\check{\beta}$, $\check{\gamma}$, and $\check{\delta}$ can be regarded as known constants obtained from model II regression analysis, x_i' and y_i' can be calculated and display the same statistical distribution as raw data x_i and y_i. Thus, the new relationships described by equation (9.18) are identical to a linear log-transformed allometry equation (9.3). By using this approximate model, we are now able to estimate the haplotype effect parameters and residual variances/correlation. The existence of the underlying QTN for more general allometry equations can be tested by formulating the corresponding hypotheses.

Example 9.1

Long et al. (2006) used the approximate model for mapping quantitative trait loci (QTLs) that control allometric scaling between growth rate and body mass in the mouse. Although they did not use a haplotyping model for sequencing this allometric relationship, their QTL application can nicely demonstrate the application of allometric mapping. For this reason, we used Long et al's (2006) results as an example for allometric mapping of biological traits. Long et al. (2006) reanalyzed Cheverud et al. (1996) data for a mouse genome project in which 19 chromosomes were mapped with 75 microsatellite markers for 535 F_2 progeny derived from two strains, Large and Small. The F_2 hybrids were weighted at 10 weekly intervals starting at age 7 days. The raw weights were corrected for the effects of each covariate due to dam, litter size at birth, parity, and sex. The growth rate at each time interval $[t, t+1]$ was calculated for each mouse by subtracting body weight at time t from body weight from time $t+1$. The mean growth rate across all the time intervals was then calculated for each mouse. Figure 9.5 illustrates the plot of mean growth rate against body weight measured at the last time points for the F_2 mice. The model-II non-linear regression was used to estimate the parameters for the equation that specifies the allometric relationship between growth rate and body mass.

By comparing the values of loss function among these equations, Robb's equation was found to be the most parsimonious. Based on Robb's equation, we estimate the three parameters underlying the allometry equation as $\tilde{\alpha} = 0.024$, $\tilde{\beta} = 0.68$, and $\tilde{\gamma} = -0.016$. It is interesting to find that the estimated β value is not significantly

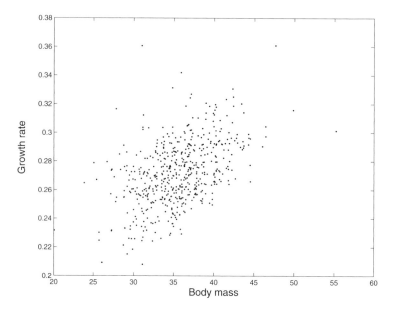

FIGURE 9.5

The plot of mean growth rate against body mass in the F_2 mice derived from two different strains.

different from 0.75, which supports the three-quarter law for a biological process to scale with body mass (West et al., 1997).

These estimated parameters from Model II regression by minimizing the loss function (9.17) were substituted to a mapping model, expressed as

$$L(\Omega|x,y) = \prod_{i=1}^{n} \sum_{j=0}^{2} \left[\omega_{j|i} f_j(x_i, y_i) \right],$$

where $\omega_{j|i}$ is the conditional probability of a QTL genotype ($j = 2$ for QQ, 1 for Qq, and 0 for qq) given the marker interval genotype of mouse i (Wu et al., 2007a) and $f_j(x_i, y_i)$ is modeled by (9.15) with parameters $(\mu, a, d, \alpha, \beta, r_1, \sigma_x^2, \sigma_y^2, R)$.

We scan all the 19 chromosomes for the existence of QTL affecting the allometric scaling relationship between growth rate and body weight. Figure 9.6 gives the profile of the log-likelihood (LR) ratio test statistics for claiming the existence of QTL across the entire mouse genome. There are two peaks for the LR profile, one (16.47) between markers D6Nds5 and D6Mit15 on chromosome 6 and the other (22.46) between D7Nds1 and D7Mit17 on chromosome 7. These two LR values are well beyond the genomewide critical threshold (9.29) at the 0.001 significance level determined on the basis of permutation tests (Churchill and Doerge, 1994). These tests suggest the existence of the two QTL that are located at the positions corresponding to the peaks of the profile.

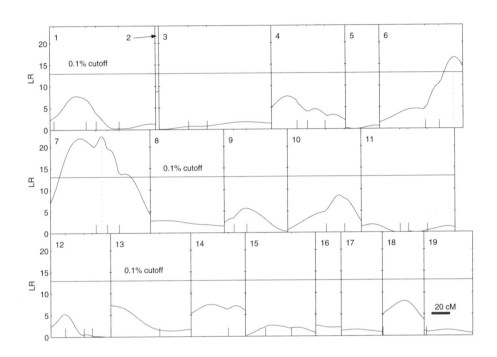

FIGURE 9.6

The LR profile between the full and reduced (no QTL) model for allometric scaling of mean growth rate to body mass across the entire genome using the linkage map constructed from microsatellite markers. The genomic positions corresponding to the peaks of the curves are the MLEs of the QTL positions. The genome-wide threshold values for claiming the existence of QTL is given as the horizonal lines. Tick marks on the *x*-axis represent the positions of markers on the linkage group.

The QTL detected on chromosome 6 affects body weight in a partial dominance manner, whereas the QTL on chromosome 7 displays a strong dominant or over-dominance effect on body weight. Our detection is broadly consistent with simple interval mapping analysis of the same material by Cheverud et al. (1996). Of 16 chromosomes observed to carry the QTL for body weight, the QTL on chromosomes 6 and 7 were detected with larger LOD scores, explaining larger percentages of the observed variation, than those on the other chromosomes.   ⧠

10

Functional Mapping of Drug Response with Allometric Scaling

Allometric scaling has been shown to be a good predictor of a variety of biological rates, times, and dimensions (including metabolic rate, lifespan, growth rate, heart rate, DNA nucleotide substitution rates, and length of aortas) between species of several orders of magnitude in body weight (Schmidt-Nielsen, 1984; West et al., 1997, 1999b; West and Brown, 2004, 2005). It is not surprising that allometric scaling principles can provide useful guidance for the scaling preclinical evaluation of drug metabolism and response (Boxenbaum, 1982; Gronert et al., 1995; Hu and Hayton, 2001; Lepist and Jusko, 2004; Zuideveld et al., 2007). Most of the work in allometric scaling of pharmacokinetic (PK) and pharmacodynamic (PD) responses was motivated to predict the time course of drug effects in man using the data observed in animals. Limited attention has been given to the use of allometric scaling to enhance an understanding of the genetic regulation of PK and PD reactions, and ultimately to predict the time courses of drug concentrations and drug effects.

Allometric scaling relationships between different organ parts can be understood from three different perspectives: static, ontogenetic, and evolutionary (Stern, 1999; Klingenberg and Zimmermann, 1992). Static allometry refers to the scaling among individuals between two different traits after growth has ceased or at a particular developmental stage. Ontogenetic allometry is the growth trajectory of one trait relative to the other (i.e., shape) during an individual's lifetime. Evolutionary (or phylogenetic) allometry is the size relationship between traits across species. Current allometric studies of pharmacological responses to make interspecific comparisons and prediction virtually has capitalized on the knowledge and technique of evolutionary allometry. Static allometry and ontogenetic allometry, on the other hand, convey an incredible amount of information about the underlying genetic and developmental mechanisms for pharmacodynamic effects of a drug.

In Chapter 9, we introduced a basic idea and procedure of integrating static allometric scaling into genetic mapping of metabolism-related traits, and formulated a number of hypotheses about the genetic control patterns of these traits. In this chapter, we will incorporate allometric principles into a process mapping framework, in a hope to better understand the genetic architecture of drug kinetics and dynamics. Statistical approaches for gene haplotyping based on allometrically scaled mechanistic PK-PD models could be used more widely to predict the pharmacodynamic response in practical clinic trials. Overall, we believe that these approaches provide a novel way of interpreting pre-clinical pharmacological responses through a genetic component which will ultimately facilitate the understanding of pharmacological responses.

10.1 Allometric Scaling of Pharmacokinetic and Pharmacodynamic Responses

Biological structures and processes ranging from cellular metabolism to population dynamics are affected by the size of the organism. Although the sizes of mammalian species span seven orders of magnitude, interspecies similarities in structural, physiological, and biochemical attributes result in an empirical power law (the allometric equation) that characterizes the dependency of biological variables on body mass,

$$B = \alpha M^{\beta}, \tag{10.1}$$

where B is the dependent biological variable of interest, α is a normalization constant known as the allometric coefficient, M is the body weight, and β is the allometric exponent.

Allometric scaling has been applied in pharmacokinetics for approximately two decades. The major interest has been prediction of pharmacokinetic parameters in man from parameter values determined in animals (Feng et al., 2000; Sinha et al., 2008). This technique is now used for gene or sequence identification of drug response.

Pharmacokinetic models are used to describe the fate of a drug in a biological system following its administration. Recalling equation (6.2), the plasma concentration (c_p) of a drug for a single dose is derived (Hochhaus and Derendorf, 1995) as

$$c_p = \begin{cases} \dfrac{d}{V_d} e^{-k_e t} & \text{for intravenous (IV) bolus} \\[2ex] \dfrac{dk_a}{V_d(k_a - k_e)}(e^{-k_a t} - e^{-k_e t}) & \text{for first-order absorption} \\[2ex] \dfrac{k_o}{k_e V_d}(e^{k_e \tau} - 1)e^{-k_e t} & \text{for zero-order absorption} \end{cases} \tag{10.2}$$

where d is the bioavailable dose, V_d is the volume of distribution, k_e is the elimination rate constant, k_a is a first-order absorption rate constant, k_o is a zero-order absorption rate constant, τ is the duration of zero-order absorption, and t is the time after the last dose was administrated. From these parameters given in equation (10.2), some interesting pharmacokinetic parameters can be derived, which include clearance, CL (the volume of plasma in the vascular compartment cleared of drug per unit time by the processes of metabolism and excretion), and phase half-time, $t_{1/2}$ (the time required to reduce the plasma concentration to one half its initial value).

The allometric equations used to relate the pharmacokinetic parameters with body weight M have been established (Hu and Hayton, 2001; Zuideveld et al., 2007) as

$$k_e = \alpha_e M^{\beta_e},\tag{10.3}$$

$$k_a = \alpha_a M^{\beta_a},\tag{10.4}$$

$$k_o = \alpha_o M^{\beta_o},\tag{10.5}$$

$$\tau = \alpha_\tau M^{\beta_\tau},\tag{10.6}$$

$$CL = \alpha_c M^{\beta_c},\tag{10.7}$$

$$t_{1/2} = \alpha_t M^{\beta_t},\tag{10.8}$$

where the normalization constant and allometric exponent are defined for different PK parameters, respectively. Depending the purpose of the experiment, biological properties of subjects, and sample size, all or part of these scaling equations may be incorporated into the functional mapping of PK processes.

Pharmacodynamic models are derived to describe the effect (E) of a drug as a function of its concentration (c). The most commonly used pharmacodynamic model is the sigmoid curve or Emax model, expressed as

$$E = E_0 + \frac{E_{\max} c^H}{EC_{50}^H + c^H},\tag{10.9}$$

where E_0 is the constant or baseline value of drug response, $E_{\max}$ is the asymptotic effect, EC_{50} is the drug concentration that results in 50% of the maximal effect, and H is the slope parameter that determines the slope of the concentration-response curve. The allometric equations have been applied to model these PD parameters (Khor et al., 2000) expressed as

$$E_0 = \alpha_0 M^{\beta_0},\tag{10.10}$$

$$E_{\max} = \alpha_{\max} M^{\beta_{\max}},\tag{10.11}$$

$$EC_50 = \alpha_{50} M^{\beta_{50}},\tag{10.12}$$

$$H = \alpha_H M^{\beta_H}.\tag{10.13}$$

Similarly, whether all or part of these power coefficients are implemented into functional mapping depends on the interest of an experimenter, etc.

10.2 Model Derivations

10.2.1 Experimental Design

We adopt a conventional design for pharmcokinetic and pharmacodynamic studies. A random sample of n subjects are collected from a Hardy–Weinberg equilibrium population. Single nucleotide polymorphoisms (SNPs) are typed for these subjects. For a putative QTN composed of multiple SNPs that form many different haplotypes through the combination of alleles at different SNPs, we may reasonably assume that one haplotype is the risk haplotype (A) in drug response distinct from the rest of haplotypes denoted as the non-risk haplotype $(\bar{A})$. The combination between the risk and non-risk haplotypes form three different composite diplotypes, AA (**2**), $A\bar{A}$ (**1**), and $\bar{A}\bar{A}$ (**0**). The construction of composite diplotypes are shown in Table 2.1.

For the collection of PK data, all subjects are dosed with a drug at a single dose. Blood samples were collected to measure the concentration of the drug at a certain number of time points (T), aimed to draw a PK concentration-time profile for each subject. Body mass was measured for all the subjects studied. For the collection of PD data, all subjects are dosed with multiple doses of a drug (C). Physiological or clinical parameters that are related to drug effects are measured after a certain amount of time following the administration of the drug at each dose level. This will allow us to construct a PD response-dose profile. The experiment can also be designed to collect both PK and PD data simultaneously.

10.2.2   Model Structure and Estimation

As always, the haplotyping model of drug response includes two types of genetic parameters, population (i.e., haplotype frequencies, allele frequencies, and linkage disequilibria) and quantitative (i.e., haplotype additive and dominance effects as well as residual variance). The closed form for the EM algorithm has been derived to estimate haplotype frequencies based on a multinomial likelihood (see Section 2.3). Haplotype genetic effects on the process of drug response are explained in terms of PK or PD parameters and can be estimated through the EM algorithm or simplex algorithm.

We assume that longitudinal PK and PD data follow a multivariate normal distribution with mean vectors for subject i specified by different composite diplotypes expressed as

$$\mathbf{u}_{Kj|i} = (u_{Kj|i}(1), \cdots, u_{Kj|i}(T)) \quad \text{for PK modeling,} \qquad (10.14)$$
$$\mathbf{u}_{Dj|i} = (u_{Dj|i}(1), \cdots, u_{Dj|i}(C)) \quad \text{for PD modeling.} \qquad (10.15)$$

For PK modeling, if the first-order absorption model is considered (equation (10.2)), we will have an elimination rate constant (k_e) and a first-order absorption rate constant (k_a) and may derive the half-time expressed as $t_{1/2} = \ln 2 / k_a$. These parameters can be modeled by allometric scaling of body mass as shown in equations (10.3), (10.4), and (10.8), respectively. Note that, if $t_{1/2}$ is modeled, i.e., $\ln 2 / k_a = \alpha_t M^{\beta_t}$, then it turns out to model k_a by

$$k_a = \frac{\ln 2}{\alpha_t M^{\beta_t}}.$$

Thus, diplotype-specific means of equation (10.14) are modeled by

$$u_{Kj|i}(t) = \frac{d\left(\frac{\ln 2}{\alpha_{tj}M_i^{\beta_{tj}}}\right)}{V_d\left(\frac{\ln 2}{\alpha_{tj}M_i^{\beta_{tj}}} - k_{ej}\right)}\left[e^{-\left(\frac{\ln 2}{\alpha_{tj}M_i^{\beta_{tj}}}\right)t} - e^{-k_{ej}t}\right], \tag{10.16}$$

where $\Omega_{u_j} = (\alpha_{tj}, \beta_{tj}, k_{ej})$ that determine the mean curve in equation (10.16) are unknown parameters being estimated if the half-time is modeled by a power equation.

The longitudinal covariance structure can be be modeled by the AR(1) or SAD(1) model. The parameters that model the covariance structure, along with Ω_{u_j}, are estimated through the EM, simplex, or Newton-Raphson algorithms, or their hybrids.

Similarly, if the PD process is modeled, we will model diplotype-specific mean vector in (equation (10.15) by a power equation. Previous PD studies suggest that the asymptotic effect of a drug well scales as body mass (Zuideveld et al., 2007). This information can be incorporated into the mean vector (10.15) through equation (10.11), i.e.,

$$u_{Dj|i}(c) = E_{0j} + \frac{\alpha_{\max j}M_i^{\beta_{\max j}}c^{H_j}}{EC_{50j}^{H_j} + c^{H_j}}, \tag{10.17}$$

where $\Omega_{u_j} = (\alpha_{\max j}, \beta_{\max j}, EC_{50j}, H_j)$ are the parameters to be estimated for composite diplotype j. The structure of the longitudinal covariance among different doses can be modeled and estimated by the AR(1) or SAD(1) model.

10.2.3 Hypothesis Testing

After an optimal risk haplotype is determined, for both PK and PD modeling, we can have two questions to be answered:

(1) Whether this risk haplotype is significant in the genetic control of drug response;

(2) Whether this risk haplotype also controls the allometric scaling relationship.

For PK modeling, if the half-time is modeled (equation (10.16)), the null hypotheses for testing the first and second questions are formulated, respectively, as

$$H_0 : (\alpha_{tj}, \beta_{tj}, k_{ej}) \equiv (\alpha_t, \beta_t, k_e),$$

and

$$H_0 : (\alpha_{tj}, \beta_{tj}) \equiv (\alpha_t, \beta_t).$$

Similarly, for PD modeling, if $E_{\max}$ is modeled (equation (10.17)), the null hypotheses for testing the first and second questions above are formulated, respectively, as

$$H_0 : (\alpha_{\max j}, \beta_{\max j}, EC_{50j}, H_j) \equiv (\alpha_{\max}, \beta_{\max}, EC_{50}, H),$$

and

$$H_0 : (\alpha_{\max j}, \beta_{\max j}) \equiv (\alpha_{\max}, \beta_{\max}).$$

Example 10.1

Lin and Wu (2005) reported on the haplotyping mapping of drug response in humans. The study was designed to use dobutamine to affect the heart rates of patients. Dobutamine is a medication that is able to treat congestive heart failure by increasing heart rate and cardiac contractility, with actions on the heart similar to the effect of exercise. About 160 patients ages 32 to 86 years old participated in this study. Dobutamine was injected into these subjects to investigate their responses in heart rate to this drug. The subjects received increasing doses of dobutamine, until they achieved a target heart rate response or predetermined maximum dose. The dose levels used were 0 (baseline), 5, 10, 20, 30, and 40 mcg/min, at each of which heart rate was measured. The time interval of 3 minutes is allowed between two successive doses for subjects to reach a plateau in response to that dose. Only those (107) in whom there were heart rate data at all the six dose levels were included for data analyses. All subjects were weighted for their body mass.

Dobutamine is a synthetic catecholamine that primarily stimulates β-adrenergic receptors (βAR), which play an important role in cardiovascular function and responses to drugs (Nabel, 2003; Johnson and Terra, 2002). Each of these subjects was typed for SNP markers at codons 49 (Ser49Gly) and 389 (Arg389Gly) within the β1AR gene and at codons 16 (Arg16Gly) and 27 (Gln27Glu) within the β2AR gene. By assuming that one haplotype is different from the rest of the haplotypes, we used a PD-based allometric mapping model through modeling the Emax (10.17) to detect an optimal risk haplotype that is associated with the PD response of dobutamine to its different doses.

At the β1AR gene, we did not find any haplotype that contributed to inter-individual difference in heart rate response. A significant effect was observed for haplotype [Gly16(G)Glu27(G)] within the β2AR gene. Table 10.1 gives the MLEs of allele frequencies and linkage disequilibrium between the two SNPs, composite-specific mean-structuring and covariance-structuring parameters. The covariance matrix was structured by modeling the dose-dependent variance as

$$\sigma^2(c) = u + vc + wc^2$$

and between-dose correlation as

$$\mathrm{corr}(c_1, c_2) = \rho^{|c_1 - c_2|}, \quad (c_1, c_2 = 1, ..., C).$$

The log-likelihood ratio (LR) test statistics by assuming that [GG] is a risk haplotype was 30.03, which is significant at $p = 0.021$ based on the critical threshold determined from 1000 permutation tests. The LR values when selecting any haplotype rather than [GG] as a risk haplotype gave no significant results ($p = 0.16–0.40$).

It is interesting to find that the power curves of the Emax scaling with body mass are remarkably different among the three composite diplotypes, [GG][GG],

TABLE 10.1

The maximum likelihood estimates (MLEs) of SNP allele frequencies and linkage disequilibria, and the parameters that model haplotype effects and the covariance structure in a sample of 107 subjects within the $\beta 2AR$ gene.

Diplotype	Parameters	Standard Estimates	Errors
	Population Genetic Parameters		
	p_G	0.6168	0.0405
	p_G	0.4019	0.0251
	D	0.1302	0.0494
	Mean-Structuring Parameters		
[GG][GG]	$\hat{\alpha}_2$	224.6300	28.1764
	$\hat{\beta}_2$	-0.1736	0.0226
	$\widehat{EC}_{50/2}$	36.8860	4.9208
	$\hat{H}_2$	1.7774	0.1535
[GG][$\overline{GG}$]	$\hat{\alpha}_1$	1.4217	0.1574
	$\hat{\beta}_1$	0.8525	0.0367
	$\widehat{EC}_{50/1}$	28.2980	3.4118
	$\hat{H}_1$	2.0566	0.1839
[$\overline{GG}$][$\overline{GG}$]	$\hat{\alpha}_0$	49.7200	3.7460
	$\hat{\beta}_0$	0.0435	0.0051
	$\widehat{EC}_{50/0}$	22.7290	1.6638
	$\hat{H}_0$	2.0173	0.1020
	Covariance-Structuring Parameters		
	$\hat{u}$	2.8361	0.1214
	$\hat{v}$	0.1205	0.0078
	$\hat{w}$	-0.0012	0.0002
	$\hat{\rho}$	0.7373	0.0246

The standard errors were estimated using an approximated approach given in Hou et al. (2005).

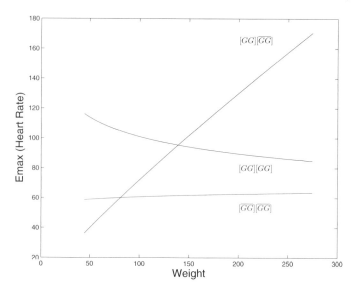

FIGURE 10.1

Different patterns of allometric scaling between the asymptotic effect of dobutamine (*E*max), measured in heart rate, and body mass for different composite diplotypes constituted by haplotype [GG] and [$\overline{GG}$].

[GG][$\overline{GG}$], and [$\overline{GG}$][$\overline{GG}$] (Fig. 10.1). The heterozygote displays much greater sensitivity to body mass in heart rate than the two homozygotes. With lighter body mass, the three composite diplotypes respond to the dose of dobutamine in a similar manner, but the slope of the responsiveness is much larger for the heterozygote than two homozygotes when body mass increases (Fig. 10.2). It is also interesting to note that the significance level to detect the QTN for drug response is increased from pure mapping of drug response ($p = 0.021$) to mapping with allometric scaling ($p = 0.002$).

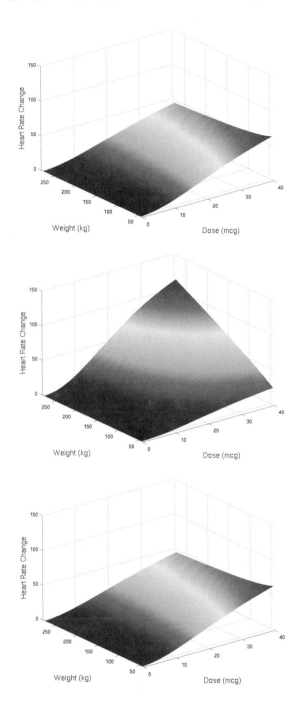

FIGURE 10.2
Heart rate as a function of dose and body mass for different composite diplotypes
constituted by haplotype [GG] and [$\overline{\text{GG}}$].

10.3   A Pleiotropic Model for Allometric Mapping

In Section 10.2, we assumed that a QTN affects the scaling relationship between the PK or PD parameters and body mass, but did not assume that this QTN also affects body mass. It is reasonable to model the genetic control of body mass as a biological trait with the same QTN. To do so, we need to remodel the mean vectors for diplotype-specific multivariate normal distributions, expressed as

$$(\mathbf{u}_{Kj|i}, u_{M_j}) = (u_{Kj|i}(1), \cdots, u_{Kj|i}(T), u_{M_j}) \quad \text{for PK modeling,} \qquad (10.18)$$

$$(\mathbf{u}_{Dj|i}, u_{M_j}) = (u_{Dj|i}(1), \cdots, u_{Dj|i}(C), u_{M_j}) \quad \text{for PD modeling.} \qquad (10.19)$$

odeling. In each case, when both null hypotheses are rejected, it is concluded that this QTN exerts a pleiotropic effect on a pharmacological process and body mass.

In these models (10.18) and (10.18), the genotypic values (u_{M_j}, $j = 0, 1, 2$) of composite diplotypes for body mass need to be estimated. In modeling the covariance structure, we will need to consider the variance for body mass (σ_M^2) and its correlation with PK responses at different times and PD responses at different doses. It is reasonable to assume that such a correlation between body mass and PK or PD responses is time- or dose-invariant, denoted as ρ_K or ρ_D. Thus, two more parameters, (σ_M^2, ρ_K) or (σ_M^2, ρ_D), are needed to model the covariance structure.

To test whether a detected QTN affects both pharmacological response and body mass, the following null hypotheses can be formulated:

$$H_0: \quad \mathbf{u}_{Kj|i} \equiv \mathbf{u}_{K|i},$$
$$H_0: \quad u_{Mj} \equiv u_M,$$

for PK modeling, and

$$H_0: \quad \mathbf{u}_{Dj|i} \equiv \mathbf{u}_{D|i},$$
$$H_0: \quad u_{Dj} \equiv u_D,$$

for PD m The model that incorporates a pleiotropic QTN for PD response and body mass into functional mapping was used to re-analyze the same data set described in Example 10.2.3, leading to the consistent results. That is, a optimal risk haplotype is [GG], there is a significant haplotype effect on differentiation in power curves, and allometric mapping can increase the power for gene detection.

10.4   Genetic Haplotyping with Developmental Allometry

The past two decades have witnessed a surge of interest in applying geometric morphometric approaches to understand how body shape changes and how such a change is associated with allometry during ontogeny (Bookstein, 1991; Rohlf, 1998; Zelditch et al., 1998). For instance, these approaches have been used to study the ontogeny of body shape change for a few number of fishes (Zelditch and Fink, 1995; Reis et al., 1998), showing that body shape changes during ontogeny are not simply the result of uniform large-scale events but that localized small-scale shape changes contribute to its ontogeny. The technique derived to model such so-called ontogenetic allometry can be used to understand the mechanisms for the process and rate of drug response. Below, we will describe statistical models for haplotypeing genes for pharmacokinetic (PK) or pharmacodynamic (PD) processes of drug response. Much of our description was derived from the work of Li et al. (2007).

10.4.1 Likelihood

Consider a Hardy–Weinberg equilibrium population from which n subjects are drawn at random. All the sampled subjects are typed for SNPs. During a time course (or a range of dose), we measure some PK- or PD-related physiological variables and observe the temporal change of the genetic control of these two variables. Again, population genetic parameters of SNPs (Ω_p) can be estimated with the EM algorithm. Our focus is on the development of a statistical model for haplotyping two allometrically related biological processes (with results reflected in Ω_q) through the recognition of risk and non-risk haplotypes (and therefore composite diplotypes, AA (2), $A\bar{A}$ (1), $\bar{A}\bar{A}$ (0)) related to the processes.

Consider two dynamic traits, $y(t)$ and $z(t)$, measured at a series of time points (say T), and two SNPs that constitute a QTN that encode these two traits. Based on Table 2.1, the likelihood for dynamic traits and SNP data ($\mathbf{S}$) is constructed as

$$\log L(\Omega_q | \mathbf{y}, \mathbf{z}, \mathbf{S}, \hat{\Omega}_p)$$

$$= \sum_{i=1}^{n_{11/11}} (2\pi)^{-\frac{T}{2}} |\mathbf{\Sigma}|^{-\frac{1}{2}} \exp\left[-\frac{1}{2}(\mathbf{y}_i - \mathbf{u}_{2|i})' \mathbf{\Sigma}^{-1} (\mathbf{y}_i - \mathbf{u}_{2|i})\right]$$

$$+ \sum_{i=1}^{n_{11/10}+n_{10/11}} (2\pi)^{-\frac{T}{2}} |\mathbf{\Sigma}|^{-\frac{1}{2}} \exp\left[-\frac{1}{2}(\mathbf{y}_i - \mathbf{u}_{1|i})' \mathbf{\Sigma}^{-1} (\mathbf{y}_i - \mathbf{u}_{1|i})\right]$$

$$+ \sum_{i=1}^{n_{10/10}} \log\left\{ \omega(2\pi)^{-\frac{T}{2}} |\mathbf{\Sigma}|^{-\frac{1}{2}} \exp\left[-\frac{1}{2}(\mathbf{y}_i - \mathbf{u}_{1|i})' \mathbf{\Sigma}^{-1} (\mathbf{y}_i - \mathbf{u}_{1|i})\right] \right.$$

$$\left. + (1-\omega)(2\pi)^{-\frac{T}{2}} |\mathbf{\Sigma}|^{-\frac{1}{2}} \exp\left[-\frac{1}{2}(\mathbf{y}_i - \mathbf{u}_{0|i})' \mathbf{\Sigma}^{-1} (\mathbf{y}_i - \mathbf{u}_{0|i})\right] \right\}$$

$$+ \sum_{i=1}^{n_{11/00}+n_{10/00}+n_{00/11}+n_{00/10}+n_{00/00}} (2\pi)^{-\frac{T}{2}} |\mathbf{\Sigma}|^{-\frac{1}{2}}$$

$$\times \exp\left[-\frac{1}{2}(\mathbf{y}_i - \mathbf{u}_{0|i})' \mathbf{\Sigma}^{-1} (\mathbf{y}_i - \mathbf{u}_{0|i})\right], \tag{10.20}$$

where $\mathbf{y}_i = (y_i(1), \cdots, y_i(T))$ is the observation vector for trait y,

$$\mathbf{u}_{j|i} = (u_{j|i}(1), \cdots, u_{j|i}(T)) \tag{10.21}$$

is the composite diplotype-specific mean vector ($j = 0, 1, 2$), and ω is the relative probability with which the double heterozygote carries one of its two possible diplotypes.

The time-dependent covariance matrix (Σ) generally follows a structure which can be mathematically modeled. A number of statistical models, such as autoregressive (Diggle et al., 2002) and antedependent (Zimmerman and Núñez-Antón, 2001), have been formulated to model such a structure. In Zimmerman and Núñez-Antón (2001), the advantage of structured antedependent (SAD) model has been extensively discussed. The structure of the first-order SAD is expressed as

$$
\Sigma = \sigma^2 \begin{bmatrix} 1 & \rho & \rho^2 & \cdots \\ \rho & 1+\rho^2 & \rho(1+\rho^2) & \cdots \\ \rho^2 & \rho(1+\rho^2) & 1+\rho^2+\rho^4 & \cdots \\ \vdots & \vdots & \vdots & \cdots \\ \rho^T & \rho^{(T-1)}(1+\rho^2) & \rho^{(T-2)}(1+\rho^2+\rho^4) & \cdots \end{bmatrix}
$$

$$
\begin{matrix} \rho^{(T-1)} & \rho^T \\ \rho^{(T-2)}(1+\rho^2) & \rho^{(T-1)}(1+\rho^2) \\ \rho^{(T-3)}(1+\rho^2+\rho^4) & \rho^{(T-2)}(1+\rho^2+\rho^4) \\ \vdots & \vdots \\ \rho(1+\rho^2+\ldots+\rho^{(2T-4)}) & 1+\rho^2+\ldots+\rho^{(2T-2)} \end{matrix}
$$

whose inverse and determinant have closed forms as follows:

$$
\Sigma^{-1} = \frac{1}{\sigma^2} \begin{bmatrix} 1+\rho^2 & -\rho & 0 & \cdots & 0 & 0 \\ -\rho & 1+\rho^2 & -\rho & \cdots & 0 & 0 \\ 0 & -\rho & 1+\rho^2 & \cdots & 0 & 0 \\ \vdots & \vdots & \vdots & \cdots & \vdots & \vdots \\ 0 & 0 & 0 & \cdots & -\rho & 1 \end{bmatrix}_{T \times T}
$$

and

$$
|\Sigma| = (\sigma^2)^T \tag{10.22}
$$

If the allometric relationship between two biological traits is controlled by a QTN, the power equation (9.1) can be used to model the genotypic mean vector in the likelihood (10.20) with diplotype-specific parameter sets (α_j, β_j). Thus, by testing the difference between the two parameter sets, we can conclude whether there is a specific QTN for allometric scaling and how the QTN controls the scaling relationship. Wu and Hou (2006) modeled the allometric scaling relationship by incorporating the genotypic vectors of the two traits, y and z, i.e.,

$$
u_{j|i}(t) = \alpha_j + \beta_j v_{j|i}(t), \quad j = 2, 1, 0 \tag{10.23}
$$

where $u_{j|i}(t)$ and $v_{j|i}(t)$ are the genotypic values of traits y and z for composite diplotype j at time point t. This treatment needs to estimate the genotypic vector of trait z based on its observations. Thus, equation (10.4.1) is now changed as a bivariate model, i.e.,

$$\mathbf{u}_{j|i} = (u_{j|i}(1),\cdots,u_{j|i}(T),v_{j|i}(1),\cdots,v_{j|i}(T)).$$

This bivariate model is needed to specify the time-dependent covariance matrix for each trait and the time-dependent covariance matrix between the two traits. All these will largely increase the number of parameters to be estimated, making the computation quickly prohibitive and the parameter estimation imprecise.

To overcome this problem, Li et al. (2007) formulated a different model for the allometric relationship. Given that the allometric change of one trait is not only by the underlying genes, but regulated by physiology- and metabolism-related characteristics that contain the influences of both genes and environments (Niklas, 2006), we model the genotypic vector of a trait with the phenotypic value of a second allometrically related trait. Thus, equation (10.4.1) is modeled by

$$u_{j|i}(t) = \alpha_j + \beta_j z_i(t), \quad j = 2,1,0 \tag{10.24}$$

which can be explained from genetic and statistical perspectives. In genetics, Equation 10.24 states how the genotypic value of trait y for QTN diplotype j scales as, or responds to, the phenotypic change of trait z during ontogeny. If parameter set (α_j,β_j) is not different between diplotypes, this means that this QTN does not determine the allometric scaling between traits y and z. In statistics, equation 10.24 constructs a linear regression model in which the change of genotypic value of trait y is specified by the phenotypic value of trait z as a covariate.

When equation (10.24) is substituted into the likelihood (10.20), the underlying unknown quantitative genetic parameters are composed of $(\{\alpha_j,\beta_j\}_{j=0}^2,\rho,\sigma^2)$.

10.4.2 Algorithm

Below, we provide the EM algorithm for estimating the diplotype-specific power parameters (α_j,β_j), and (ρ,σ^2) that model the covariance matrix structure with the SAD(1) model. Taking the log-likelihood function of (10.20), we will have

$$\mathbf{u}_{j|i} = (\alpha_j + \beta_j z_i(1), \alpha_j + \beta_j z_i(2),\ldots,\alpha_j + \beta_j z_i(T))'.$$

We define the E-step by expressing the posterior probabilities of subject i to be diplotype j as

$$\Psi_{j|i} = \frac{\omega_{j|i}(2\pi)^{-\frac{T}{2}}|\Sigma|^{-\frac{1}{2}}\exp\left[-\frac{1}{2}(\mathbf{y}_i - \mathbf{u}_{j|i})'\Sigma^{-1}(\mathbf{y}_i - \mathbf{u}_{j|i})\right]}{\sum_{j'=0}^{2}\omega_{j'|i}(2\pi)^{-\frac{T}{2}}|\Sigma|^{-\frac{1}{2}}\exp\left[-\frac{1}{2}(\mathbf{y}_i - \mathbf{u}_{j'|i})'\Sigma^{-1}(\mathbf{y}_i - \mathbf{u}_{j'|i})\right]}$$

The M step is derived by solving the log-likelihood equations:

$$0 = \frac{\partial \log L}{\partial \alpha_j} = \sum_{i=1}^{n} \Psi_{j|i}(\mathbf{z}_i - \mathbf{u}_{j|i})' \Sigma^{-1} \frac{\partial \mathbf{u}_{j|i}}{\partial \alpha_j}$$

$$0 = \frac{\partial \log L}{\partial \beta_j} = \sum_{i=1}^{n} \Psi_{j|i}(\mathbf{z}_i - \mathbf{u}_{1|i})' \Sigma^{-1} \frac{\partial \mathbf{u}_{j|i}}{\partial \beta_j}$$

$$0 = \frac{\partial \log L}{\partial \rho} = -\frac{1}{2\sigma^2}\left[\Psi_{1|i}(\mathbf{z}_i - \mathbf{u}_{1|i})'(\rho \mathbf{B}_2 - \mathbf{B}_1)(\mathbf{z}_i - \mathbf{u}_{1|i}) \right.$$
$$\left. + \Psi_{2|i}(\mathbf{z}_i - \mathbf{u}_{2|i})'(\rho \mathbf{B}_2 - \mathbf{B}_1)(\mathbf{z}_i - \mathbf{u}_{2|i}) \right]$$

$$0 = \frac{\partial \log L}{\partial \sigma^2} = \frac{1}{2\sigma^4} \sum_{i=1}^{n} \left[-T\sigma^2 + \Psi_{1|i}(\mathbf{z}_i - \mathbf{u}_{1|i})'\mathbf{A}(\mathbf{z}_i - \mathbf{u}_{1|i}) \right.$$
$$\left. + \Psi_{2|i}(\mathbf{z}_i - \mathbf{u}_{2|i})'\mathbf{A}(\mathbf{z}_i - \mathbf{u}_{2|i}) \right]$$

where

$$\mathbf{A} = \sigma^2 \Sigma^{-1}$$

$$\mathbf{B}_1 = \begin{bmatrix} 0 & 1 & 0 & \ldots & 0 & 0 \\ 1 & 0 & 1 & \ldots & 0 & 0 \\ 0 & 1 & 0 & \ldots & 0 & 0 \\ \vdots & \vdots & \vdots & \ldots & \vdots & \vdots \\ 0 & 0 & 0 & \ldots & 1 & 0 \end{bmatrix}$$

$$\mathbf{B}_2 = \mathbf{diag}\{2, ..., 2, 0\}$$

and $\mathbf{B} = \frac{\partial \mathbf{A}}{\partial \rho} = \rho \mathbf{B}_2 - \mathbf{B}_1$, which leads to

$$\hat{\alpha}_1 = \frac{\sum_{i=1}^{n} \Psi_{1|i} \mathbf{z}_i' \Sigma^{-1} \frac{\partial \mathbf{u}_{1|i}}{\partial \alpha_1}}{\sum_{i=1}^{n} \Psi_{1|i} \frac{\partial \mathbf{u}_{1|i}'}{\partial \alpha_1} \Sigma^{-1} \frac{\partial \mathbf{u}_{1|i}}{\partial \alpha_1}}$$

$$\hat{\alpha}_2 = \frac{\sum_{i=1}^{n} \Psi_{2|i} \mathbf{z}_i' \Sigma^{-1} \frac{\partial \mathbf{u}_{2|i}}{\partial \alpha_2}}{\sum_{i=1}^{n} \Psi_{2|i} \frac{\partial \mathbf{u}_{2|i}'}{\partial \alpha_2} \Sigma^{-1} \frac{\partial \mathbf{u}_{2|i}}{\partial \alpha_2}}$$

$$\hat{\rho} = \frac{\sum_{i=1}^{n} [\Psi_{1|i}(\mathbf{z}_i - \mathbf{u}_{1|i})'\mathbf{B}_1(\mathbf{z}_i - \mathbf{u}_{1|i}) + \Psi_{2|i}(\mathbf{z}_i - \mathbf{u}_{2|i})'\mathbf{B}_1(\mathbf{z}_i - \mathbf{u}_{2|i})]}{\sum_{i=1}^{n} [\Psi_{1|i}(\mathbf{z}_i - \mathbf{u}_{1|i})'\mathbf{B}_2(\mathbf{z}_i - \mathbf{u}_{1|i}) + \Psi_{2|i}(\mathbf{z}_i - \mathbf{u}_{2|i})'\mathbf{B}_2(\mathbf{z}_i - \mathbf{u}_{2|i})]}$$

$$\hat{\sigma}^2 = \frac{1}{nT} \sum_{i=1}^{n} [\Psi_{1|i}(\mathbf{z}_i - \mathbf{u}_{1|i})'\mathbf{A}(\mathbf{z}_i - \mathbf{u}_{1|i}) + \Psi_{2|i}(\mathbf{z}_i - \mathbf{u}_{2|i})'\mathbf{A}(\mathbf{z}_i - \mathbf{u}_{2|i})]$$

For parameters β_j, it is not possible to derive a closed form for their solution. Instead, the Newton-Raphson algorithm is implemented to estimate these parameters. The τth step in the Newton-Raphson algorithm is expressed as

$$\beta_1^{\{\tau\}} = \beta_1^{\{\tau-1\}} - \left[\left(\frac{\partial^2 \log L}{\partial \beta_1^2} \right)^{-1} \left(\frac{\partial \log L}{\partial \beta_1} \right) \right] \Big|_{\beta_1 = \beta_1^{\{\tau-1\}}},$$

$$\beta_2^{\{\tau\}} = \beta_2^{\{\tau-1\}} - \left[\left(\frac{\partial^2 \log L}{\partial \beta_2^2}\right)^{-1} \left(\frac{\partial \log L}{\partial \beta_2}\right)\right]\bigg|_{\beta_2=\beta_2^{\{\tau-1\}}}.$$

The E and M steps are iteratively repeated until the estimates of parameters are stable. These stable estimates are regarded as the MLEs of parameters. The estimates of the sampling variances for the MLEs are obtained using the Louis (1982) approach.

10.4.3 Hypothesis Testing

As shown in earlier publications (Ma et al., 2002; Wu et al., 2002a, 2004b), functional mapping is advantageous for the tests of biologically meaningful hypotheses regarding genetic actions and organ development. Here, we outline several important hypotheses for the genetic control of allometric scaling. The first hypothesis is about the existence of QTN, which can be tested by formulating the null hypothesis,

$$H_0: \ (\alpha_j, \beta_j) \equiv (\alpha, \beta), \ j = 2, 1, 0.$$

H_1 : At least one of the equalities in H_0 does not hold.

The log-likelihood ratio between the null and alternative hypothesis is calculated as

$$\text{LR} = -2[\ln L_0(\tilde{\alpha}, \tilde{\beta}, \tilde{\rho}, \tilde{\sigma}^2 | \mathbf{y}, \mathbf{z}) - \ln L_1(\hat{r}_1 \text{ or } \hat{r}_2, \hat{\alpha}_j, \hat{\beta}_j, \hat{\rho}, \hat{\sigma}^2 | \mathbf{y}, \mathbf{z}, \mathbf{S})],$$

where the tildes and hats are the MLEs of parameters under the null and hypotheses, respectively. The plug-in LR value is then compared with the critical threshold determined from permutation tests to test the significance of the QTN hypothesized.

We are also interested in the genetic cause for the differentiation in ontogenetic allometric scaling. This can be investigated by testing the normalization (α) and exponent constant (β) individually. While α is suggested to be a characteristic of species or populations (Niklas, 1994), a recent survey by Niklas and Enquist (2001) finds that all plants have a similar normalization constant, and thus follow a single allometric formula. By testing whether α equals to a specific constant for different plant species, this debate can be addressed.

In practice, the exponent coefficient β can be treated as a constant if the allometric relationship of the two traits studied is known. For example, body length scales as the $1/4$-power of body mass (West et al., 1997, 1999a,b). In this case, $\beta = 1/4$ can be directly substituted in Equation 5 to obtain estimates for the remainder of the unknown parameters. Owing to the reduced number of the unknowns to be estimated, such a substitution can potentially increase the precision and power of parameter estimation.

All the tests for α and β can be performed by calculating a likelihood ratio statistic which asymptotically follows a chi-square distribution with the corresponding degrees of freedom. In actual data analyses, an empirical approach based on simulation studies can be used to determine the threshold for these tests.

Example 10.2
Li et al. (2007) used their developmental allometry model to map quantitative trait loci (QTLs) that regulate the allometric scaling between main stems and whole-plant

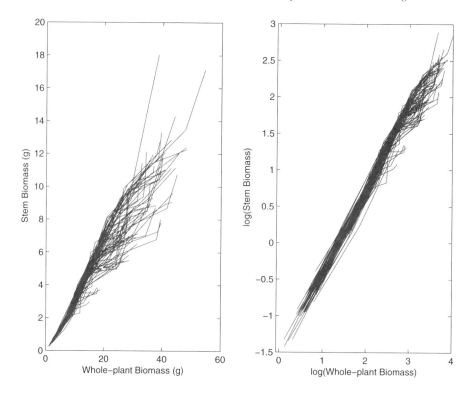

FIGURE 10.3

The allometric scaling relationship between stem biomass and whole-plant biomass for the RILs from a soybean mapping population.

biomass repeatedly measured at multiple different ages during a growing season in soybeans. The data were from a soybean genome project in which a total of 184 recombinant inbred lines (RILs) derived from two inbred lines, Kefeng No. 1 and Nannong 1138-2, was used as a mapping population (Fig. 10.3). All the RILs were typed for 488 molecular markers (restricted fragment length polymorphisms, simple sequence repeats and amplified fragment length polymorphsim) that construct a linkage map with 25 linkage groups covering 4,151.2 cM of the soybean genome (Zhang et al., 2004) (Fig. 10.4).

The RILs were planted in a simple lattice design with multiple replicates in a plot and harvested to measure their above- and under-ground biomass for 8 times with the first time at the 28th day after emergence and successive 7 times every 10 days thereafter. For the same RIL, the phenotypic values measured for different times correspond to successive measurements on a time scale. The analysis of Li et al. (2007) was based on the ontogenetic allometric scaling relationship between stem and whole-plant biomass. As shown by a subset of mapping RILs in Fig. 10.3, stem

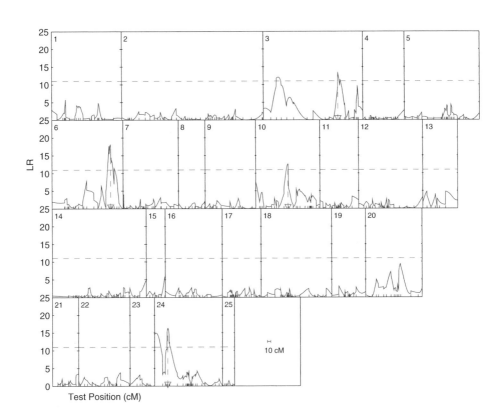

FIGURE 10.4

The LR profile of the likelihoods under the null (there is no QTL) and alternative hypothesis (there is a QTL) across the lengths of 25 linkage groups for the allometric scaling relationship between stem and whole-plant biomass growth trajectories in a soybean RIL population. The 5% significance critical threshold (10.98) determined from 1000 permutation tests is indicated by the broken horizonal line. The arrowed broken vertical line indicates the MLE of the QTL location.

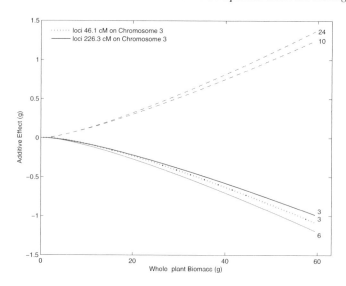

FIGURE 10.5
Body size-dependent additive genetic effects calculated from ontogenetic allometry
curves for two different genotypes at each of the five QTL detected on linkage groups
3, 6, 10, and 24.

biomass scales as a power function of whole-plant biomass. The Pearson correlation
coefficients between the two log-transformed traits from the samples for all subjects
are all close to 1, ranging from 0.9381 to 0.9992.

Based on the modeling strategy of equation (10.24), five significant allometry
QTLs were detected, two located on linkage group 3, and one on linkage groups 6,
10, and 24, respectively (Fig. 10.4), two located between markers GMKF082c and
GMKF168b and at marker A520T on chromosome 3, one located between markers
GMKF059a and satt319 on chromosome 6, one located between markers Satt372
and Satt154 on chromosome 10, and one between markers GMKF082b and satt331
on chromosome 24. The model provided the MLEs of genotype-specific curve parameters and covariance-structuring SAD parameters when each of the significant
QTL was detected. The estimated genotypic power curve parameters are used to
calculate additive genetic effects, $a(t)$, at each QTL that vary with time-dependent
whole-plant biomass by

$$a(t) = \frac{1}{2} \left\{ \exp\left[\alpha_1 + \beta_1 z(t)\right] - \exp\left[\alpha_2 + \beta_2 z(t)\right] \right\}$$

for an RIL design. The positive value of $a(t)$ implies that parent Kefeng No. 1 contributes favorable alleles to increased stem biomass, whereas the negative value corresponds to the favorable contribution made by parent Nannong 1138-2. As shown
by Fig. 10.5, the additive effects of each QTL on stem biomass change with whole-

plant biomass. Based on their signs, it is suggested that at the two QTLs on chromosomes 10 and 24 favorable alleles for increased stem biomass are contributed by parent Kefeng No. 1, whereas the inverse pattern is true for the three QTLs on chromosomes 3 and 6.   ☐

11

Joint Functional Mapping of Drug Efficacy and Toxicity

The administration of a medicine to patients would produce two different responses, beneficial therapeutic effects (efficacy) and adverse drug reactions (ADRs, or toxicity). According to the information provided by Marsh and McLeod (2006), 7% of patients were affected by ADRs in the UK in 1994, whereas this number is estimated as 2,216,000 hospital patients, leading to approximately 106,000 patient deaths and making ADR fatalities the fourth to sixth leading cause of death in the US in the same year. There is considerable variability in the pattern of drug response among different individuals (Fig. 11.1), which can be categorized into four types: (1) There is a benefit and also toxicity, (2) there is a benefit, but no toxicity, (3) there is no benefit, but there is toxicity, and (4) there is neither benefit nor toxicity.

Because of such variation, the currently used "one dose fits all" regime is not ideal for patients, and therefore is not cost-effective in the health care field. The identification of the pharmacogenetic basis for interindividual differences in the pattern of drug response will help to make a personalized therapy strategy to improve treatment efficacy and, meanwhile, reduce the incidence of ADRs (Evans and Johnson, 2001; Evans and McLeod, 2003; Johnson, 2003; Weinshilboum, 2003; Marsh, 2005; Marsh and McLeod, 2006). However, the genetic control of both efficacy and toxicity is typically complex, in which multiple genes interact with various biochemical, developmental, and environmental factors in coordinated ways to determine the overall phenotypes (Watters and McLeod, 2003). With the advent of recent genomic technologies, interindividual differences in drug response can now be attributed to sequence variants in genes that encode the metabolism and disposition of drugs and the targets of drug therapy (such as receptors) (Evans and Relling, 1999; Evans and Johnson, 2001). To understand more comprehensively the genetic architecture of drug response, sophisticated approaches are needed with which genes for drug efficacy and genes for drug toxicity can be identified.

Taking advantage of sequence-based association studies and functional mapping, the DNA sequence structure of drug response can be estimated and the genetic differences between efficacy and toxicity can be compared at the single DNA base level. In this chapter, we will introduce basic principles and procedures for joint sequence mapping of drug efficacy and toxicity within the framework of functional mapping. We formulate a series of hypotheses about the difference in genetic control of these two processes. An example will be shown for practical usefulness of the joint functional mapping.

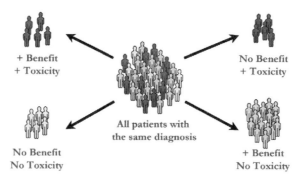

FIGURE 11.1
Substantial inter-patient variation occurs in the pattern of responsiveness to a medicine. Adapted from Marsh and McLeod (2006).

11.1   A Joint Model

11.1.1   Genetic Design

An experiment is launched to haplotype drug efficacy and drug toxicity via modeling pharmacokinetic and pharmacodynamic processes of drug response. The experiment is based on a random sample of *n* subjects drawn randomly from a natural population at Hardy–Weinberg equilibrium. Many toxicogenomic studies are conducted with an animal as model system. The model being described for haplotyping drug efficacy and drug toxicity can be used in an experimental cross derived from inbred strains.

The sampled subjects are typed for different single nucleotide polymorphisms (SNPs), at each of which two alleles are denoted as 1 and 0. The model should allow the characterization of any number SNPs for haplotyping, but for the sake of simplicity, we will focus on two SNPs that constitute four haplotypes [11], [10], [01], and [00] that may be responsible for drug efficacy and toxicity. The haplotype frequencies are expressed as p_{11}, p_{10}, p_{01}, and p_{00}. The four haplotypes generate 10 distinct diplotypes expressed as [11][11], [11][10], $\cdots$, [00][00] which are sorted into nine genotypes 11/11, 11/10, $\cdots$, 00/00 (Table 2.1). The double heterozygotic genotype 10/10 contains two possible diplotypes [11][00] and [10][01]. We hope to characterize the relationship between diplotypes and drug response (rather than relationship between genotypes and drug response, as detected by conventional mapping approaches) although these diplotypes cannot always be distinguished clearly from genotype data.

Let $n_{r_1 r'_1 / r_2 r'_2}$ $(r_1 \geq r'_1, r_2 \geq r'_2 = 0, 1)$ be the observations of various genotypes (**S**). Based on these observations and the expected genotype frequencies expressed in terms of haplotype frequencies (Table 2.1), we can construct a multinomial likelihood. The EM algorithm is implemented to provide a closed form solution for four haplotype frequencies (see Section 2.3). To estimate haplotype effects on drug response, we will need to assume that one of these haplotypes is different from the rest in trait expression. The former is called the risk haplotype denoted as A, and the latter is called the non-risk haplotype denoted as $\bar{A}$. The risk and non-risk haplotypes are combined to generate three possible composite diplotypes, AA (2), $A\bar{A}$ (1), and $\bar{A}\bar{A}$ (0). The model will be able to detect the differences in trait phenotypes among these three diplotypes, again although they cannot be seen clearly.

11.1.2   Clinical Design

Efficacy and toxicity describe how patients respond to different doses or concentrations of drugs. Statistically, these represent a longitudinal problem whose underlying genetic determinants can be mapped using the functional mapping strategy. To integrate sequence mapping and functional mapping to directly characterize DNA sequence variants that are responsible for efficacy and toxicity processes, we need not only SNP data as described above, but also a series of measures for patient's pharmacological response to a drug at multiple times and/or multiple doses of the drug. To better quantify pharmacological responses, we will measure both pharmacokinetic (PK) and pharmacodynamic (PD) parameters in a clinical trial.

Following the adminstration of a drug, blood or urine samples will be collected from each subject to test the following response variables: plasma concentration (PK), and PD-related drug efficacy and toxicity parameters in a course of T time points. Let x, y, and z be the serial measurements of plasma concentration, drug efficacy, and drug toxicity, respectively. A number of mathematical models derived from zero-order or first-order differentiate equations are proposed to describe the PK response for one, two or multiple compartments. Here, as an example, we will use a standard PK one-compartment open model with first order absorption and elimination to specify the time-dependent change of plasma concentration, expressed as

$$c_p(t) = \frac{dk_a}{V_d(k_a - k_e)}(e^{-k_a t} - e^{-k_e t}) \qquad (11.1)$$

where d is the bioavailable dose, V_d is the volume of distribution, k_e is the elimination rate constant, k_a is a first-order absorption rate constant, and t is the time after the last dose was administrated.

The efficacy and toxicity of a drug is modeled by specifying their relationships with plasma concentration. The sigmoid Emax model for drug efficacy and the power function for drug toxicity (McClish and Roberts, 2003) will be used for PD modeling, expressed as

$$E(c) = E_0 + \frac{E_{max}c^H}{EC_{50}^H + c^H}, \qquad (11.2)$$

where E_0 is the baseline value for the drug response parameter, E_{max} is the asymptotic effect, EC_{50} is the drug concentration that results in 50% of the maximal effect, and

H is the slope parameter that determines the slope of the concentration–response curve, and

$$\mathscr{T}(c) = \alpha c^\beta, \tag{11.3}$$

where β determines the shape of the dose-response relationship, and α adjusts the dose–related gain to which that shape conforms.

11.1.3 Statistical Design

At a particular time t $(t = 1,...,T)$, the relationship between the observations of plasma concentration, drug efficacy, and drug toxicity and their expected genotypic means can be described by linear regression models, i.e.,

$$
\begin{aligned}
x_i(t) &= \sum_{j_1=0}^{2} \xi_{ij_1} \mu_{j_1 x}(t) + e_{ix}(t), \\
y_i(t) &= \sum_{j_2=0}^{2} \zeta_{ij_2} \mu_{j_2 y}(t) + e_{iy}(t), \\
z_i(t) &= \sum_{j_3=0}^{2} \eta_{ij_3} \mu_{j_3 z}(t) + e_{iz}(t),
\end{aligned}
\tag{11.4}
$$

where ξ_{ij_1}, ζ_{ij_2}, or η_{ij_3} are the indicator variables denoted as 1 if a particular composite diplotype j_1, j_2, or j_3 is considered for subject i in terms of traits x, y, or z, respectively, and 0 otherwise, $\mu_{j_1 x}(t)$, $\mu_{j_1 y}(t)$, and $\mu_{j_1 z}(t)$ are the genotypic values of different composite diplotypes for the three traits, respectively, and $e_{ix}(t)$, $e_{iy}(t)$, and $e_{iz}(t)$ are the residual errors that are iid normal with the mean of zero and the variances of $\sigma_x^2(t)$, $\sigma_y^2(t)$, and $\sigma_z^2(t)$, respectively.

The errors at two different times, t_1 and t_2 $(t_1,t_2 = 1,...,T)$, are correlated with the covariances of $\sigma_x(t_1,t_2)$ for plasma concentration, $\sigma_x(t_1,t_2)$ for drug efficacy, and $\sigma_z(t_1,t_2)$ for drug toxicity, and $\sigma_{xy}(t)$ and $\sigma_{xy}(t_1,t_2)$ between plasma concentration and efficacy, $\sigma_{xz}(t)$ and $\sigma_{xz}(t_1,t_2)$ between plasma concentration and toxicity, and $\sigma_{yz}(t)$ and $\sigma_{yz}(t_1,t_2)$ between efficacy and toxicity. All these variances and covariances comprise the structure of a large covariance matrix Σ, expressed as

$$
\Sigma = \begin{pmatrix} \Sigma_x & \Sigma_{xy} & \Sigma_{xz} \\ \Sigma_{yx} & \Sigma_y & \Sigma_{yz} \\ \Sigma_{zx} & \Sigma_{zy} & \Sigma_z \end{pmatrix}, \tag{11.5}
$$

where Σ_x is composed of $\sigma_x^2(t)$ and $\sigma_x(t_1,t_2)$, Σ_y composed of $\sigma_y^2(t)$ and $\sigma_y(t_1,t_2)$, Σ_z composed of $\sigma_z^2(t)$ and $\sigma_z(t_1,t_2)$, Σ_{xy} composed of $\sigma_{xy}(t)$ and $\sigma_{xy}(t_1,t_2)$, Σ_{xz} composed of $\sigma_{xz}(t)$ and $\sigma_{xz}(t_1,t_2)$, and Σ_{yz} composed of $\sigma_{yz}(t)$ and $\sigma_{yz}(t_1,t_2)$.

11.1.4 The Haplotyping Framework

While a drug is expected to display favorable effects, it may also be toxic. Also, the PK and PD responses are physiologically not independent of each other. For these reasons, the haplotyping model should integrate the genetic control mechanisms of these related (although different) processes. Let $\mathbf{x}_i = [x_i(1), \cdots, x_i(T)]$ be the plasma concentration observation vector, $\mathbf{y}_i = [y_i(1), \cdots, y_i(T)]$ be the efficacy observation vector, and $\mathbf{z}_i = [z_i(1), \cdots, z_i(T)]$ be the toxicity observation vector for a drug administrated to subject i. Because the phenotypic measurements of drug efficacy and toxicity are continuously variable, it is reasonable to assume that joint vector $(\mathbf{x}_i, \mathbf{y}_i, \mathbf{z}_i)$ follows a multivariate normal distribution, as used in general quantitative genetic studies.

To jointly haplotype the PK, drug efficacy, and drug toxicity, two log-likelihood functions as described in equation (2.2) should be constructed for observed SNP markers (**S**) and longitudinal phenotypic data ($\mathbf{x}, \mathbf{y}, \mathbf{z}$). One likelihood based on a multinomial distribution is formulated to estimate haplotype frequencies ($\Omega_p = (p_{11}, p_{10}, p_{01}, p_{00})$). The estimated Ω_p is used to construct a mixture model-based likelihood with observations ($\mathbf{x}, \mathbf{y}, \mathbf{z}, \mathbf{S}$), aimed to estimate the parameters (Ω_q) that are associated with haplotype effects and covariance structure for these PK and PD variables.

We can assume that the same risk haplotype is affecting the three physiological variables, but we use a more general model to model these three variables by assuming different risk haplotypes specific to each variable. Assume that the risk haplotypes of plasma concentration, drug efficacy, and drug toxicity are [11], [10], and [00], respectively. Thus, we will need to model the haplotype effects by specifying different composite diplotypes for the three variables, j_1 for plasma concentration, j_2 for drug efficacy, and j_3 for drug toxicity ($j_1, j_2, j_3 = 2, 1, 0$). Based on equations (11.1), (11.2), and (11.3), PK and PD parameters are specified in terms of different composite diplotypes as $(V_{dj_1}, v_{aj_1}, v_{ej_1})$, $(E_{\max j_2}, EC_{50j_2}, H_{j_2})$, and $(\alpha_{j_3}, \beta_{j_3})$, respectively. Table 11.1 provides information about diplotypes, genotypes and composite diplotypes for plasma concentration, drug efficacy, and drug toxicity.

The mixture-based likelihood is constructed as

$$
\begin{aligned}
\log L(\Omega_q | \mathbf{x}, \mathbf{y}, \mathbf{z}, \mathbf{S}, \hat{\Omega}_p) \\
= \sum_{i=1}^{n_{11/11}} \log f_{200}(\mathbf{x}_i, \mathbf{y}_i, \mathbf{z}_i) + \sum_{i=1}^{n_{11/10}} \log f_{110}(\mathbf{x}_i, \mathbf{y}_i, \mathbf{z}_i) \\
+ \sum_{i=1}^{n_{11/00}} \log f_{120}(\mathbf{x}_i, \mathbf{y}_i, \mathbf{z}_i) + \sum_{i=1}^{n_{10/11}} \log f_{101}(\mathbf{x}_i, \mathbf{y}_i, \mathbf{z}_i) \\
+ \sum_{i=1}^{n_{10/10}} \log [\omega f_{100}(\mathbf{x}_i, \mathbf{y}_i, \mathbf{z}_i) + (1 - \omega) f_{011}(\mathbf{x}_i, \mathbf{y}_i, \mathbf{z}_i)] \\
+ \sum_{i=1}^{n_{10/00}} \log f_{010}(\mathbf{x}_i, \mathbf{y}_i, \mathbf{z}_i) + \sum_{i=1}^{n_{00/11}} \log f_{002}(\mathbf{x}_i, \mathbf{y}_i, \mathbf{z}_i) \\
+ \sum_{i=1}^{n_{00/10}} \log f_{001}(\mathbf{x}_i, \mathbf{y}_i, \mathbf{z}_i) + \sum_{i=1}^{n_{00/00}} \log f_{000}(\mathbf{x}_i, \mathbf{y}_i, \mathbf{z}_i),
\end{aligned}
\tag{11.6}
$$

TABLE 11.1

Possible diplotype configurations of nine genotypes at two SNPs that affect plasma concentration (PC), drug efficacy (DE), and drug toxicity (DT).

Genotype	Diplotype Configuration	Diplotype Frequency	Haplotype Composition 11	10	01	00	Observation	Genotypic Mean Vector	Parameters for Genotypic Mean Vector PC	DE	DT
11/11	[11][11]	p_{11}^2	1	0	0	0	$n_{11/11}$	u_{200}	(V_{d2}, v_{a2}, v_{e2})	$(E_{max0}, EC_{50/0}, H_0)$	(α_0, β_0)
11/10	[11][10]	$2p_{11}p_{10}$	$\frac{1}{2}$	$\frac{1}{2}$	0	0	$n_{11/10}$	u_{11}	(r_{M2}, q_{M2})	(r_{P0}, q_{P0})	(h_0, m_0, τ_0)
11/00	[10][10]	p_{10}^2	0	1	0	0	$n_{11/00}$	u_{12}	(r_{M2}, q_{M2})	(r_{P0}, q_{P0})	(h_0, m_0, τ_0)
10/11	[11][01]	$2p_{11}p_{01}$	$\frac{1}{2}$	0	$\frac{1}{2}$	0	$n_{10/11}$	u_{10}	(r_{M2}, q_{M2})	(r_{P0}, q_{P0})	(h_0, m_0, τ_0)
10/10	$\Big\{$ [11][00]	$\Big\{$ $2p_{11}p_{00}$	$\frac{1}{2}\omega$	$\frac{1}{2}(1-\omega)$	$\frac{1}{2}(1-\omega)$	$\frac{1}{2}\omega$	$n_{10/10}$	$\Big\{$ u_{10}	$\Big\{$ (r_{M2}, q_{M2})	$\Big\{$ (r_{P0}, q_{P0})	$\Big\{$ (h_0, m_0, τ_0)
	[10][01]	$2p_{10}p_{01}$						u_{01}	(r_{M2}, q_{M2})	(r_{P0}, q_{P0})	(h_0, m_0, τ_0)
10/00	[10][00]	$2p_{10}p_{00}$	0	$\frac{1}{2}$	0	$\frac{1}{2}$	$n_{10/00}$	u_{01}	(r_{M2}, q_{M2})	(r_{P0}, q_{P0})	(h_0, m_0, τ_0)
00/11	[01][01]	p_{01}^2	0	0	1	0	$n_{00/11}$	u_{00}	(r_{M2}, q_{M2})	(r_{P0}, q_{P0})	(h_0, m_0, τ_0)
00/10	[01][00]	$2p_{01}p_{00}$	0	0	$\frac{1}{2}$	$\frac{1}{2}$	$n_{00/10}$	u_{00}	(r_{M2}, q_{M2})	(r_{P0}, q_{P0})	(h_0, m_0, τ_0)
00/00	[00][00]	p_{00}^2	0	0	0	1	$n_{00/00}$	u_{00}	(r_{M2}, q_{M2})	(r_{P0}, q_{P0})	(h_0, m_0, τ_0)

$\omega = \dfrac{p_{11}p_{00}}{p_{11}p_{00} + p_{10}p_{01}}$ where p_{11}, p_{10}, p_{01} and p_{00} are the haplotype frequencies of [11], [10], [01], and [00], respectively. Different risk haplotypes are assumed for PK and PD parameters, [11] for plasma concentration (PC), [10] for drug efficacy (DE), and [01] for drug toxicity (DT).

where
$$(\mathbf{x}_i, \mathbf{y}_i, \mathbf{z}_i) = (x_i(1), y_i(1), z_i(1), ..., x_i(T), y_i(T), z_i(T))$$

is the $3T$-dimensional vector for three longitudinal physiological traits measured at all times for patient i, and $f_{j_1 j_2 j_3}(\mathbf{x}_i, \mathbf{y}_i, \mathbf{z})$ is a multivariate normal distribution.

Functional mapping intends to model the mean vector and covariance matrix in $f_{j_1 j_2 j_3}(\mathbf{x}_i, \mathbf{y}_i, \mathbf{z}_i)$. For this particular question, we model the mean vector by

$$\mathbf{u}_{j_1 j_2 j_3} = [\mu_{x j_1}(1), \cdots, \mu_{x j_1}(T), \mu_{y j_2}(1), \cdots, \mu_{y j_2}(T), \mu_{z j_3}(1), \cdots, \mu_{z j_3}(T)],$$

modeled by equations (11.1), (11.2), and (11.3), respectively, i.e.,

$$\mu_{x j_1}(t) = \frac{d k_{a j_1}}{V_{d j_1}(k_{a j_1} - k_{e j_1})}(e^{-k_{a j_1} t} - e^{-k_{e j_1} t}), \quad j_1 = 2, 1, 0 \qquad (11.7)$$

$$\mu_{x j_2}(t) = E_{0 j_2} + \frac{E_{\max j_2} c^{H_{j_2}}}{E C_{50 j_2}^{H_{j_2}} + c^{H_{j_2}}}, \quad j_2 = 2, 1, 0 \qquad (11.8)$$

$$\mu_{x j_3}(t) = \alpha_{j_3} c^{\beta_{j_3}} \quad j_3 = 2, 1, 0. \qquad (11.9)$$

All the parameters that are associated with haplotype effects are contained in $\Omega_q = \{V_{d j_1}, v_{a j_1}, v_{e j_1}, E_{\max j_2}, E C_{50 j_2}, H_{j_2}, \alpha_{j_3}, \beta_{j_3}\}_{j_1, j_2, j_3 = 0}^2$.

11.1.5 Covariance Structure

Many statistical approaches have been proposed to model the structure of the covariance matrix for longitudinal traits measured at multiple time points (Diggle et al., 2002). Here, we use the structured antedependence (SAD) (Zimmerman and Núñez-Antón, 2001) to model the covariance structure.

According to the SAD model, an observation at a particular dosage c depends on the previous ones, with the degree of dependence decaying with time lag. For plasma concentration, drug efficacy, and toxicity, the residual errors for subject i in equation (11.4) can be expressed, in terms of the first-order SAD (SAD(1)) model, as

$$e_{xi}(t) = \phi_x e_{xi}(t-1) + \phi_{x \leftarrow y} e_{yi}(t-1) + \phi_{x \leftarrow z} e_{zi}(t-1) + \varepsilon_{xi}(t),$$
$$e_{yi}(t) = \phi_y e_{yi}(t-1) + \phi_{y \leftarrow x} e_{xi}(t-1) + \phi_{y \leftarrow z} e_{zi}(t-1) + \varepsilon_{yi}(t), \qquad (11.10)$$
$$e_{zi}(t) = \phi_z e_{zi}(t-1) + \phi_{z \leftarrow x} e_{xi}(t-1) + \phi_{z \leftarrow y} e_{yi}(t-1) + \varepsilon_{zi}(t),$$

where ϕ_x, ϕ_y, and ϕ_z are the unrestricted antedependence parameters induced by trait x, y, and z itself, respectively, $\phi_{k \leftarrow k'}$ is the the unrestricted antedependence parameter for trait k induced by trait k' ($k, k' = x, y, x$), respectively, and $\varepsilon_{xi}(t)$, $\varepsilon_{yi}(t)$, and $\varepsilon_{zi}(t)$ are the "innovation" errors for the three traits, respectively, normally distributed as $N(0, v_x^2(t))$, $N(0, v_y^2(t))$, and $N(0, v_z^2(c))$ (Zimmerman and Núñez-Antón, 2001).

The trivariate SAD(1) model for subject i described by equation (11.10) can be expressed in matrix notation as

$$\mathbf{e}_i = \Phi^{-1} \varepsilon_i, \qquad (11.11)$$

where

$$\mathbf{e}_i = [e_{xi}(1), \cdots, e_{xi}(T), e_{yi}(1), \cdots, e_{yi}(T), e_{zi}(1), \cdots, e_{zi}(T)]^{\mathrm{T}},$$
$$\boldsymbol{\varepsilon}_i = [\varepsilon_{xi}(1), \cdots, \varepsilon_{xi}(T), \varepsilon_{yi}(1), \cdots, \varepsilon_{yi}(T), \varepsilon_{zi}(1), \cdots, \varepsilon_{zi}(T)]^{\mathrm{T}},$$

and

$$\boldsymbol{\Phi} = \begin{pmatrix} \boldsymbol{\Phi}_x & \boldsymbol{\Phi}_{xy} & \boldsymbol{\Phi}_{xz} \\ \boldsymbol{\Phi}_{yx} & \boldsymbol{\Phi}_y & \boldsymbol{\Phi}_{zy} \\ \boldsymbol{\Phi}_{zx} & \boldsymbol{\Phi}_{zy} & \boldsymbol{\Phi}_z \end{pmatrix},$$

$$\boldsymbol{\Phi}_k = \begin{pmatrix} 1 & 0 & 0 & \cdots & 0 \\ -\phi_k & 1 & 0 & \cdots & 0 \\ 0 & -\phi_k & 1 & \cdots & 0 \\ 0 & 0 & -\phi_k & \cdots & 0 \\ \vdots & \vdots & \vdots & \ddots & \vdots \\ 0 & 0 & 0 & \cdots & 1 \end{pmatrix},$$

$$\boldsymbol{\Phi}_{kk'} = \begin{pmatrix} 0 & 0 & 0 & \cdots & 0 \\ -\phi_{k \leftarrow k'} & 0 & 0 & \cdots & 0 \\ 0 & -\phi_{k \leftarrow k'} & 0 & \cdots & 0 \\ 0 & 0 & -\phi_{k \leftarrow k'} & \cdots & 0 \\ \vdots & \vdots & \vdots & \ddots & \vdots \\ 0 & 0 & 0 & \cdots & 0 \end{pmatrix},$$

for $k, k' = x, y, z$.

The covariance matrix of $\mathbf{e}_i$ under the trivariate SAD(1) model (11.10) can be obtained as

$$\boldsymbol{\Sigma} = \boldsymbol{\Phi}^{-1} \boldsymbol{\Sigma}_\varepsilon (\boldsymbol{\Phi}^{-1})^{\mathrm{T}}, \tag{11.12}$$

where

$$\boldsymbol{\Sigma}_\varepsilon = \begin{pmatrix} \boldsymbol{\Sigma}_{\varepsilon x} & \boldsymbol{\Sigma}_{\varepsilon xy} & \boldsymbol{\Sigma}_{\varepsilon xz} \\ \boldsymbol{\Sigma}_{\varepsilon yx} & \boldsymbol{\Sigma}_{\varepsilon y} & \boldsymbol{\Sigma}_{\varepsilon yz} \\ \boldsymbol{\Sigma}_{\varepsilon zx} & \boldsymbol{\Sigma}_{\varepsilon zy} & \boldsymbol{\Sigma}_{\varepsilon z} \end{pmatrix},$$

$$\boldsymbol{\Sigma}_{\varepsilon k} = \begin{pmatrix} v_k^2(1) & 0 & 0 & \cdots & 0 \\ 0 & v_k^2(2) & 0 & \cdots & 0 \\ \vdots & \vdots & \vdots & \ddots & \vdots \\ 0 & 0 & 0 & \cdots & v_k^2(C) \end{pmatrix},$$

and

$$\boldsymbol{\Sigma}_{\varepsilon kk'} = \boldsymbol{\Sigma}_{\varepsilon k'k} = \begin{pmatrix} v_k v_{k'} \rho_{kk'}(1) & 0 & 0 & \cdots & 0 \\ 0 & v_k v_{k'} \rho_{kk'}(2) & 0 & \cdots & 0 \\ \vdots & \vdots & \vdots & \ddots & \vdots \\ 0 & 0 & 0 & \cdots & v_k v_{k'} \rho_{kk'}(T) \end{pmatrix}.$$

In the covariance modeling here, the innovative variance is assumed to be constant over different dosages. As shown by Jaffrézic et al. (2003), the SAD(1) model with this assumption can still allow for both the variance and correlation to change with the dose level. In addition, we assume that the correlation ($\rho_{kk'}$) in innovative errors between the two traits k and k' is stable over time course. With these assumptions, we need to estimate an array of parameters contained in $\{v_k^2, \phi_k, \phi_{k \leftarrow k'}, \rho_{kk'}\}_{k,k'=x}^z$ that model the structure of the covariance matrix under the SAD model.

As can be proven, in the context of the trivariate SAD model, the cross-correlation functions can be asymmetrical, i.e., $\text{Corr}_{kk'}(t_{1k}, t_{2k'}) \neq \text{Corr}_{k'k}(t_{1k'}, t_{2k}), t_{1k}, t_{1k'}, t_{2k}, t_{2k'} = 1, ..., T$. This favorable feature of the bivariate SAD model makes it useful for understanding the genetic correlation between different traits. In practice, innovative variance and correlation can be modeled by a polynomial and exponential function, respectively (Zimmerman and Núñez-Antón, 2001).

11.1.6 Algorithm and Determination of Risk Haplotypes

The EM algorithm is implemented to obtain the MLEs of two groups of unknown parameters in the integrated sequence and function mapping model (11.6), that is, the curve parameters that model the mean vector, and the parameters that model the structure of the covariance matrix. Zhao et al. (2004b) implemented the simplex method as advocated by Nelder and Mead (1965) to the estimation process of functional mapping, which can strikingly increase computational efficiency. The simplex algorithm can be embedded in the EM algorithm to provide simultaneous estimation of haplotype frequencies and mean- and covariance-structuring parameters.

A search for optimal risk haplotypes, each for a different group of circadian parameters, is important, although it is a tedious task. For each parameter group, there are four possibilities to choose a risk haplotype from two SNPs. Thus, three parameter groups will need to choose 4^3 combinations of risk haplotypes. The largest likelihood among these 64 combinations corresponds to the optimal combination of risk haplotypes for the three groups of circadian parameters.

11.2 Hypothesis Testing

The model described to jointly haplotype PK and PD processes has many significant advantages for asking biologically meaningful questions and addressing them at the interplay between gene actions and the process and rate of PK and PD response. These questions can be addressed by formulating a series of hypothesis tests. These hypotheses include the overall genetic control of drug response and the temporal pattern of haplotype effects on different types of drug responses.

11.2.1 Existence of Risk Haplotypes

Whether there are risk haplotypes for PK and PD responses (i.e., plasma concentration, drug efficacy, and drug toxicity in this example) is the most important question that needs to be first tested. The genetic control over drug response can be tested by formulating the following hypotheses:

$$\begin{cases} H_0: & (V_{dj_1}, v_{aj_1}, v_{ej_1}) \equiv (V_d, v_a, v_e), \\ & (E_{\max j_2}, EC_{50j_2}, H_{j_2}) \equiv (E_{\max}, EC_{50}, H), \\ & (\alpha_{j_3}, \beta_{j_3}) \equiv (\alpha, \beta) \\ H_1: & \text{at least one of the equalities above does not hold,} \end{cases} \quad (11.13)$$

for $j_1, j_2, j_3 = 2, 1, 0$. The H_0 states that there are no risk haplotypes that affect each PK and PD variable (the reduced model), whereas the H_1 proposes that such risk hapltypes do exist (the full model). The test statistic for testing the hypotheses (11.13) is calculated as the log-likelihood ratio of the reduced to the full model:

$$\text{LR} = -2[\ln L(\tilde{\Omega}_q | \mathbf{x}, \mathbf{y}, \mathbf{z}) - \ln L(\hat{\Omega}_q | \mathbf{x}, \mathbf{y}, \mathbf{z}, \mathbf{S}, \hat{\Omega}_p,)], \quad (11.14)$$

where the tildes and hats denote the MLEs of the unknown parameters under H_0 and H_1, respectively, i.e.,

$$\tilde{\Omega}_q = (\tilde{V}_d, \tilde{v}_a, \tilde{v}_e, \tilde{E}_{\max}, \widetilde{EC}_{50}, \tilde{H}, \tilde{\alpha}, \tilde{\beta}, \{\tilde{v}_k^2, \tilde{\phi}_k, \tilde{\phi}_{k \leftarrow k'}, \tilde{\rho}_{kk'}\}_{k,k'=x}^z),$$

and

$$\hat{\Omega}_q = (\{\hat{V}_{dj_1}, \hat{v}_{aj_1}, \hat{v}_{ej_1}\}_{j_1=0}^2, \{\hat{E}_{\max j_2}, \widehat{EC}_{50j_2}, \hat{H}_{j_2}\}_{j_2=0}^2, \{\hat{\alpha}_{j_3}, \hat{\beta}_{j_3}\}_{j_3=0}^2,$$

$$\{\hat{v}_k^2, \hat{\phi}_k, \hat{\phi}_{k \leftarrow k'}, \hat{\rho}_{kk'}\}_{k,k'=x}^z).$$

The LR is asymptotically χ^2-distributed with the number of degrees of freedom that is the difference in unknown parameters between the two hypotheses.

11.2.2 Different Risk Haplotypes for PK and PD

After the existence of significant risk haplotypes that affect PK and PD responses is confirmed, the next test is to detect whether these haplotypes affect jointly or separately these two processes. The hypothesis for testing the effect of the risk haplotypes on the PK response is formulated as

$$\begin{aligned} H_0: & (V_{dj_1}, v_{aj_1}, v_{ej_1}) \equiv (V_d, v_a, v_e), \\ H_1: & \text{At least one of the equalities above does not hold.} \end{aligned} \quad (11.15)$$

In a similar way, a test is formulated for detecting the effect of the risk haplotypes on drug efficacy and drug toxicity, expressed as

$$\begin{aligned} H_0: & (E_{\max j_2}, EC_{50j_2}, H_{j_2}) \equiv (E_{\max}, EC_{50}, H), \\ H_1: & \text{At least one of the equalities above does not hold.} \end{aligned} \quad (11.16)$$

and

$$H_0 : (\alpha_{j_3}, \beta_{j_3}) \equiv (\alpha, \beta),$$

H_1 : At least one of the equalities above does not hold. $\qquad(11.17)$

The log-likelihood values under the H_0 and H_1 of hypotheses (11.15), (11.16), and (11.17) are calculated from which the LRs are then calculated and compared with the critical values obtained from a χ^2-distribution table. But for these two hypothesis tests, an empirical approach based on simulation studies can also be used to determine the critical threshold. If all the three null hypotheses of (11.15), (11.16), and (11.17) are rejected, this means that we have detected a QTN that pleiotropically affects PK, drug efficacy and drug toxicity in a haplotype form.

Another interesting question is whether the risk haplotype detected for the PK (or PD) response also affects the PD (or PK) response. The hypotheses for such a test can be formulated as

$$H_0 : (E_{\max j_1}, EC_{50 j_1}, H_{j_1}) \equiv (E_{\max}, EC_{50}, H), (\alpha_{j_1}, \beta_{j_1}) \equiv (\alpha, \beta)$$

H_1 : At least one of the equalities above does not hold, $\qquad(11.18)$

$$H_0 : (V_{d j_2}, V_{a j_2}, V_{e j_2}) \equiv (V_d, V_a, V_e),$$

H_1 : At least one of the equalities above does not hold, $\qquad(11.19)$

or

$$H_0 : (V_{d j_3}, V_{a j_3}, V_{e j_3}) \equiv (V_d, V_a, V_e),$$

H_1 : At least one of the equalities above does not hold. $\qquad(11.20)$

If the null hypotheses of (11.15) and (11.18) are both rejected, this indicates that risk haplotype [11] affects both the PK and PD (including drug efficacy and drug toxicity). If both null hypotheses of (11.16) and (11.19) are rejected, and if both null hypotheses of (11.17) and (11.20) are rejected, then we can say that risk haplotypes [10] and [01] are both pleiotropic to PK and PD responses.

11.2.3 Different Risk Haplotypes for Drug Efficacy and Drug Toxicity

Similar tests can be made to detect how the same risk haplotype affects drug efficacy and drug toxicity. The results from this test will be of great significance in practical chemotherapy. If both null hypotheses of (11.16) and (11.17) are rejected, then the QTN detected pleiotropically affects drug efficacy and drug toxicity. If the null hypotheses of the following formulations

$$H_0 : (E_{\max j_3}, EC_{50 j_3}, H_{j_3}) \equiv (E_{\max}, EC_{50}, H),$$

H_1 : At least one of the equalities above does not hold,

and

$$H_0 : (\alpha_{j_2}, \beta_{j_2}) \equiv (\alpha, \beta),$$

H_1 : At least one of the equalities above does not hold,

are both rejected, then this indicates that the same risk haplotypes [10] and [01] jointly affect drug efficacy and drug toxicity.

11.2.4 Risk Haplotypes Responsible for Individual Curve Parameters

The shapes of PK and PD responses are defined by various parameters. Different parameters determine different aspects of the curve shape. For example, the Hill coefficient H is a parameter that determines the steepness of the Emax curve for PD responses. As shown in Fig. 6.2, a larger H value leads to a steeper sigmoid curve than its smaller counterpart. A variety of hypotheses can be made to test how risk haplotypes control these parameters individually.

Example 11.1

Congestive heart failure (CHF) is a pervasive and insidious clinical syndrome that most commonly results from ischemic heart disease and hypertension. It is estimated that almost 5 million Americans are affected by CHF. Dobutamine is primarily an agonist at β-adrenergic receptors (βARs) that predominate in the heart. It is used to relieve symptoms in patients with CHF by increasing stroke volume in a dose-dependent manner. Consequently, systolic blood pressure (SBP) is generally elevated while diastolic blood pressure (DBP) usually remains unchanged or slightly decreased. The understanding of the association between βAR polymorphisms and the inter-patient variability in blood pressure responses (SBP and DBP) to dobutamine will provide an objective genetic basis for individualization in treating CHF. Lin et al. (2006) used a bivariate model to study the relationship between βAR SNP genotype and blood pressure responses to dobutamine. In their study, the haplotype-based DNA sequence variants associated with SBP and DBP responses were identified.

This study contains about 160 subjects in ages from 32 to 86 years old. The subjects were typed for two single nucleotide polymorphisms (SNPs) at codon 16 (with two alleles A and G) and 27 (with two alleles C and G) with β_2ARand and injected by dobutamine, aimed to investigate the genetic control of patients' responses in SBP and DBP to this drug. The subjects received increasing doses of dobutamine, until they achieved a target heart rate response or predetermined maximum dose. The dose levels used were 0 (baseline), 5, 10, 20, 30, and 40 mcg/min, at each of which heart rate was measured. The time interval of 3 minutes is allowed between two successive doses for subjects to reach a plateau in response to that dose.

Figure 11.2A and 11.2C show response curves of SBP and DBP for different subjects to changing doses. The Emax model (6.1) was used to model dose-dependent SBP and DBP curves (Lin et al., 2005). It is possible that SBP and DBP have different risk haplotypes, so their composite diplotypes should be treated differently. Let

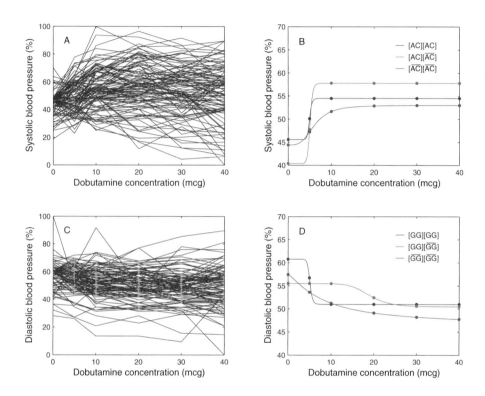

FIGURE 11.2

Response curves for systolic (SBP) (**A**) and diastolic blood pressure (DBP) (**C**) to dobutamine in a pharmacogenomic study. From these curves, Lin et al. (2005) detected significant risk haplotypes [AC] (**B**) for SBP and [GG] for DBP (**D**) that form three composite genotypes for blood pressure.

TABLE 11.2

Likelihood ratios for 16 possible combinations of assumed risk haplotypes for SBP and DBP within $\beta2AR$ gene.

SBP	DBP			
	[GC]	[GG]	[AC]	[AG]
[GC]	15.1335	17.3061	11.2845	8.6332
[GG]	16.6025	17.7130	13.9015	12.1757
[AC]	16.5969	**21.5680**	13.0381	10.5079
[GG]	18.9051	21.5001	13.9138	10.8167

The maximum likelihood ratio value is detected when [AC] and [GG] are used as the risk haplotypes for SBP and DBP, respectively.

j_1 and j_2 ($j_1, j_2 = 2, 1, 0$) denote the composite diplotype associated with SBP and DBP curves, respectively. The eight unknown parameters to be estimated for curve fitting can be arrayed as $(E^S_{0j_1}, E^S_{\max j_1}, EC^S_{50j_1}, H^S_{j_1}, E^D_{0j_2}, E^D_{\max j_2}, EC^D_{50j_2}, H^D_{j_2})$.

By comparing all possible combinations of risk haplotype between SBP and DBP, H. Lin et al. (2005) found significant risk haplotypes [AC] for SBP response and [GG] for DBP response. The log-likelihood ratio (LR) test statistics for the combination between these two risk haplotypes is 21.6, which is statistically significant based on the critical threshold determined from 1000 permutation tests and is also greater than the LR values for any other combinations (Table 11.2).

The maximum likelihood estimates (MLEs) of population and quantitative genetic parameters were given in Table 11.3. Using the estimated response parameters, the profiles of SBP and DBP response were drawn to increasing dose levels of dobutamine for three composite diplotypes (Fig. 11.2B and 11.2D). As shown, the three composite diplotypes displayed different curves across all dose levels for each blood pressure. We used area under curves to test in which gene action mode (additive or dominant) the haplotypes affect drug response curves for blood pressure. The testing results suggest that both additive and dominant effects are important in determining the shape of the response curve (Table 11.4).

Lin et al. (2006) also performed simulation studies to investigate the statistical properties of bivariate haplotyping model. The data were simulated by mimicking the example used in order to determine the reliability of our estimates in this real application. One haplotype was assumed to be different from the other three for each trait. The data simulated under this assumption were subject to statistical analyses, pretending that haplotype distinction is unknown. There is reasonably high power to detect the correct risk haplotypes. The parameters associated with haplotype effects can be accurately and precisely estimated.

☐

TABLE 11.3

Maximum likelihood estimates (MLEs) of population genetic parameters (allele frequencies and linkage disequilibria) for two SNPs as well as quantitative genetic parameters (drug response and matrix-structuring parameters) within $\beta 2AR$ genes.

	Population Genetic Parameters		
	p_G	p_G	D
	0.6191	0.5952	0.0484

Composite	Curve Parameters: SBP			
Diplotype	$\hat{E}_0^S$	$\hat{E}_{max}^S$	$\widehat{EC}_{50}^S$	$\hat{H}^S$
[AC][AC]	0.4560	0.0890	4.9920	18.3690
[AC][$\overline{AC}$]	0.4038	0.1743	5.0934	16.4025
[$\overline{AC}$][$\overline{AC}$]	0.4442	0.0858	6.1354	3.4347

Composite	Curve Parameters: DBP			
Diplotype	$\hat{E}_0^D$	$\hat{E}_{max}^D$	$\widehat{EC}_{50}^D$	$\hat{H}^D$
[GG][GG]	0.6075	-0.0978	5.1087	16.6864
[GG][$\overline{GG}$]	0.5550	-0.0505	18.9088	8.0138
[$\overline{GG}$][$\overline{GG}$]	0.5745	-0.1091	8.0254	1.2959

	Matrix-structuring Parameters		
σ_S^2	σ_D^2	ρ_S	ρ_D
0.0262	0.0133	0.8348	0.8384

The risk haplotypes for SBP and DBP are [AC] and [GG], respectively.

TABLE 11.4

Testing results for additive and dominant effects for SBP and DBP based on AUC under the optimal haplotype model.

Test	Additive^S	Dominance^S	Additive^D	Dominant^D
LR	7.7843	10.1180	13.5510	14.4160
p-value	<0.05	<0.05	<0.05	<0.05

11.3 Closed Forms for the SAD Structure

In this section, we give the structure of the covariance matrix fit by a SAD(1) model and the closed forms of the determinant and inverse of the matrix (Zhao et al., 2005). Our focus is on a bivariate SAD model which can be extended to any dimensional SAD model. Consider two longitudinal traits $\mathbf{x}_i = (x_i(1), \ldots, x_i(T))$ and $\mathbf{y}_i = (y_i(1), \ldots, y_i(T))$. We need to model the structure of their residual covariance ($\mathbf{\Sigma}$).

According to Newton (1988) (p. 359), $\mathbf{\Sigma}$ is positive definite if and only if there exists a unique low triangle matrix, $\mathbf{L}$, and a diagonal matrix, $\mathbf{\Sigma}_\varepsilon$, such that $\mathbf{L\Sigma L}' = \mathbf{\Sigma}_\varepsilon$ (Jaffrézic et al., 2003), where

$$\mathbf{L} = \begin{pmatrix} \mathbf{I} & 0 & \cdots & 0 & 0 \\ -\mathbf{V} & \mathbf{I} & \cdots & 0 & 0 \\ 0 & -\mathbf{V} & \cdots & 0 & 0 \\ \vdots & \vdots & \ddots & \vdots & \vdots \\ 0 & 0 & \cdots & -\mathbf{V} & \mathbf{I} \end{pmatrix},$$

$$\mathbf{\Sigma}_\varepsilon = \begin{pmatrix} \mathbf{\Sigma}_\epsilon(1) & 0 & \cdots & 0 \\ 0 & \mathbf{\Sigma}_\epsilon(2) & \cdots & 0 \\ \vdots & \vdots & \ddots & \vdots \\ 0 & 0 & \cdots & \mathbf{\Sigma}_\epsilon(T) \end{pmatrix}$$

with

$$\mathbf{V} = \begin{pmatrix} \phi_x & \phi_{x \leftarrow y} \\ \phi_{y \leftarrow x} & \phi_y \end{pmatrix} \text{ and } \mathbf{I} = \begin{pmatrix} 1 & 0 \\ 0 & 1 \end{pmatrix},$$

Given that matrix $\mathbf{L}$ is non-singular and matrix $\mathbf{\Sigma}_\varepsilon$ is positive semi-definite, it can be shown that

$$|\mathbf{\Sigma}|^{-1/2} = |\mathbf{\Sigma}_\varepsilon|^{-1/2}$$

$$= \prod_{t=1}^{T} \left[\sqrt{1 - \rho^2(t)} \delta_x(t) \delta_y(t) \right]^{-1},$$

and

$$\mathbf{\Sigma}^{-1} = \mathbf{L}' \mathbf{\Sigma}_\varepsilon^{-1} \mathbf{L},$$

$$= \begin{pmatrix} \mathbf{\Sigma}_\varepsilon^{-1}(1) + \mathbf{V}' \mathbf{\Sigma}_\varepsilon^{-1}(2) \mathbf{V} & -\mathbf{V}' \mathbf{\Sigma}_\varepsilon^{-1}(1) & \cdots & 0 & 0 \\ -\mathbf{\Sigma}_\varepsilon^{-1}(2) \mathbf{V} & \mathbf{\Sigma}_\varepsilon^{-1}(2) + \mathbf{V}' \mathbf{\Sigma}_\varepsilon^{-1}(3) \mathbf{V} & \cdots & 0 & 0 \\ 0 & -\mathbf{\Sigma}_\varepsilon^{-1}(3) \mathbf{V} & \cdots & 0 & 0 \\ \vdots & \vdots & \ddots & \vdots & \vdots \\ 0 & 0 & \cdots & -\mathbf{\Sigma}_\varepsilon^{-1}(T-1) \mathbf{V} & \mathbf{\Sigma}_\varepsilon^{-1}(T) \end{pmatrix}$$

$$= \mathbf{\Lambda}_1 + \mathbf{\Lambda}_2 + \mathbf{\Lambda}_3 + \mathbf{\Lambda}_3'$$

where

$$\mathbf{\Lambda}_1 = \begin{pmatrix} \mathbf{\Sigma}_\epsilon^{-1}(1) & 0 & \cdots & 0 \\ 0 & \mathbf{\Sigma}_\epsilon^{-1}(2) & \cdots & 0 \\ \vdots & \vdots & \ddots & \vdots \\ 0 & 0 & \cdots & \mathbf{\Sigma}_\epsilon^{-1}(T) \end{pmatrix},$$

$$\mathbf{\Lambda}_2 = \begin{pmatrix} \mathbf{V}'\mathbf{\Sigma}_\epsilon^{-1}(2)\mathbf{V} & 0 & \cdots & 0 & 0 \\ 0 & \mathbf{V}'\mathbf{\Sigma}_\epsilon^{-1}(3)\mathbf{V} & \cdots & 0 & 0 \\ \vdots & \vdots & \ddots & \vdots & \vdots \\ 0 & 0 & \cdots & \mathbf{V}'\mathbf{\Sigma}_\epsilon^{-1}(T)\mathbf{V} & 0 \\ 0 & 0 & \cdots & 0 & 0 \end{pmatrix},$$

$$\mathbf{\Lambda}_3 = \begin{pmatrix} 0 & 0 & \cdots & 0 & 0 \\ -\mathbf{\Sigma}_\epsilon^{-1}(2)\mathbf{V} & 0 & \cdots & 0 & 0 \\ 0 & -\mathbf{\Sigma}_\epsilon^{-1}(3)\mathbf{V} & \cdots & 0 & 0 \\ \vdots & \vdots & \ddots & \vdots & \vdots \\ 0 & 0 & \cdots & -\mathbf{\Sigma}_\epsilon^{-1}(T)\mathbf{V} & 0 \end{pmatrix}.$$

Combine the two longitudinal vectors $\mathbf{x}_i = (x_i(1),\ldots,x_i(T))$ and $\mathbf{y}_i = (y_i(1),\ldots,y_i(T))$ into one vector. We have

$$\mathbf{z}_i = (\underbrace{x_i(1),y_i(1)}_{\mathbf{z}_i(1)},\underbrace{x_i(2),y_i(2)}_{\mathbf{z}_i(2)},\ldots,\underbrace{x_i(T),y_i(T)}_{\mathbf{z}_i(T)})$$

The quadratic form $\mathbf{z}_i \mathbf{\Sigma}^{-1} \mathbf{z}_i'$ can then be divided into three parts

$$\mathbf{z}_i \mathbf{\Sigma}^{-1} \mathbf{z}_i' = \mathbf{z}_i \mathbf{\Lambda}_1 \mathbf{z}_i' + \mathbf{z}_i \mathbf{\Lambda}_2 \mathbf{z}_i' + \mathbf{z}_i \mathbf{\Lambda}_3 \mathbf{z}_i' + \mathbf{z}_i \mathbf{\Lambda}_3' \mathbf{z}_i',$$

where

$$\mathbf{z}_i \mathbf{\Lambda}_1 \mathbf{z}_i' = \sum_{t=1}^{T} \mathbf{z}_i(t) \mathbf{\Sigma}_\epsilon^{-1}(t) \mathbf{z}_i'(t),$$

$$\mathbf{z}_i \mathbf{\Lambda}_2 \mathbf{z}_i' = \sum_{t=1}^{T-1} \mathbf{z}_i(t) \mathbf{V} \mathbf{\Sigma}_\epsilon^{-1}(t+1) \mathbf{V}' \mathbf{z}_i'(t),$$

$$\mathbf{z}_i \mathbf{\Lambda}_3 \mathbf{z}_i' = -\sum_{t=1}^{T-1} \mathbf{z}_i(t+1) \mathbf{\Sigma}_\epsilon^{-1}(t+1) \mathbf{V}' \mathbf{z}_i'(t).$$

11.4 Allometric Mapping of Drug Efficacy and Drug Toxicity

The genetic mapping of drug efficacy and drug toxicity can be conducted by incorporating the principle of allometric scaling. This can be done in two ways. First,

212 Statistical and Computational Pharmacogenomics

the genetic control of body mass itself is not interesting, but the inclusion of body mass can increase biological relevance of haplotyping results and statistical power of the model. Second, the genetic control mechanism of body mass can be incorporated into the haplotyping model, allowing the test of whether body mass shares the common genetic basis with PK, drug efficacy and toxicity. For the second way, we can assume a different risk haplotype that is associated with body mass. Recall Table 11.1. If [00] is a risk haplotype for body mass (w), the likelihood (11.6) will be changed as

$$
\begin{aligned}
&\log L(\Omega_q | \mathbf{x}, \mathbf{y}, \mathbf{z}, w, \mathbf{S}, \hat{\Omega}_p) \\
&= \sum_{i=1}^{n_{11/11}} \log f_{2000}(\mathbf{x}_i, \mathbf{y}_i, \mathbf{z}_i, w_i) + \sum_{i=1}^{n_{11/10}} \log f_{1100}(\mathbf{x}_i, \mathbf{y}_i, \mathbf{z}_i, w_i) \\
&+ \sum_{i=1}^{n_{11/00}} \log f_{1200}(\mathbf{x}_i, \mathbf{y}_i, \mathbf{z}_i, w_i) + \sum_{i=1}^{n_{10/11}} \log f_{1010}(\mathbf{x}_i, \mathbf{y}_i, \mathbf{z}_i, w_i) \\
&+ \sum_{i=1}^{n_{10/10}} \log[\omega f_{1001}(\mathbf{x}_i, \mathbf{y}_i, \mathbf{z}_i, w_i) + (1-\omega) f_{0110}(\mathbf{x}_i, \mathbf{y}_i, \mathbf{z}_i, w_i)] \\
&+ \sum_{i=1}^{n_{10/00}} \log f_{0101}(\mathbf{x}_i, \mathbf{y}_i, \mathbf{z}_i, w_i) + \sum_{i=1}^{n_{00/11}} \log f_{0020}(\mathbf{x}_i, \mathbf{y}_i, \mathbf{z}_i, w_i) \\
&+ \sum_{i=1}^{n_{00/10}} \log f_{0011}(\mathbf{x}_i, \mathbf{y}_i, \mathbf{z}_i, w_i) + \sum_{i=1}^{n_{00/00}} \log f_{0002}(\mathbf{x}_i, \mathbf{y}_i, \mathbf{z}_i, w_i).
\end{aligned}
$$

$$(11.21)$$

A model selection procedure is needed to find the best combination of risk haplotypes for PK, drug efficacy, drug toxicity, and body mass. A number of hypothesis tests can be formulated to address many interesting biological questions about pharmacological response.

12

Modeling Epistatic Interactions in Drug Response

Interactions between different genes, coined the *epistasis*, have long been recognized to play a central role in shaping the genetic architecture of a quantitative trait (Whitlock et al., 1995; Wolf, 2000; Cheverud, 2006). Recent genetic studies from vast quantities of molecular data have also indicated that epistasis is of paramount importance in the pathogenesis of most common human diseases, such as cancer or cardiovascular disease (Cordell, 2002; Moore, 2003; Talkowski et al., 2008), and patients' responsiveness to a medicine (Lin et al., 2007). The evidence for this is the nonlinear relationship detected between genotype and phenotype. The decipherment of interconnected networks of genes and their associations with disease susceptibility has become a pressing demand for a detailed understanding of the genetic basis for disease processes (Moore et al., 2008).

The most common and powerful approaches for detecting genome-wide epistasis are based on genetic mapping that associates phenotypic variation of a trait with a linkage map constructed by polymorphic markers (Lander and Botstein, 1989). The epistasis between different quantitative trait loci (QTLs) is considered to occur if the effect of one QTL depends on the expression of other QTL. While traditional genetic mapping can only make an indirect inference about QTL actions and interactions, genetic haplotyping based on the haplotype map (HapMap) constructed from single nucleotide polymorphisms (SNPs) (Gibbs et al., 2003, 2005; Frazer et al., 2007) can directly characterize specific DNA sequence variants for a phenotype (Liu et al., 2004; Lin and Wu, 2006b). More precise than chromosomal segments identified by QTL mapping, the haplotyping model can probe concrete nucleotoxic sites, i.e., quantitative trait nucleotides (QTNs), that contribute to variation in a quantitative trait, and thus provide a way to push epistasis identification at the DNA sequence level.

In this chapter, we will first provide an analytical approach of quantifying epistasis the formulation of Mather and Jinks (1982), and then describe a statistical model for detecting epistatic interactions that control dynamic traits including drug response. This model is constructed through integrating functional mapping into a genetic mapping or haplotyping framework for epistatic interactions (Wu et al., 2004b; Lin et al., 2007). We will provide a computational procedure for testing the effects of genetic interactions on the pattern and process of dynamic traits and use an example to demonstrate the interpretation of the epistatic model.

TABLE 12.1

An example showing the type of epistasis.

	AB	Ab	aB	ab
No epistasis	20	10	10	0
Synergistic epistasis	30	10	10	0
Antagonistic epistasis	10	10	10	0

TABLE 12.2

Testing the existence and type of epistasis.

Trait values	Type of epistasis
$AB = Ab + aB - ab$	No epistasis
$AB > Ab + aB - ab$	Synergistic epistasis
$AB < Ab + aB - ab$	Antagonistic epistasis

12.1 Quantitative Genetic Models for Epistasis

12.1.1 Definition and Type

Epistasis is defined as the expression of an allele at one gene dependent on alleles at one or several other genes (Bateson, 1909). If this gene-dependent expression is more favorable than the independent expression of all genes considered, such epistasis is called the synergistic (positive) epistasis. If epistasis leads to unfavorable expression, it is the antagonistic (or negative) epistasis. Consider a haploid organism in which two genes **A** and **B** forms four genotypes *AB*, *Ab*, *aB*, and *ab*. Each genotype is assigned as a metric value for a phenotypic trait. The epistasis between the two genes and its types are shown in Table 12.1.

If there is no epitasis, the overall genotypic value for the two genes is equal to the sum of genotypic values at individual genes. The genes with no epistasis are called the additive genes. Thus, we can determine whether a gene is additive or epistatic and what type of epistasis is true (if there is epistasis) using the relationships shown in Table 12.2.

Although the concept epistasis was originally proposed to describe the interactions among individual genes in the setting of Mendelian genetics, where epistasis gives rise to distorted Mendelian ratios of genotypes, it has now been well quantified by a statistical measure because of the pioneering work of Fisher (1918) . Quantitative epistasis has been instrumental for many areas of biology including population and quantitative genetics, evolutionary biology, plant and animal breeding, and human genetics (Lynch and Walsh, 1998).

TABLE 12.3

Additive, dominance, and epistatic compositions of two-gene genotypic values.

	BB	*Bb*	*bb*
AA	$\mu_{AABB} =$ $\mu + a_1 + a_2 + i_{aa}$	$\mu_{AABb} =$ $\mu + a_1 + d_2 + i_{ad}$	$\mu_{AAbb} =$ $\mu + a_1 - a_2 - i_{aa}$
Aa	$\mu_{AaBB} =$ $\mu + d_1 + a_2 + i_{da}$	$\mu_{AaBb} =$ $\mu + d_1 + d_2 + i_{dd}$	$\mu_{Aabb} =$ $\mu + d_1 - a_2 - i_{da}$
aa	$\mu_{aaBB} =$ $\mu - a_1 + a_2 - i_{aa}$	$\mu_{aaBb} =$ $\mu - a_1 + d_2 - i_{ad}$	$\mu_{aabb} =$ $\mu - a_1 - a_2 + i_{aa}$

12.1.2 Quantifying Epistasis

Traditional epistatic modeling of different genes is based on the combination of genotypes at different loci. Among many ways to model epistasis, Mather and Jinks's (1982) formulation, which will be followed in this chapter, seems to be the most popular and easiest to understood. For example, there is a pair of genetic loci **A** (with two alleles *A* and *a*) and **B** (with two alleles *B* and *b*) that interact with each other to determine the phenotype of a trait in a diploid population. The two loci form a total of nine two-gene genotypes *AABB*, *AABb*, *AAbb*, *AaBB*, *AaBb*, *Aabb*, *aaBB*, *aaBb*, and *aabb*, whose genotypic values are dissolved into different components: μ, the overall mean; a_1 and a_2, the additive effects of loci **A** and **B**; d_1 and d_2, the dominance effects of loci **A** and **B**; and i_{aa}, i_{aa}, i_{aa}, and i_{aa}, the additive × additive, additive × dominance, dominance × additive, and dominance × dominance epistatic effects between the two loci (Table 12.3).

Lin and Wu (2006a) extended the idea of Mather and Jinks (1982) to model epistatic interactions between different QTNs at the haplotype level. Because phenotypic variation in a complex disease can be explained by haplotype diversity, a particular haplotype can be assumed to be different from other haplotypes for a given phenotype (Bader, 2001). Such a distinct haplotype is defined as the *risk haplotype*, and all the others defined as the non-risk haplotype. The risk and non-risk haplotypes are combined to produce composite diplotypes. For two different QTNs **A** and **B**, let A, $\bar{A}$ and B, $\bar{B}$ be the risk haplotype and non-risk haplotypes, respectively. These two QTNs form nine different composite diplotypes expressed as $AABB$, $AAB\bar{B}$, $AA\bar{B}\bar{B}$, $A\bar{A}BB$, $A\bar{A}B\bar{B}$, $A\bar{A}\bar{B}\bar{B}$, $\bar{A}\bar{A}BB$, $\bar{A}\bar{A}B\bar{B}$, and $\bar{A}\bar{A}\bar{B}\bar{B}$. We used Mather and Jinks's formulation (Table 12.3) to model the genetic effects of the composite diplotypes. The genotypic value ($\mu_{j_1 j_2}$) of a joint composite diplotype at the two QTNs can be decomposed into nine different components as follows:

$$\mu_{j_1 j_2} = \qquad \mu \qquad\qquad \text{Overall mean}$$
$$+ (j_1 - 1)a_1 + (j_2 - 1)a_2 \quad \text{Additive effects}$$
$$+ j_1 d_1 + j_2 d_2 \qquad \text{Dominant effects}$$
$$+ (j_1 - 1)(j_2 - 1)i_{aa} \quad \text{Additive} \times \text{additive effect}$$

$$+ (j_1 - 1)j_2 i_{ad} \qquad \text{Additive} \times \text{dominance effect}$$
$$+ j_1(j_2 - 1)i_{da} \qquad \text{Dominance} \times \text{additive effect}$$
$$+ (1 - j_1)(1 - j_2)i_{dd} \quad \text{Dominance} \times \text{dominance effect,}$$

$$(12.1)$$

where

$$j_1, j_2 = \begin{cases} 2 & \text{for } AA \text{ or } BB \\ 1 & \text{for } A\bar{A} \text{ or } B\bar{B} \\ 0 & \text{for } \bar{A}\bar{A} \text{ or } \bar{B}\bar{B} \end{cases}$$

stand for the composite diplotypes at QTNs **A** and **B**, respectively, a_1, a_2, d_1, d_2, and i_{aa}, i_{ad}, i_{da}, and i_{dd} are defined as above (12.3) but at the haplotype level.

Different types of genetic actions and interactions can be expressed in terms of genotypic values by solving a group of regular equations (12.1). This lets us describe the overall mean, additive, dominance, and four kinds of epistatic effects between two QTNs by

$$\mu = \frac{1}{4}(\mu_{\bar{A}\bar{A}\bar{B}\bar{B}} + \mu_{AA\bar{B}\bar{B}} + \mu_{\bar{A}\bar{A}BB} + \mu_{AABB})$$

$$a_1 = \frac{1}{4}(\mu_{AABB} - \mu_{\bar{A}\bar{A}\bar{B}\bar{B}} + \mu_{AA\bar{B}\bar{B}} - \mu_{\bar{A}\bar{A}BB})$$

$$a_2 = \frac{1}{4}(\mu_{\bar{A}\bar{A}BB} - \mu_{\bar{A}\bar{A}\bar{B}\bar{B}} - \mu_{AA\bar{B}\bar{B}} + \mu_{AABB})$$

$$d_1 = \frac{1}{4}(2\mu_{A\bar{A}\bar{B}\bar{B}} - \mu_{\bar{A}\bar{A}\bar{B}\bar{B}} - \mu_{AA\bar{B}\bar{B}} - \mu_{\bar{A}\bar{A}BB} - \mu_{AABB} + 2\mu_{A\bar{A}BB})$$

$$d_2 = \frac{1}{4}(2\mu_{\bar{A}\bar{A}B\bar{B}} - \mu_{\bar{A}\bar{A}\bar{B}\bar{B}} - \mu_{AA\bar{B}\bar{B}} - \mu_{\bar{A}\bar{A}BB} - \mu_{AABB} + 2\mu_{AAB\bar{B}})$$

$$i_{aa} = \frac{1}{4}(\mu_{AABB} - \mu_{AA\bar{B}\bar{B}} - \mu_{\bar{A}\bar{A}BB} + \mu_{\bar{A}\bar{A}\bar{B}\bar{B}})$$

$$i_{ad} = \frac{1}{4}(2\mu_{AAB\bar{B}} - \mu_{AABB} - 2\mu_{\bar{A}\bar{A}B\bar{B}} + \mu_{\bar{A}\bar{A}\bar{B}\bar{B}} - \mu_{AA\bar{B}\bar{B}} + \mu_{\bar{A}\bar{A}BB})$$

$$i_{da} = \frac{1}{4}(2\mu_{A\bar{A}BB} - 2\mu_{A\bar{A}\bar{B}\bar{B}} + \mu_{\bar{A}\bar{A}\bar{B}\bar{B}} + \mu_{AA\bar{B}\bar{B}} - \mu_{\bar{A}\bar{A}BB} - \mu_{AABB})$$

$$i_{dd} = \frac{1}{4}(4\mu_{A\bar{A}B\bar{B}} + \mu_{\bar{A}\bar{A}\bar{B}\bar{B}} + \mu_{AA\bar{B}\bar{B}} + \mu_{\bar{A}\bar{A}BB} + \mu_{AABB} - 2\mu_{A\bar{A}\bar{B}\bar{B}}$$
$$- 2\mu_{A\bar{A}BB} - 2\mu_{\bar{A}\bar{A}B\bar{B}} - 2\mu_{AAB\bar{B}}).$$

$$(12.2)$$

Thus, by testing the significance of i_{aa}, i_{ad}, i_{da}, and i_{dd}, we judge whether there is epistasis and how the epistasis affects a phenotypic trait.

12.1.3 From Static to Dynamic

In current developmental genetic and pharmacogenetic studies, there is an increasing interest to explore the relationship between epistasis and trait development and disease progression. This will need to model the temporal or longitudinal pattern of epistatic effects in a time course. If we have estimated a set of curve parameters $(\Omega_{j_1 j_2})$ that define the dynamic change of a composite diplotype $j_1 j_2$, the time-dependent genotypic value of this diplotype can be expressed as

$$\mu_{j_1 j_2}(t) = g(t : \Omega_{j_1 j_2}), \quad t = 1, ..., T.$$

The time-dependent overall mean,

$$\mu = (\mu(1), ..., \mu(t), ..., \mu(T)),$$

time-dependent additive effects for QTNs **A** and **B**,

$$\mathbf{a}_1(t) = (a_1(1), ..., a_1(t), ..., a_1(T)), \tag{12.3}$$
$$\mathbf{a}_2(t) = (a_2(1), ..., a_2(t), ..., a_2(T)), \tag{12.4}$$

time-dependent dominance effects for QTNs **A** and **B**,

$$\mathbf{d}_1(t) = (d_1(1), ..., d_1(t), ..., d_1(T)), \tag{12.5}$$
$$\mathbf{d}_2(t) = (d_2(1), ..., d_2(t), ..., d_2(T)), \tag{12.6}$$

and time-dependent additive × additive, additive × dominance, dominance × additive, and dominance × dominance epistatic effects between the two QTNs,

$$\mathbf{i}_{aa}(t) = (i_{aa}(1), ..., i_{aa}(t), ..., i_{aa}(T)), \tag{12.7}$$
$$\mathbf{i}_{ad}(t) = (i_{ad}(1), ..., i_{ad}(t), ..., i_{ad}(T)), \tag{12.8}$$
$$\mathbf{i}_{da}(t) = (i_{da}(1), ..., i_{da}(t), ..., i_{da}(T)), \tag{12.9}$$
$$\mathbf{i}_{dd}(t) = (i_{dd}(1), ..., i_{dd}(t), ..., i_{dd}(T)), \tag{12.10}$$

can be estimated using equation (12.2).

12.2 Haplotyping Epistasis

12.2.1 Population Genetic Structure

Consider a natural human population at Hardy–Weinberg equilibrium from which a random sample is drawn for pharmacogenomic studies. In order to identify DNA sequences responsible for a complex disease, we genotype a number of SNPs genome-wide and construct a haplotype map. Recent molecular surveys suggest that the human genome contains many discrete haplotype blocks that are sites of closely located SNPs (Daly et al., 2001; Patil et al., 2001; Gabriel et al., 2002). Each block may have a few common haplotypes which account for a large proportion of chromosomal variation. Between adjacent blocks there are large regions, called *hotspots*, in which recombination events occur with high frequencies. Several algorithms have been developed to identify a minimal subset of SNPs, i.e., "tagging" SNPs, that can characterize the most common haplotypes (Zhang et al., 2002; Kimmel and Shamir, 2005). We assume that the number and type of tagging SNPs within each haplotype block has been determined. The rationale behind the epistatic model to be described is that the effect of a given DNA sequence in one haplotype block on a complex disease is masked or enhanced by one or more sequences in other blocks. We assume that each of these blocks affects a trait in a unit of QTN.

12.2.2   Genetic Design

Suppose there are R and S $(R, S > 1)$ tagging SNPs for two arbitrary haplotype blocks or QTNs **A** and **B**, respectively. Let two alleles of a tagging SNP r or s at QTN **A** or **B** be denoted by $U_{k_r}^r$ $(k_r = 1, 2; r = 1, \cdots R)$ and $V_{l_s}^s$ $(l_s = 1, 0; s = 1, \cdots S)$, respectively. We use $p_{k_r}^A$ and $p_{l_s}^B$ to denote allele frequencies at the corresponding SNP. All the SNPs within each of the two blocks form 2^R or 2^S possible haplotypes expressed as $U_{k_1}^1 U_{k_2}^2 \cdots U_{k_R}^R$ and $V_{l_1}^1 V_{l_2}^2 \cdots V_{l_S}^S$, respectively. The corresponding haplotype frequencies within each block are denoted by $p_{k_1 k_2 \cdots k_R}^A$ and $p_{l_1 l_2 \cdots l_S}^B$, which are composed of allele frequencies at each SNP and linkage disequilibria of different orders among SNPs (Lynch and Walsh, 1998). A general expression for the relationships between haplotype frequencies and allele frequencies and linkage disequilibria was given by Bennett (1954).

The random combination of maternal and paternal haplotypes generates $2^{R-1}(2^R + 1)$ *diplotypes* expressed as $[U_{k_1}^1 U_{k_2}^2 \cdots U_{k_R}^R] [U_{k_1'}^1 U_{k_2'}^2 \cdots U_{k_R'}^R]$ $(k_1 \geq k_1', k_2 \geq k_2', ..., k_R \geq k_R' = 1, 0)$ for QTN **A** and $2^{S-1}(2^S + 1)$ diplotypes expressed as $[V_{l_1}^1 V_{l_2}^2 \cdots V_{l_S}^S][V_{l_1'}^1 V_{l_2'}^2 \cdots V_{l_S'}^S]$ $(l_1 \geq l_1', k_2 \geq l_2', ..., l_S \geq l_S' = 1, 0)$ for QTN **B**. We use the brackets to separate maternal (former) and paternal haplotypes (latter) for a given diplotype. Unless there are two or more SNPs that are heterozygous, observable *zygotic genotypes* will be the same as diplotypes. Thus, the numbers of zygotic genotypes, 3^R or 3^S, will be less than the number of diplotypes and the difference between these two numbers is statistically viewed as *missing data*. The observed zygotic genotypes are expressed as

$$U_{k_1}^1 U_{k_1'}^1 / U_{k_2}^2 U_{k_2'}^2 / \cdots / U_{k_R}^R U_{k_R'}^R, \quad (k_1 \geq k_1', k_2 \geq k_2', ..., k_R \geq k_R' = 1, 0)$$

or

$$V_{l_1}^1 V_{l_1'}^1 / V_{l_2}^2 A_{l_2'}^2 / \cdots / V_{l_S}^S V_{l_S'}^S, \quad (l_1 \geq l_1', k_2 \geq l_2', ..., l_S \geq l_S' = 1, 0)$$

for the two QTNs, respectively.

Let $n_{(k_1 k_1' / k_2 k_2' / \cdots / k_R k_R')(l_1 l_1' / l_2 l_2' / \cdots / l_S l_S')}$ (which sums to a total sample size of n subjects) be the observation of a typical joint-SNP genotype at two different QTNs. For each subject, a phenotypic trait related to drug response is measured at different time points or dose levels. For simplicity, we assume that repeated measurements are equally spaced, although the haplotyping model is able to handle unequally-spaced longitudinal data.

12.2.3 Population Genetic Model

For two different haplotype blocks (or QTNs) **A** and **B**, between which no linkage disequilibria exist (Daly et al., 2001; Gabriel et al., 2002), across-block haplotype frequencies can be calculated as the product of the corresponding haplotype frequencies from a different block, expressed as

$$P_{(k_1k_2\cdots k_R)(l_1l_2\cdots l_S)} = p^{\mathbf{A}}_{k_1k_2\cdots k_R} p^{\mathbf{B}}_{l_1l_2\cdots l_S}, \tag{12.11}$$

where the parentheses are used to separate two different blocks for a given across-block haplotype. But in the model being described, we assume that the two QTNs are not independent of each other. This treatment will cover the situation of equation (12.11). In any case, with across-block haplotype frequencies, expected across-block diplotype frequencies and across-block genotype frequencies can be calculated, respectively, under Hardy–Weinberg equilibrium.

To simplify our presentation, we will first assume two tagging SNPs for each QTN. Thus, we will have a total of 16 across-block haplotypes and 81 observed SNP genotypes. Using the above notation, the expected frequencies of these genotypes are expressed in terms of across-block haplotypes as shown in Table 12.4, in which genotypic observations are also given. For a given four-SNP haplotype, its frequency $(p_{(k_1k_2)(l_1l_2)})$ will be decomposed into the terms given in Table 12.5. All these linkage disequilibria (LD) of different orders among these SNPs can be estimated and tested.

12.2.4 Likelihood for Estimating Across-Block Haplotype Frequencies

Based on the information provided by Table 12.4, a multinomual likelihood is constructed to estimate the across-block haplotype frequencies whose MLEs are given as

$$
\begin{aligned}
P_{(k_1k_2)(l_1l_2)} = \frac{1}{2n}\big[& 2n_{(k_1k_1/k_2k_2)(l_1l_1/l_2l_2)} \\
& + n_{(k_1k_1/k_2k_2)(l_1l_1/l_2l_2')}, \quad l_2' < l_2 \\
& + n_{(k_1k_1/k_2k_2)(l_1l_1'/l_2l_2)}, \quad l_1' < l_1 \\
& + n_{(k_1k_1/k_2k_2')(l_1l_1/l_2l_2)}, \quad k_2' < k_2 \\
& + n_{(k_1k_1'/k_2k_2)(l_1l_1/l_2l_2)}, \quad k_1' < k_1 \\
& + \phi_1 n_{(k_1k_1/k_2k_2)(l_1l_1'/l_2l_2')}, \quad l_1' < l_1, l_2' < l_2 \\
& + \phi_2 n_{(k_1k_1/k_2k_2')(l_1l_1/l_2l_2')}, \quad k_2' < k_2, l_2' < l_2 \\
& + \phi_3 n_{(k_1k_1'/k_2k_2)(l_1l_1/l_2l_2')}, \quad k_1' < k_1, l_2' < l_2 \\
& + \phi_4 n_{(k_1k_1/k_2k_2')(l_1l_1'/l_2l_2)}, \quad k_2' < k_2, l_1' < l_1 \\
& + \phi_5 n_{(k_1k_1'/k_2k_2)(l_1l_1'/l_2l_2)}, \quad k_1' < k_1, l_1' < l_1 \\
& + \phi_6 n_{(k_1k_1'/k_2k_2')(l_1l_1/l_2l_2)}, \quad k_1' < k_1, k_2' < k_2 \\
& + \psi_1 n_{(k_1k_1/k_2k_2')(l_1l_1'/l_2l_2')}, \quad k_2' < k_2, l_1' < l_1, l_2' < l_2
\end{aligned}
$$

$$+ \psi_2 n_{(k_1 k_1' / k_2 k_2)(l_1 l_1' / l_2 l_2')}, \quad k_1' < k_1, l_1' < l_1, l_2' < l_2$$

$$+ \psi_3 n_{(k_1' k_1 / k_2' k_2)(l_1 l_1 / l_2 l_2')}, \quad k_1' < k_1, k_2' < k_2, l_2' < l_2$$

$$+ \psi_4 n_{(k_1' k_1 / k_2' k_2)(l_1 l_1' / l_2 l_2)}, \quad k_1' < k_1, k_2' < k_2, l_1' < l_1$$

$$+ \varphi n_{(k_1 k_1' / k_2 k_2')(l_1 l_1' / l_2 l_2')}, \quad k_1' < k_1, k_2' < k_2, l_1' < l_1, l_2' < l_2,$$

(12.12)

where

$$\phi_1 = \frac{P_{(k_1 k_2)(l_1 l_2)} P_{(k_1 k_2)(l_1' l_2')}}{P_{(k_1 k_2)(l_1 l_2)} P_{(k_1 k_2)(l_1' l_2')} + P_{(k_1 k_2)(l_1 l_2')} P_{(k_1 k_2)(l_1' l_2)}},$$

$$\phi_2 = \frac{P_{(k_1 k_2)(l_1 l_2)} P_{(k_1 k_2')(l_1 l_2')}}{P_{(k_1 k_2)(l_1 l_2)} P_{(k_1 k_2')(l_1 l_2')} + P_{(k_1 k_2)(l_1 l_2')} P_{(k_1 k_2')(l_1 l_2)}},$$

$$\phi_3 = \frac{P_{(k_1 k_2)(l_1 l_2)} P_{(k_1' k_2)(l_1 l_2')}}{P_{(k_1 k_2)(l_1 l_2)} P_{(k_1' k_2)(l_1 l_2')} + P_{(k_1 k_2)(l_1 l_2')} P_{(k_1' k_2)(l_1 l_2)}},$$

$$\phi_4 = \frac{P_{(k_1 k_2)(l_1 l_2)} P_{(k_1 k_2')(l_1' l_2)}}{P_{(k_1 k_2)(l_1 l_2)} P_{(k_1 k_2')(l_1' l_2)} + P_{(k_1 k_2)(l_1' l_2)} P_{(k_1 k_2')(l_1 l_2)}},$$

$$\phi_5 = \frac{P_{(k_1 k_2)(l_1 l_2)} P_{(k_1' k_2)(l_1' l_2)}}{P_{(k_1 k_2)(l_1 l_2)} P_{(k_1' k_2)(l_1' l_2)} + P_{(k_1 k_2)(l_1' l_2)} P_{(k_1' k_2)(l_1 l_2)}},$$

$$\phi_6 = \frac{P_{(k_1 k_2)(l_1 l_2)} P_{(k_1' k_2')(l_1 l_2)}}{P_{(k_1 k_2)(l_1 l_2)} P_{(k_1' k_2')(l_1 l_2)} + P_{(k_1 k_2')(l_1 l_2)} P_{(k_1' k_2)(l_1 l_2)}},$$

$$\psi_1 = [P_{(k_1 k_2)(l_1 l_2)} P_{(k_1 k_2')(l_1' l_2')}] / [P_{(k_1 k_2)(l_1 l_2)} P_{(k_1 k_2')(l_1' l_2')} + P_{(k_1 k_2)(l_1 l_2')} P_{(k_1 k_2')(l_1' l_2)}$$
$$+ P_{(k_1 k_2)(l_1' l_2)} P_{(k_1 k_2')(l_1 l_2')} + P_{(k_1 k_2)(l_1' l_2')} P_{(k_1 k_2')(l_1 l_2)}],$$

$$\psi_2 = [P_{(k_1 k_2)(l_1 l_2)} P_{(k_1' k_2)(l_1' l_2')}] / [P_{(k_1 k_2)(l_1 l_2)} P_{(k_1' k_2)(l_1' l_2')} + P_{(k_1 k_2)(l_1 l_2')} P_{(k_1' k_2)(l_1' l_2)}$$
$$+ P_{(k_1 k_2)(l_1' l_2)} P_{(k_1' k_2)(l_1 l_2')} + P_{(k_1 k_2)(l_1' l_2')} P_{(k_1' k_2)(l_1 l_2)}],$$

$$\psi_3 = [P_{(k_1 k_2)(l_1 l_2)} P_{(k_1' k_2')(l_1 l_2')}] / [P_{(k_1 k_2)(l_1 l_2)} P_{(k_1' k_2')(l_1 l_2')} + P_{(k_1 k_2)(l_1 l_2')} P_{(k_1' k_2')(l_1 l_2)}$$
$$+ P_{(k_1 k_2)(l_1 l_2')} P_{(k_1' k_2')(l_1 l_2)} + P_{(k_1 k_2')(l_1 l_2)} P_{(k_1' k_2)(l_1 l_2')}],$$

$$\psi_4 = [P_{(k_1 k_2)(l_1 l_2)} P_{(k_1' k_2')(l_1' l_2)}] / [P_{(k_1 k_2)(l_1 l_2)} P_{(k_1' k_2')(l_1' l_2)} + P_{(k_1 k_2)(l_1' l_2)} P_{(k_1' k_2')(l_1 l_2)}$$
$$+ P_{(k_1 k_2')(l_1' l_2)} P_{(k_1' k_2)(l_1 l_2)} + P_{(k_1 k_2)(l_1' l_2)} P_{(k_1' k_2')(l_1 l_2)}],$$

$$\varphi = [P_{(k_1 k_2)(l_1 l_2)} P_{(k_1' k_2')(l_1' l_2')}] / p_\varphi,$$

(12.13)

where

$$p_\varphi = P_{(k_1 k_2)(l_1 l_2)} P_{(k_1' k_2')(l_1' l_2')} + P_{(k_1 k_2)(l_1 l_2')} P_{(k_1' k_2')(l_1' l_2)} + P_{(k_1 k_2')(l_1 l_2)} P_{(k_1' k_2)(l_1' l_2')}$$

$$+ P_{(k_1' k_2)(l_1 l_2)} P_{(k_1 k_2')(l_1' l_2')} + P_{(k_1 k_2)(l_1' l_2')} P_{(k_1' k_2')(l_1 l_2)} + P_{(k_1 k_2')(l_1 l_2')} P_{(k_1' k_2)(l_1' l_2)}$$

$$+ P_{(k_1' k_2)(l_1 l_2')} P_{(k_1 k_2')(l_1' l_2)} + P_{(k_1' k_2)(l_1' l_2)} P_{(k_1 k_2')(l_1 l_2')}].$$

TABLE 12.4
Observed 81 SNP genotypes and their frequencies described in terms of their haplotype/diplotype compositions.

No.	Genotype		Observation	Frequency
	QTN A	QTN B		
1	$U_1^1 U_1^1 / U_1^2 U_1^2$	$V_1^1 V_1^1 / V_1^2 V_1^2$	$n_{(11/11)(11/11)}$	$p_{(11)(11)}^2$
2	$U_1^1 U_1^1 / U_1^2 U_1^2$	$V_1^1 V_1^1 / V_1^2 V_0^2$	$n_{(11/11)(11/10)}$	$2p_{(11)(11)}P_{(11)(10)}$
3	$U_1^1 U_1^1 / U_1^2 U_1^2$	$V_1^1 V_1^1 / V_0^2 V_0^2$	$n_{(11/11)(11/00)}$	$2p_{(11)(10)}^2$
4	$U_1^1 U_1^1 / U_1^2 U_1^2$	$V_1^1 V_0^1 / V_1^2 V_1^2$	$n_{(11/11)(10/11)}$	$2p_{(11)(11)}P_{(11)(01)}$
5	$U_1^1 U_1^1 / U_1^2 U_1^2$	$V_1^1 V_0^1 / V_1^2 V_0^2$	$n_{(11/11)(10/10)}$	$2p_{(11)(11)}P_{(11)(00)} + 2p_{(11)(10)}P_{(11)(01)}$
6	$U_1^1 U_1^1 / U_1^2 U_1^2$	$V_1^1 V_0^1 / V_0^2 V_0^2$	$n_{(11/11)(10/00)}$	$2p_{(11)(10)}P_{(11)(00)}$
7	$U_1^1 U_1^1 / U_1^2 U_1^2$	$V_0^1 V_0^1 / V_1^2 V_1^2$	$n_{(11/11)(00/11)}$	$p_{(11)(01)}^2$
8	$U_1^1 U_1^1 / U_1^2 U_1^2$	$V_0^1 V_0^1 / V_1^2 V_0^2$	$n_{(11/11)(00/10)}$	$2p_{(11)(01)}P_{(11)(00)}$
9	$U_1^1 U_1^1 / U_1^2 U_1^2$	$V_0^1 V_0^1 / V_0^2 V_0^2$	$n_{(11/11)(00/00)}$	$2p_{(11)(00)}^2$
10	$U_1^1 U_1^1 / U_1^2 U_0^2$	$V_1^1 V_1^1 / V_1^2 V_1^2$	$n_{(11/10)(11/11)}$	$2p_{(11)(11)}P_{(10)(11)}$
11	$U_1^1 U_1^1 / U_1^2 U_0^2$	$V_1^1 V_1^1 / V_1^2 V_0^2$	$n_{(11/10)(11/10)}$	$2p_{(11)(11)}P_{(10)(10)} + 2p_{(10)(11)}P_{(11)(10)}$
12	$U_1^1 U_1^1 / U_1^2 U_0^2$	$V_1^1 V_1^1 / V_0^2 V_0^2$	$n_{(11/10)(11/00)}$	$2p_{(11)(10)}P_{(10)(10)}$
13	$U_1^1 U_1^1 / U_1^2 U_0^2$	$V_1^1 V_0^1 / V_1^2 V_1^2$	$n_{(11/10)(10/11)}$	$2p_{(11)(11)}P_{(10)(01)} + 2p_{(10)(11)}P_{(11)(01)}$
14	$U_1^1 U_1^1 / U_1^2 U_0^2$	$V_1^1 V_0^1 / V_1^2 V_0^2$	$n_{(11/10)(10/10)}$	$2p_{(11)(11)}P_{(10)(00)} + 2p_{(11)(10)}P_{(10)(01)} + 2p_{(10)(11)}P_{(11)(00)} + 2p_{(10)(10)}P_{(11)(01)}$
15	$U_1^1 U_1^1 / U_1^2 U_0^2$	$V_1^1 V_0^1 / V_0^2 V_0^2$	$n_{(11/10)(10/00)}$	$2p_{(11)(10)}P_{(10)(00)} + 2p_{(11)(00)}P_{(10)(10)}$
16	$U_1^1 U_1^1 / U_1^2 U_0^2$	$V_0^1 V_0^1 / V_1^2 V_1^2$	$n_{(11/10)(00/11)}$	$2p_{(11)(01)}P_{(10)(01)}$
17	$U_1^1 U_1^1 / U_1^2 U_0^2$	$V_0^1 V_0^1 / V_1^2 V_0^2$	$n_{(11/10)(00/10)}$	$2p_{(11)(01)}P_{(10)(00)} + 2p_{(11)(00)}P_{(10)(01)}$
18	$U_1^1 U_1^1 / U_1^2 U_0^2$	$V_0^1 V_0^1 / V_0^2 V_0^2$	$n_{(11/10)(00/00)}$	$2p_{(11)(00)}P_{(10)(00)}$
19	$U_1^1 U_1^1 / U_0^2 U_0^2$	$V_1^1 V_1^1 / V_1^2 V_1^2$	$n_{(11/00)(11/11)}$	$p_{(10)(11)}^2$
20	$U_1^1 U_1^1 / U_0^2 U_0^2$	$V_1^1 V_1^1 / V_1^2 V_0^2$	$n_{(11/00)(11/10)}$	$2p_{(10)(11)}P_{(10)(10)}$

21	$U^1_1 U^1_1 / U^2_0 U^2_0$	$V^1_1 V^1_1 / V^2_0 V^2_0$	$n_{(11/00)(11/00)}$	$2p^2_{(10)(10)}$
22	$U^1_1 U^1_1 / U^2_0 U^2_0$	$V^1_1 V^1_1 / V^2_2 V^2_-$	$n_{(11/00)(10/11)}$	$2P_{(10)(11)}P_{(10)(01)}$
23	$U^1_1 U^1_1 / U^2_0 U^2_0$	$V^1_1 V^1_0 / V^2_2 V^2_0$	$n_{(11/00)(10/10)}$	$2P_{(10)(11)}P_{(1C)(00)} + P_{(10)(10)}P_{(10)(01)}$
24	$U^1_1 U^1_1 / U^2_0 U^2_0$	$V^1_1 V^1_0 / V^2_0 V^2_0$	$n_{(11/00)(10/00)}$	$2P_{(10)(10)}P_{(1C)(00)}$
25	$U^1_1 U^1_1 / U^2_0 U^2_0$	$V^1_0 V^1_0 / V^2_2 V^2_-$	$n_{(11/00)(00/11)}$	$P^2_{(10)(01)}$
26	$U^1_1 U^1_1 / U^2_0 U^2_0$	$V^1_0 V^1_1 / V^2_0 V^2_2$	$n_{(11/00)(00/10)}$	$2P_{(10)(01)}P_{(10)(00)}$
27	$U^1_1 U^1_1 / U^2_0 U^2_0$	$V^1_0 V^1_0 / V^2_0 V^2_2$	$n_{(11/00)(00/00)}$	$2p^2_{(10)(00)}$
28	$U^1_1 U^1_0 / U^2_- U^2_-$	$V^1_1 V^1_1 / V^2_2 V^2_2$	$n_{(10/11)(11/11)}$	$2P_{(11)(11)}P_{(01)(11)}$
29	$U^1_1 U^1_0 / U^2_- U^2_-$	$V^1_1 V^1_1 / V^2_2 V^2_0$	$n_{(10/11)(11/10)}$	$2P_{(11)(11)}P_{(01)(10)} + 2P_{(01)(11)}P_{(11)(10)}$
30	$U^1_1 U^1_0 / U^2_- U^2_-$	$V^1_1 V^1_1 / V^2_0 V^2_0$	$n_{(10/11)(11/00)}$	$2P_{(11)(10)}P_{(01)(10)}$
31	$U^1_1 U^1_0 / U^2_- U^2_-$	$V^1_1 V^1_0 / V^2_2 V^2_2$	$n_{(10/11)(10/11)}$	$2P_{(11)(11)}P_{(01)(01)} + 2P_{(01)(11)}P_{(11)(10)}$
32	$U^1_1 U^1_0 / U^2_- U^2_-$	$V^1_1 V^1_0 / V^2_1 V^2_0$	$n_{(10/11)(10/10)}$	$2P_{(11)(11)}P_{(01)(00)} + 2P_{(11)(10)}P_{(01)(01)} + 2P_{(01)(11)}P_{(01)(10)}P_{(11)(01)}$ $+ 2P_{(01)(11)}P_{(11)(00)} + 2P_{(10)(10)}P_{(01)(10)}$
33	$U^1_1 U^1_0 / U^2_2 U^2_2$	$V^1_1 V^1_0 / V^2_0 V^2_0$	$n_{(10/11)(10/00)}$	$2P_{(11)(10)}P_{(00)(10)} + 2P_{(10)(10)}P_{(01)(10)}$
34	$U^1_1 U^1_0 / U^2_- U^2_-$	$V^1_0 V^1_0 / V^2_- V^2_-$	$n_{(10/11)(00/11)}$	$2P_{(11)(01)}P_{(C1)(01)}$
35	$U^1_1 U^1_0 / U^2_- U^2_-$	$V^1_0 V^1_0 / V^2_- V^2_-$	$n_{(10/11)(00/10)}$	$2P_{(11)(01)}P_{(01)(00)} + 2P_{(01)(01)}P_{(11)(00)}$
36	$U^1_1 U^1_0 / U^2_2 U^2_2$	$V^1_0 V^1_0 / V^2_0 V^2_2$	$n_{(10/11)(00/00)}$	$2P_{(11)(00)}P_{(01)(00)}$
37	$U^1_1 U^1_0 / U^2_- U^2_-$	$V^1_1 V^1_1 / V^2_2 V^2_2$	$n_{(10/10)(11/11)}$	$2P_{(11)(11)}P_{(00)(11)} + 2P_{(10)(11)}P_{(01)(11)}$
38	$U^1_1 U^1_0 / U^2_- U^2_0$	$V^1_1 V^1_1 / V^2_2 V^2_0$	$n_{(10/10)(11/10)}$	$2P_{(11)(11)}P_{(00)(10)} + 2P_{(10)(11)}P_{(01)(10)}$ $+ 2P_{(11)(10)}P_{(00)(11)} + 2P_{(10)(10)}P_{(01)(11)}$
39	$U^1_1 U^1_0 / U^2_0 U^2_2$	$V^1_1 V^1_1 / V^2_0 V^2_0$	$n_{(10/10)(11/00)}$	$2P_{(11)(10)}P_{(00)(10)} + 2P_{(10)(10)}P_{(01)(10)}$
40	$U^1_1 U^1_0 / U^2_0 U^2_0$	$V^1_1 V^1_0 / V^2_0 V^2_1$	$n_{(10/10)(10/11)}$	$2P_{(11)(11)}P_{(30)(01)} + 2P_{(10)(11)}P_{(01)(01)}$ $+ 2P_{(11)(01)}P_{(00)(11)} + 2P_{(10)(01)}P_{(01)(11)}$

41	$U_1^1U_0^1/U_1^2U_0^2$	$V_1^1V_0^1/V_1^2V_0^2$	$n_{(10/10)(10/10)}$	$2p_{(11)(11)}P(00)(00) + 2p_{(11)(10)}P(00)(01)$ $+2p_{(10)(10)}P(01)(01) + 2p_{(10)(10)}P(01)(01)$ $+2p_{(01)(11)}P(10)(00) + 2p_{(01)(10)}P(10)(01)$ $+2p_{(00)(10)}P(11)(01) + 2p_{(00)(10)}P(11)(01)$
42	$U_1^1U_0^1/U_1^2U_0^2$	$V_1^1V_0^1/V_0^2V_0^2$	$n_{(10/10)(10/00)}$	$2p_{(11)(10)}P(00)(00) + 2p_{(10)(10)}P(01)(00)$ $+2p_{(00)(10)}P(11)(00) + 2p_{(01)(10)}P(10)(00)$
43	$U_1^1U_0^1/U_1^2U_0^2$	$V_0^1V_0^1/V_1^2V_1^2$	$n_{(10/10)(00/11)}$	$2p_{(11)(01)}P(00)(01) + 2p_{(10)(01)}P(01)(01)$
44	$U_1^1U_0^1/U_1^2U_0^2$	$V_0^1V_0^1/V_1^2V_1^2$	$n_{(10/10)(00/10)}$	$2p_{(11)(01)}P(00)(00) + 2p_{(11)(00)}P(00)(01)$ $+2p_{(10)(01)}P(01)(00) + 2p_{(10)(00)}P(01)(01)$
45	$U_1^1U_0^1/U_1^2U_0^2$	$V_0^1V_0^1/V_1^2V_0^2$	$n_{(10/10)(00/00)}$	$2p_{(11)(00)}P(00)(00) + 2p_{(10)(00)}P(01)(00)$
46	$U_1^1U_0^1/U_0^2U_0^2$	$V_1^1V_1^1/V_1^2V_1^2$	$n_{(10/00)(11/11)}$	$2p_{(10)(11)}P(00)(11)$
47	$U_1^1U_0^1/U_0^2U_0^2$	$V_1^1V_1^1/V_1^2V_0^2$	$n_{(10/00)(11/10)}$	$2p_{(10)(11)}P(00)(10) + 2p_{(00)(11)}P(10)(10)$
48	$U_1^1U_0^1/U_0^2U_0^2$	$V_1^1V_1^1/V_0^2V_0^2$	$n_{(10/00)(11/00)}$	$2p_{(10)(10)}P(00)(10)$
49	$U_1^1U_0^1/U_0^2U_0^2$	$V_1^1V_0^1/V_1^2V_1^2$	$n_{(10/00)(10/11)}$	$2p_{(10)(11)}P(00)(01) + 2p_{(00)(11)}P(10)(01)$
50	$U_1^1U_0^1/U_0^2U_0^2$	$V_1^1V_0^1/V_1^2V_0^2$	$n_{(10/00)(10/10)}$	$2p_{(10)(11)}P(00)(00) + 2p_{(10)(10)}P(00)(01)$ $+2p_{(00)(11)}P(10)(00) + 2p_{(10)(10)}P(00)(01)$
51	$U_1^1U_0^1/U_0^2U_0^2$	$V_1^1V_0^1/V_0^2V_0^2$	$n_{(10/00)(10/00)}$	$2p_{(10)(10)}P(00)(00) + 2p_{(10)(00)}P(00)(10)$
52	$U_1^1U_0^1/U_0^2U_0^2$	$V_0^1V_0^1/V_1^2V_1^2$	$n_{(10/00)(00/11)}$	$2p_{(10)(01)}P(00)(01)$
53	$U_1^1U_0^1/U_0^2U_0^2$	$V_0^1V_0^1/V_1^2V_0^2$	$n_{(10/00)(00/10)}$	$2p_{(10)(01)}P(00)(00) + 2p_{(10)(00)}P(00)(01)$
54	$U_0^1U_0^1/U_1^2U_1^2$	$V_0^1V_0^1/V_1^2V_0^2$	$n_{(10/00)(00/00)}$	$2p_{(10)(00)}P(00)(00)$
55	$U_0^1U_0^1/U_1^2U_1^2$	$V_1^1V_1^1/V_1^2V_1^2$	$n_{(00/11)(11/11)}$	$p_{(01)(11)}^2$
56	$U_0^1U_0^1/U_1^2U_1^2$	$V_1^1V_1^1/V_1^2V_0^2$	$n_{(00/11)(11/10)}$	$2p_{(01)(11)}P(01)(10)$
57	$U_0^1U_0^1/U_1^2U_0^2$	$V_0^1V_0^1/V_0^2V_0^2$	$n_{(00/11)(11/00)}$	$2p_{(01)(10)}^2$
58	$U_0^1U_0^1/U_1^2U_0^2$	$V_1^1V_0^1/V_1^2V_1^2$	$n_{(00/11)(10/11)}$	$2p_{(01)(11)}P(01)(01)$
59	$U_0^1U_0^1/U_1^2U_0^2$	$V_1^1V_0^1/V_1^2V_0^2$	$n_{(00/11)(10/10)}$	$2p_{(01)(11)}P(01)(00) + P(01)(10)P(01)(01)$

60	$U_1^1 U_1^1 / U_1^2 U_1^2$	$V_1^1 V_1^1 / V_1^2 V_1^2$	$n(00/11)(10/00)$	$2P(01)(10)P(01)(00)$
61	$U_1^1 U_0^1 / U_1^2 U_1^2$	$V_1^1 V_0^1 / V_1^2 V_1^2$	$n(00/11)(00/11)$	$P^2_{(01)(01)}$
62	$U_1^1 U_0^1 / U_1^2 U_1^2$	$V_0^1 V_0^1 / V_1^2 V_0^2$	$n(00/11)(00/10)$	$2P(01)(01)P(01)(00)$
63	$U_0^1 U_0^1 / U_1^2 U_1^2$	$V_0^1 V_0^1 / V_1^2 V_0^2$	$n(00/11)(00/00)$	$2P^2_{(01)(00)}$
64	$U_1^1 U_1^1 / U_1^2 U_0^2$	$V_1^1 V_1^1 / V_1^2 V_1^2$	$n(00/10)(11/11)$	$2P(01)(11)P(00;(11)$
65	$U_1^1 U_0^1 / U_1^2 U_0^2$	$V_1^1 V_0^1 / V_1^2 V_1^2$	$n(00/10)(11/10)$	$2P(01)(11)P(00)(10) + 2P(00)(11)P(01)(10)$
66	$U_1^1 U_0^1 / U_1^2 U_0^2$	$V_1^1 V_0^1 / V_1^2 V_0^2$	$n(00/10)(11/00)$	$2P(01)(10)P(00)(10)$
67	$U_1^1 U_1^1 / U_1^2 U_0^2$	$V_1^1 V_0^1 / V_1^2 V_0^2$	$n(00/10)(10/11)$	$2P(01)(11)P(00)(01) + 2P(00)(11)P(01)(01)$
68	$U_0^1 U_0^1 / U_1^2 U_0^2$	$V_1^1 V_0^1 / V_1^2 V_0^2$	$n(00/10)(10/10)$	$2P(01)(11)P(0C)(00) + 2P(01)(10)P(00)(01) + 2P(00)(11)P(01)(00)(00) + 2P(00)(10)P(01)(01)$
69	$U_0^1 U_0^1 / U_1^2 U_0^2$	$V_1^1 V_1^1 / V_2^2 V_2^2$	$n(00/10)(10/00)$	$2P(01)(10)P(00)(00) + 2P(01)(00)P(00)(10)$
70	$U_0^1 U_0^1 / U_1^2 U_0^2$	$V_0^1 V_0^1 / V_1^2 V_1^2$	$n(00/10)(00/11)$	$2P(01)(01)P(00)(01)$
71	$U_0^1 U_0^1 / U_1^2 U_0^2$	$V_0^1 V_1^1 / V_1^2 V_2^2$	$n(00/10)(00/10)$	$2P(01)(01)P(00)(00) + 2P(01)(00)P(00)(01)$
72	$U_0^1 U_0^1 / U_1^2 U_0^2$	$V_0^1 V_0^1 / V_1^2 V_0^2$	$n(00/10)(00/00)$	$2P(01)(00)P(00)(00)$
73	$U_1^1 U_1^1 / U_0^2 U_0^2$	$V_1^1 V_1^1 / V_2^2 V_1^2$	$n(00/00)(11/11)$	$P^2_{(00)(11)}$
74	$U_1^1 U_0^1 / U_0^2 U_0^2$	$V_1^1 V_1^1 / V_2^2 V_0^2$	$n(00/00)(11/10)$	$2P(00)(11)P(00)(10)$
75	$U_0^1 U_0^1 / U_0^2 U_0^2$	$V_1^1 V_1^1 / V_0^2 V_0^2$	$n(00/00)(11/00)$	$2P^2_{(00)(10)}$
76	$U_1^1 U_1^1 / U_0^2 U_0^2$	$V_1^1 V_0^1 / V_2^2 V_1^2$	$n(00/00)(10/11)$	$2P(00)(11)P(11)(01)$
77	$U_1^1 U_0^1 / U_0^2 U_0^2$	$V_1^1 V_0^1 / V_2^2 V_0^2$	$n(00/00)(10/10)$	$2P(00)(11)P(00)(00) + P(00)(10)P(00)(01)$
78	$U_0^1 U_0^1 / U_0^2 U_0^2$	$V_1^1 V_0^1 / V_0^2 V_0^2$	$n(00/00)(10/00)$	$2P(00)(10)P(00)(00)$
79	$U_0^1 U_0^1 / U_0^2 U_0^2$	$V_0^1 V_0^1 / V_2^2 V_1^2$	$n(00/00)(00/11)$	$P^2_{(00)(01)}$
80	$U_0^1 U_1^1 / U_0^2 U_0^2$	$V_1^1 V_0^1 / V_2^2 V_0^2$	$n(00/00)(00/10)$	$2P(00)(01)P(01)(00)$
81	$U_0^1 U_0^1 / U_0^2 U_0^2$	$V_0^1 V_0^1 / V_0^2 V_0^2$	$n(00/00)(00/00)$	$2P^2_{(00)(00)}$

TABLE 12.5
Disequilibrium compositions of four-gene haplotype frequencies.

Term.	Composition	Remark
(1)	$p_{k_1}^1 p_{k_2}^2 p_{l_1}^1 p_{l_2}^2$	No LD
(2)	$(-1)^{k_1+k_2} p_{l_1}^1 p_{l_2}^2 D_{\mathbf{A}_1\mathbf{A}_2}$	Digenic LD within **A**
(3)	$(-1)^{l_1+l_2} p_{k_1}^1 p_{k_2}^2 D_{\mathbf{B}_1\mathbf{B}_2}$	Digenic LD within **B**
(4)	$(-1)^{k_2+l_1} p_{k_1}^1 p_{l_2}^2 D_{\mathbf{A}_2\mathbf{B}_1}$	Digenic LD between SNP 2 of **A** and SNP 1 of **B**
(5)	$(-1)^{k_2+l_2} p_{k_1}^1 p_{l_1}^1 D_{\mathbf{A}_2\mathbf{B}_2}$	Digenic LD between SNP 2 of **A** and SNP 2 of **B**
(6)	$(-1)^{k_1+l_1} p_{k_2}^2 p_{l_2}^2 D_{\mathbf{A}_1\mathbf{B}_1}$	Digenic LD between SNP 1 of **A** and SNP 1 of **B**
(7)	$(-1)^{k_1+l_2} p_{k_2}^2 p_{l_1}^1 D_{\mathbf{A}_1\mathbf{B}_2}$	Digenic LD between SNP 1 of **A** and SNP 2 of **B**
(8)	$(-1)^{k_1+k_2+l_1} p_{l_2}^2 D_{\mathbf{A}_1\mathbf{A}_2\mathbf{B}_1}$	Trigenic LD between **A** and SNP 1 of **B**
(9)	$(-1)^{k_1+k_2+l_2} p_{l_1}^1 D_{\mathbf{A}_1\mathbf{A}_2\mathbf{B}_2}$	Trigenic LD between **A** and SNP 2 of **B**
(10)	$(-1)^{k_1+l_1+l_2} p_{k_2}^2 D_{\mathbf{A}_1\mathbf{B}_1\mathbf{B}_2}$	Trigenic LD between SNP 1 of **A** and **B**
(11)	$(-1)^{k_2+l_1+l_1} p_{k_1}^1 D_{\mathbf{A}_2\mathbf{B}_1\mathbf{B}_2}$	Trigenic LD between SNP 2 of **A** and **B**
(12)	$(-1)^4 (-1)^{k_1+k_2+l_1+l_2} D_{\mathbf{A}_1\mathbf{A}_2\mathbf{B}_1\mathbf{B}_2}$	Quadrigenic LD between **A** and **B**

Equations (12.12) and (12.13) construct an iterative loop of the EM algorithm. By giving initiate values for $p_{(k_1k_2)(l_1l_2)}$, we can calculate the proportion of a diplotype within a double, triple or quadruple heterozygote with equation (12.13), and then use those proportions to estimate across-block haplotype frequencies $\Omega_p = \{p_{(k_1k_2)(l_1l_2)}\}_{k_1,k_2,l_1,l_2=0}^1$. Using Table 12.5, allele frequencies and linkage disequilibria of different orders can be estimated.

12.2.5 Likelihood for Haplotype-Haplotype Interaction Effects

We need to detect how diplotypes are associated with phenotypic variation in a trait (y). We first assume haplotypes $[U_1^1 U_1^2]$ and $[V_1^1 V_1^2]$ are the risk haplotypes at QTNs **A** and **B**, respectively. As shown above, this leads to nine different across-block composite diplotypes. Based on Table 12.4, we formulate a mixture-based likelihood for quantitative genetic parameters (Ω_q) given observed phenotypes (y), SNP genotypes (**S**), and the estimated haplotype frequencies ($\hat{\Omega}_p$), expressed as

$$\log L(\Omega_q | y, \mathbf{S}, \hat{\Omega}_p) =$$

$$\sum_{i=1}^{n_{(11/11)(11/11)}} \log f_{AABB}(y_i) + \sum_{i=1}^{n_{(11/11)(\bullet)}} \log f_{AAB\bar{B}}(y_i) + \sum_{i=1}^{n_{(11/11)(\bullet\bullet)}} \log f_{AA\bar{B}\bar{B}}(y_i)$$

$$+ \sum_{i=1}^{n_{(11/11)(10/10)}} \log[\omega_{\mathbf{B}} f_{AAB\bar{B}}(y_i) + (1-\omega_{\mathbf{B}}) f_{AA\bar{B}\bar{B}}(y_i)]$$

$$+ \sum_{i=1}^{n_{(\bullet)(11/11)}} \log f_{A\bar{A}BB}(y_i) + \sum_{i=1}^{n_{(\bullet)(\bullet)}} \log f_{A\bar{A}B\bar{B}}(y_i) + \sum_{i=1}^{n_{(\bullet)(\bullet\bullet)}} \log f_{A\bar{A}\bar{B}\bar{B}}(y_i)$$

$$+ \sum_{i=1}^{n_{(\bullet)(10/10)}} \log[\omega_{\mathbf{B}} f_{A\bar{A}B\bar{B}}(y_i) + (1-\omega_{\mathbf{B}}) f_{A\bar{A}\bar{B}\bar{B}}(y_i)]$$

$$+ \sum_{i=1}^{n_{(\bullet\bullet)(11/11)}} \log f_{\bar{A}\bar{A}BB}(y_i) + \sum_{i=1}^{n_{(\bullet\bullet)(\bullet)}} \log f_{\bar{A}\bar{A}B\bar{B}}(y_i) + \sum_{i=1}^{n_{(\bullet\bullet)(\bullet\bullet)}} \log f_{\bar{A}\bar{A}\bar{B}\bar{B}}(y_i)$$

$$+ \sum_{i=1}^{n_{(\bullet\bullet)(10/10)}} \log[\omega_{\mathbf{B}} f_{\bar{A}\bar{A}B\bar{B}}(y_i) + (1-\omega_{\mathbf{B}}) f_{\bar{A}\bar{A}\bar{B}\bar{B}}(y_i)]$$

$$+ \sum_{i=1}^{n_{(10/10)(11/11)}} \log[\omega_{\mathbf{A}} f_{A\bar{A}BB}(y_i) + (1-\omega_{\mathbf{A}}) f_{\bar{A}\bar{A}BB}(y_i)]$$

$$+ \sum_{i=1}^{n_{(10/10)(\bullet)}} \log[\omega_{\mathbf{A}} f_{A\bar{A}B\bar{B}}(y_i) + (1-\omega_{\mathbf{A}}) f_{\bar{A}\bar{A}B\bar{B}}(y_i)]$$

$$+ \sum_{i=1}^{n_{(10/10)(\bullet\bullet)}} \log[\omega_{\mathbf{A}} f_{A\bar{A}\bar{B}\bar{B}}(y_i) + (1-\omega_{\mathbf{A}}) f_{\bar{A}\bar{A}\bar{B}\bar{B}}(y_i)]$$

$$+ \sum_{i=1}^{n_{(10/10)(10/10)}} \log[\omega_{A\bar{A}B\bar{B}}^{(10/10)(10/10)} f_{A\bar{A}B\bar{B}}(y_i) + \omega_{A\bar{A}\bar{B}\bar{B}}^{(10/10)(10/10)} f_{A\bar{A}\bar{B}\bar{B}}(y_i)$$

$$+ \omega_{\bar{A}\bar{A}B\bar{B}}^{(10/10)(10/10)} f_{\bar{A}\bar{A}B\bar{B}}(y_i) + \omega_{\bar{A}\bar{A}B\bar{B}}^{(10/10)(10/10)} f_{\bar{A}\bar{A}B\bar{B}}(y_i)],$$

$$(12.14)$$

where

$$n_{(11/11)(\bullet)} = n_{(11/11)(11/10)} + n_{(11/11)(10/11)},$$

$$n_{(11/11)(\bullet\bullet)} = n_{(11/11)(11/00)} + n_{(11/11)(10/00)} + n_{(11/11)(00/11)} + n_{(11/11)(00/10)}$$
$$+ n_{(11/11)(00/00)},$$

$$n_{(\bullet)(11/11)} = n_{(11/10)(11/11)} + n_{(10/11)(11/11)},$$

$$n_{(\bullet\bullet)(11/11)} = n_{(11/00)(11/11)} + n_{(10/00)(11/11)} + n_{(00/11)(11/11)} + n_{(00/10)(11/11)}$$
$$+ n_{(00/00)(11/11)},$$

$$n_{(\bullet)(\bullet)} = n_{(11/10)(11/10)} + n_{(11/10)(10/11)} + n_{(10/11)(11/10)} + n_{(10/11)(10/11)},$$

$$n_{(\bullet)(\bullet\bullet)} = n_{(11/10)(11/00)} + n_{(11/10)(10/00)} + n_{(11/10)(00/11)} + n_{(11/10)(00/10)} +$$
$$n_{(11/10)(00/00)} + n_{(10/11)(11/00)} + n_{(10/11)(10/00)} + n_{(10/11)(00/11)} +$$
$$n_{(10/11)(00/10)} + n_{(10/11)(00/00)},$$

$$n_{(\bullet\bullet)(\bullet)} = n_{(11/00)(11/10)} + n_{(10/00)(11/10)} + n_{(00/11)(11/10)} + n_{(00/10)(11/10)} +$$
$$n_{(00/00)(11/10)} + n_{(11/00)(10/11)} + n_{(10/00)(10/11)} + n_{(00/11)(10/11)} +$$
$$n_{(00/10)(10/11)} + n_{(00/00)(10/11)},$$

$$n_{(\bullet\bullet)(\bullet\bullet)} = n_{(11/00)(11/00)} + n_{(11/00)(10/00)} + n_{(11/00)(00/11)} + n_{(11/00)(00/10)} +$$
$$n_{(11/00)(00/00)} + n_{(10/00)(11/00)} + n_{(10/00)(10/00)} + n_{(10/00)(00/11)} +$$
$$n_{(10/00)(00/10)} + n_{(10/00)(00/00)} + n_{(00/11)(11/00)} + n_{(00/11)(10/00)} +$$
$$n_{(00/11)(00/11)} + n_{(00/11)(00/10)} + n_{(00/11)(00/00)} + n_{(00/10)(11/00)} +$$
$$n_{(00/10)(10/00)} + n_{(00/10)(00/11)} + n_{(00/10)(00/10)} + n_{(00/10)(00/00)} +$$
$$n_{(00/00)(11/00)} + n_{(00/00)(10/00)} + n_{(00/00)(00/11)} + n_{(00/00)(00/10)} +$$
$$n_{(00/00)(00/00)},$$

$$n_{(\bullet)(10/10)} = n_{(11/10)(10/10)} + n_{(10/11)(10/10)},$$

$$n_{(10/10)(\bullet)} = n_{(10/10)(11/10)} + n_{(10/10)(10/11)},$$

$$n_{(\bullet\bullet)(10/10)} = n_{(11/00)(10/10)} + n_{(10/00)(10/10)} + n_{(00/11)(10/10)} + n_{(00/10)(10/10)} +$$
$$n_{(00/00)(10/10)},$$

$$n_{(10/10)(\bullet\bullet)} = n_{(10/10)(11/00)} + n_{(10/10)(10/00)} + n_{(10/10)(00/11)} + n_{(10/10)(00/10)} +$$
$$n_{(10/10)(00/00)},$$

$$\omega_{\mathbf{A}} = \frac{\hat{p}_{11}^{\mathbf{A}} \hat{p}_{00}^{\mathbf{A}}}{\hat{p}_{11}^{\mathbf{A}} \hat{p}_{00}^{\mathbf{A}} + \hat{p}_{10}^{\mathbf{A}} \hat{p}_{01}^{\mathbf{A}}}, \quad \hat{p}_{k_1 k_2}^{\mathbf{A}} = \sum_{l_1=0}^{1} \sum_{l_2=0}^{1} \hat{p}_{(k_1 k_2)((l_1 l_2)},$$

$$\omega_{\mathbf{B}} = \frac{\hat{p}_{11}^{\mathbf{B}} \hat{p}_{00}^{\mathbf{B}}}{\hat{p}_{11}^{\mathbf{B}} \hat{p}_{00}^{\mathbf{B}} + \hat{p}_{10}^{\mathbf{B}} \hat{p}_{01}^{\mathbf{B}}}, \quad \hat{p}_{l_1 l_2}^{\mathbf{B}} = \sum_{k_1=0}^{1} \sum_{k_2=0}^{1} \hat{p}_{(k_1 k_2)((l_1 l_2)},$$

$$\omega_{A\bar{A}B\bar{B}}^{(10/10)(10/10)} = \frac{P_{(11)(11)}P_{(00)(00)} + P_{(11)(00)}P_{(00)(11)}}{p_{\varphi}},$$

$$\omega_{A\bar{A}\bar{B}\bar{B}}^{(10/10)(10/10)} = \frac{P_{(11)(10)}P_{(00)(01)} + P_{(11)(01)}P_{(00)(10)}}{p_{\varphi}},$$

$$\omega_{\bar{A}\bar{A}B\bar{B}}^{(10/10)(10/10)} = \frac{P_{(10)(11)}P_{(01)(00)} + P_{(01)(11)}P_{(10)(00)}}{p_{\varphi}},$$

$$\omega_{\bar{A}\bar{A}\bar{B}\bar{B}}^{(10/10)(10/10)} = \frac{P_{(10)(10)}P_{(01)(01)} + P_{(10)(01)}P_{(01)(10)}}{p_{\varphi}}.$$

In likelihood (12.14), we model $f....(y_i)$ by a normal distribution with diplotype-specific mean $\mu....$ and variance σ^2. Lin and Wu (2006a) derived a closed form for the EM algorithm to estimate these means and variances. An optimal combination between risk haplotypes at QTNs **A** and **B** needs to be selected from all 16 possible combinations. The combination that gives the largest likelihood is considered as the best risk-haplotype combination. Under such an optimal combination, we estimate genotypic values of the composite diplotypes including μ_{AABB}, $\mu_{AAB\bar{B}}$, $\mu_{AA\bar{B}\bar{B}}$, $\mu_{A\bar{A}BB}$, $\mu_{A\bar{A}B\bar{B}}$, $\mu_{A\bar{A}\bar{B}\bar{B}}$, $\mu_{\bar{A}\bar{A}BB}$, $\mu_{\bar{A}\bar{A}B\bar{B}}$, and $\mu_{\bar{A}\bar{A}\bar{B}\bar{B}}$. From the estimated $\mu....$ values, we can then solve the additive, dominance, and epistatic effects at two QTNs **A** and **B** using equations (12.2).

12.2.6   Hypothesis Tests

Two kinds of major hypotheses can be made in the following sequence: (1) The association between different SNPs within each QTN and between two QTNs by testing their linkage disequilibria LD), and (2) The significance of an assumed risk haplotype for its effect on the trait studied. The LD between four SNPs from two QTNs can be tested using the two hypotheses as follows:

$$\begin{cases} H_0: & D_{\mathbf{A}_1\mathbf{A}_2} = D_{\mathbf{B}_1\mathbf{B}_2} = D_{\mathbf{A}_1\mathbf{B}_1} = D_{\mathbf{A}_1\mathbf{B}_2} = D_{\mathbf{A}_2\mathbf{B}_1} = D_{\mathbf{A}_2\mathbf{B}_1} \\ & = D_{\mathbf{A}_1\mathbf{A}_2\mathbf{B}_1} = D_{\mathbf{A}_1\mathbf{A}_2\mathbf{B}_2} = D_{\mathbf{A}_1\mathbf{B}_1\mathbf{B}_2} = D_{\mathbf{A}_2\mathbf{B}_1\mathbf{B}_2} \\ & = D_{\mathbf{A}_1\mathbf{A}_2\mathbf{B}_1\mathbf{B}_2} = 0 \\ H_1: & \text{at least one of the LD above is not equal to zero.} \end{cases} \quad (12.15)$$

The log-likelihood ratio test statistic for the significance of LD is calculated by comparing the likelihood values under the H_1 (full model) and H_0 (reduced model) using

$$\mathrm{LR}_D = -2[\log L(\hat{p}_{k_1}^{\mathbf{A}}, \hat{p}_{k_2}^{\mathbf{A}}, \hat{p}_{l_1}^{\mathbf{B}}, \hat{p}_{l_2}^{\mathbf{B}}, \mathrm{LD} = 0|\mathbf{S}) - \log L(\hat{\boldsymbol{\Omega}}_p, \hat{\boldsymbol{\Omega}}_q|y, \mathbf{S})] \quad (12.16)$$

where $\hat{p}_{k_1}^{\mathbf{A}}$, $\hat{p}_{k_2}^{\mathbf{A}}$, $\hat{p}_{l_1}^{\mathbf{B}}$, and $\hat{p}_{l_2}^{\mathbf{B}}$ are the MLEs of allele frequencies at two QTNs. The LR_D calculated under the H_0 and H_1 hypotheses is considered to asymptotically follow a χ^2 distribution with 11 degrees of freedom.

It is also interesting to test whether the linkage disequilibria of a different particular order are significant. This can be done by formulating the null hypotheses:

$$\begin{cases} H_0: D_{\mathbf{A}_1\mathbf{A}_2} = D_{\mathbf{B}_1\mathbf{B}_2} = D_{\mathbf{A}_1\mathbf{B}_1} = D_{\mathbf{A}_1\mathbf{B}_2} = D_{\mathbf{A}_2\mathbf{B}_1} = D_{\mathbf{A}_2\mathbf{B}_1} = 0 \\ H_1: \text{at least one of the LD above is not equal to zero,} \end{cases} \quad (12.17)$$

for the digenic LD,

$$\begin{cases} H_0 : \ D_{\mathbf{A}_1\mathbf{A}_2\mathbf{B}_1} = D_{\mathbf{A}_1\mathbf{A}_2\mathbf{B}_2} = D_{\mathbf{A}_1\mathbf{B}_1\mathbf{B}_2} = D_{\mathbf{A}_2\mathbf{B}_1\mathbf{B}_2} = 0 \\ H_1 : \ \text{at least one of the LD above is not equal to zero,} \end{cases} \quad (12.18)$$

for the trigenic LD, and

$$\begin{cases} H_0 : \ D_{\mathbf{A}_1\mathbf{A}_2\mathbf{B}_1\mathbf{B}_2} = 0 \\ H_1 : \ D_{\mathbf{A}_1\mathbf{A}_2\mathbf{B}_1\mathbf{B}_2} \neq 0, \end{cases} \quad (12.19)$$

for the quadrigenic LD.

Diplotype or haplotype effects on a complex trait can be tested using the following hypotheses expressed as

$$\begin{cases} H_0 : \ \mu_{AABB} = \mu_{AAB\bar{B}} = \mu_{AA\bar{B}\bar{B}} = \mu_{A\bar{A}BB} = \mu_{A\bar{A}B\bar{B}} \\ \qquad = \mu_{A\bar{A}\bar{B}\bar{B}} = \mu_{\bar{A}\bar{A}BB} = \mu_{\bar{A}\bar{A}B\bar{B}} = \mu_{\bar{A}\bar{A}\bar{B}\bar{B}} = \mu \\ H_1 : \ \text{at least one equality in } H_0 \text{ does not hold} \end{cases} \quad (12.20)$$

The log-likelihood ratio test statistic (LR_E) under these two hypotheses can be similarly calculated. The LR_E may asymptotically follow a χ^2 distribution with eight degrees of freedom. However, the approximation of a χ^2 distribution may be inappropriate when some regularity conditions, such as normality and uncorrelated residuals, are violated. The permutation test approach (Churchill and Doerge, 1994), which does not rely upon the distribution of the LR_E, may be used to determine the critical threshold for determining the existence of QTNs.

Different genetic effects, such as the additive, dominance, and additive × additive, additive × dominance, dominance × additive, and dominance × dominance effects between two QTNs **A** and **B** can also be tested individually, with respective null hypotheses formulated as

$$H_0 : \ a_1 = 0, \qquad (12.21)$$
$$H_0 : \ a_2 = 0, \qquad (12.22)$$
$$H_0 : \ d_1 = 0, \qquad (12.23)$$
$$H_0 : \ d_2 = 0, \qquad (12.24)$$
$$H_0 : \ i_{aa} = 0, \qquad (12.25)$$
$$H_0 : \ i_{ad} = 0, \qquad (12.26)$$
$$H_0 : \ i_{da} = 0, \qquad (12.27)$$
$$H_0 : \ i_{dd} = 0. \qquad (12.28)$$

The parameter estimation under each of these null hypotheses can be obtained using the same EM algorithm as described for the alternative hypothesis (full model) of equation (12.20), with a constraint given by equation (12.2). The critical thresholds for these individual effects (12.21)–(12.28) can be determined on the basis of simulation studies.

Example 12.1

Numerous genes have been investigated as potential obesity-susceptibility genes (Mason et al., 1999; Chagnon et al., 2003). The $\beta1AR$ and $\beta2AR$ genes are two such examples (Green et al., 1995; Large et al., 1997) in each of which there are several polymorphisms common in the population. Two common polymorphisms are identified at codons 49 and 389 for the $\beta1AR$ gene on chromosome 10 and at codons 16 and 27 for the $\beta2AR$ gene on chromosome 5, respectively. The polymorphisms in each of these two receptor genes are in linkage disequilibrium, which suggests the importance of taking into account haplotypes, rather than a single polymorphism, when defining biologic function. This study attempts to detect haplotype variants within these candidate genes which determine human obesity traits.

To determine whether sequence variants at the two polymorphisms from one gene interact with those from the other gene to affect obesity phenotypes, a group of 155 men and women were investigated with ages from 32 to 86 years old with a large variation in body fat mass. Each of these patients was determined for their genotypes at codon 49 with two alleles, Ser49 (A) and Gly49 (G), and codon 389 with two alleles, Arg389 (C) and Gly389 (G), within the $\beta1AR$ gene, as well as at codon 16 with two alleles, Arg16 (A) and Gly16 (G), and codon 27 with two alleles, Gln27 (C) and Glu27 (G), within the $\beta2AR$ gene, and measured for body mass index (BMI). Two SNPs from each gene theoretically form 81 across-gene genotypes, but, because these two genes are independent, the frequencies of these genotypes can be expressed as the product of the genotype frequencies from each gene (see equation (12.11)). An integrative EM algorithm based on the likelihood function (12.14) allows for the estimates of four haplotype frequencies and the resulting allele frequencies and linkage disequilibrium at each gene (Table 12.6). Highly significant LD was detected between two SNPs for each gene ($p < 0.001$).

By assuming that one haplotype is different from the rest of the haplotypes at each gene, the epistatic model can detect the risk haplotypes that display significant main and interaction effects on the BMI trait. Using haplotypes [AC], [AG], [GC], and [GG] as a risk haplotype at the $\beta1AR$ gene, respectively, in conjunction with a risk haplotype selected from [AC], [AG], [GC], or [GG] at the $\beta2AR$ gene, Lin and Wu (2006a) calculated the corresponding log-likelihood-ratio (LR) test statistics (0.36–17.90) using equation (12.20) (Table 12.6). Based on the critical threshold value of 15.05 at the 5% significance level determined from 1000 permutation tests, the two maximal LR values, 15.1 and 17.9 for across-gene risk haplotypes (GC)(GG) and (GC)(GC), respectively,
are thought to trigger significant haplotype effects on BMI. However, because these two risk haplotypes form different numbers of composite genotypes, with all nine for (GC)(GG) and eight for (GC)(GC) (Table 12.6), an optimal risk-haplotype combination should be selected on the basis of the AIC criterion. We found that across-gene risk haplotype (GC)(GC) is the best for explaining the BMI data in this example.

The missing of one composite diplotype generated by across-gene risk haplotype (GC)(GC) prevents the estimation of all the additive, dominance, and epistatic effects with equations (12.21)–(12.28) because of inadequate degrees of freedom. As an example, we used across-gene risk haplotype (GC)(GG) to demonstrate how each

TABLE 12.6

Maximum likelihood estimates of SNP population genetic parameters (allele frequencies and linkage disequilibrium) and quantitative genetic parameters associated with phenotypic variation in BMI for 145 patients when different across-gene risk haplotypes are assumed.

Parameters	Across-gene risk haplotype combinations															
	(AC)(AC)	(AC)(AG)	(AC)(GC)	(AC)(GG)	(AG)(AC)	(AG)(AG)	(AG)(GC)	(AG)(GG)	(GC)(AC)	(GC)(AG)	(GC)(GC)	(GC)(GG)	(GG)(AC)	(GG)(AG)	(GG)(GC)	(GG)(GG)
LR_1	5.00	4.23	6.64	1.84	1.99	1.36	2.29	0.36	9.24	3.44	17.90	2.28	10.77	10.36	**15.06**	8.46
P value															**0.05**	
Population genetic parameters																
$p_1^{\beta 1}$	0.39	0.39	0.39	0.39	0.39	0.39	0.39	0.39	0.61	0.61	0.61	0.61	0.61	0.61	**0.61**	0.61
$p_2^{\beta 1}$	0.63	0.63	0.63	0.63	0.37	0.37	0.37	0.37	0.63	0.63	0.63	0.63	0.37	0.37	**0.37**	0.37
$D_{\beta 1}$	0.13	0.13	0.13	0.13	-0.13	-0.13	-0.13	-0.13	-0.13	-0.13	-0.13	-0.13	0.13	0.13	**0.13**	0.13
$p_1^{\beta 2}$	0.85	0.85	0.15	0.15	0.85	0.85	0.85	0.15	0.85	0.85	0.15	0.15	0.85	0.85	**0.15**	0.15
$p_2^{\beta 2}$	0.74	0.26	0.74	0.26	0.26	0.26	0.74	0.74	0.74	0.26	0.74	0.74	0.74	0.26	**0.74**	0.26
$D_{\beta 2}$	-0.04	0.04	0.04	-0.04	-0.04	0.04	0.04	-0.04	-0.04	0.04	0.04	-0.04	-0.04	0.04	**0.04**	-0.04

Quantitative genetic parameters

μ_{AABB}	32.72	35.97	21.68	-				-	32.16	-	-	-	29.73	24.41	**24.18**	-
$\mu_{AAB\bar{B}}$	29.07	29.74	29.79	33.21				-	34.65	29.58	27.71	22.66	29.46	29.58	**29.28**	-
$\mu_{AA\bar{B}\bar{B}}$	31.71	30.78	31.97	30.96	-			-	23.11	32.95	32.81	31.42	24.33	29.02	**28.94**	28.79
$\mu_{\bar{A}ABB}$	29.36	31.08	-		28.55	26.09	-	-	28.62	31.31	36.35	-	28.48	26.69	**36.35**	-
$\mu_{\bar{A}AB\bar{B}}$	32.41	29.55	34.03	24.06	20.75	27.48	30.69	32.76	33.19	29.66	40.12	45.72	27.53	27.11	**27.43**	24.77
$\mu_{\bar{A}A\bar{B}\bar{B}}$	30.70	31.62	29.53	30.82	31.22	27.96	24.91	27.57	33.47	31.93	28.99	31.30	27.60	28.66	**27.84**	27.93
$\mu_{\bar{A}\bar{A}BB}$	29.23	25.84	30.27	-	29.90	30.21	27.40	-	30.42	28.85	22.93	-	31.18	**33.80**	21.68	-
$\mu_{\bar{A}\bar{A}B\bar{B}}$	29.29	28.12	28.74	23.90	30.67	29.07	31.52	24.16	28.49	28.59	28.23	27.41	33.87	31.03	**36.19**	29.34
$\mu_{\bar{A}\bar{A}\bar{B}\bar{B}}$	26.26	29.61	28.72	28.81	29.26	30.69	29.69	30.11	28.00	29.41	29.51	29.05	32.30	32.88	**31.40**	32.44
σ	8.75	8.78	8.70	8.85	8.84	8.86	8.83	8.89	8.62	8.80	8.34	8.83	8.57	8.59	**8.43**	8.64

Note: "-" denotes the missing of composite genotypes under the corresponding across-gene reference haplotypes.

of these genetic effects is tested. It is found that the additive and dominance effects exerted by (GC)(GG) are not significant. Of the four kinds of epistasis between the two genes, only dominance $\times$ dominance genetic effect is significant at the 5% significance level. This type of epistasis reduces, by about 20%, the BMI of the patient who carries diplotypes [GG][$\overline{GG}$] at the β1AR gene and [GC][$\overline{GC}$] at the β2AR gene, compared to the other across-gene diplotypes. Lin and Wu (2006a) estimated the standard errors of the MLEs of the population and quantitative genetic parameters based on Louis (1982) observed information matrix, suggesting that all MLEs have reasonable estimation precision although the estimates of quantitative genetic parameters are not as precise as those of population genetic parameters due to a small sample size used.   ☐

12.3   Haplotyping Epistasis of Drug Response

12.3.1   Introduction

Given a recent upsurge of interest in pharmacogenetics and pharmacogenomics, there is a pressing need for the development of statistical models for unraveling the genetic etiology of pharmacological variation. Genetic haplotyping, by superimposing real biological phenotypes on genome sequence, structural polymorphisms, and gene expression data, can provide an unbiased view of the network of gene actions and interactions that build a complex phenotype like drug response. Through an integrated approach, studies can move to characterize the best drug, the best dosage of the drug and the best injection time for individual patients based on their genetic structure. By incorporating genetic tests into the prescription process, physicians might improve outcomes for patients.

Genetically, drug response is a complex trait, involving multiple biochemical pathways each controlled by different interacting genes (Marchini et al., 2005). Traditional genetic mapping can probe the important chromosomal segments for drug response, but is difficult to identify the causal DNA sequences of these segments. The completion of the human genome project makes it possible to genome-wide associate DNA sequence variants with complex phenotypes. Below, we will introduce a statistical model that has capacity to characterize the effects of genetic factors and their epistatic effects on drug response at the DNA sequence level.

12.3.2   Model and Estimation

As described above, genetic haplotyping is constructed within a mixture model in which the proportions of each mixture component (i.e., the probabilities of composite diplotypes) are fit and estimated by SNP data. Phenotypic observations of a trait given each component are modeled by a probability distribution function with diplotype-specific means and variance. When the trait studied is a longitudinal one like drug response, we need to model the longitudinal pattern of diplotype-specific means by a biologically and clinically meaningful mathematical equation and the longitudinal covariance by stationary or non-stationary models. This idea founds the basic framework of functional mapping (Ma et al., 2002).

In the pharmacodynamic study of drug response, we use the Hill or sigmoid Emax equation to model the response-concentration profile expressed as

$$E = \frac{E_{\max} c^H}{EC_{50}^H + c^H},$$ (12.29)

which is different from equation (6.1). Here, no same baseline is included, although the other parameters $E_{\max}$ (asymptotic effect), EC_{50} (drug concentration that results in 50% of the maximal effect), and H (slope parameter that determines the slope of the response-concentration curve) preserve the same meanings.

When equation (12.29) is used in the functional mapping of drug response, we first adjust the raw data to remove the differences in baseline among all subjects. This can be done by subtracting the baseline measurement of drug effect from the measurements at other higher doses for each subject. By assuming risk haplotypes for two different QTNs **A** and **B**, we construct the composite diplotypes. Then, composite diplotype-specific mean vectors in the mixture model from dose 1 to C is modeled by

$$
\begin{aligned}
\mathbf{u}_{j_1 j_2} &= [u_{j_1 j_2}(1), \ldots, u_{j_1 j_2}(C)] \\
&= \left\{ \left[\frac{E_{\max j_1 j_2}}{EC_{50 j_1 j_2}^{H_{j_1 j_2}} + 1^{H_{j_1 j_2}}} \right], \ldots, \left[\frac{E_{\max j_1 j_2} C^{H_{j_1 j_2}}}{EC_{50 j_1 j_2}^{H_{j_1 j_2}} + C^{H_{j_1 j_2}}} \right] \right\},
\end{aligned}
$$ (12.30)

where $\Omega_{j_1 j_2} = (E_{\max j_1 j_2}, EC_{50 j_1 j_2}, H_{j_1 j_2})$ are the mathematical parameters that describe the drug response profile for composite diplotype $j_1 j_2$. Thus, based on equation (12.30), our estimates will be concentrated on $\Omega_{j_1 j_2}$ rather than on $\mathbf{u}_{j_1 j_2}$. This modeling of the mean vectors has two advantages: (1) Clinically meaningful curves are used in genetic mapping so that the results will be closer to biological reality, and (2) The number of parameters to be estimated is reduced, thus increasing the power of the model to detect significant QTN and their interactions.

It is not parsimonious to estimate all the elements in the within-subject covariance matrix among different dose levels because some structure exists for dosage-dependent variances and correlations. The structure of the residual covariance matrix in the multivariate normal distribution can be modeled by (Zimmerman and Núñez-Antón, 2001) first-order structured antedependence (SAD(1)) model that contains parameters ϕ and v^2.

12.3.3 Hypothesis Tests

With our epistatic model, we can make a number of hypothesis tests regarding the genetic control of overall drug response to a spectrum of dosages and other clinically important events. Here, we only provide the procedure for testing the significance of an assumed across-QTN risk haplotype for its effect on drug response. The genetic effects of diplotypes on drug response can be tested using the following hypotheses, expressed as

$$\begin{cases} H_0 : \Theta_{j_1 j_2} = \Theta \;\; (j_1, j_2 = 2, 1, 0) \\ H_1 : \text{at least one equality in } H_0 \text{ does not hold} \end{cases} \tag{12.31}$$

The log-likelihood ratio test statistic (LR_E) under these two hypotheses can be calculated. The LR_E may asymptotically follow a χ^2 distribution with 24 degrees of freedom. The permutation test approach proposed by Churchill and Doerge (1994), which does not rely upon the distribution of the LR_E, may be used to determine the critical threshold for determining the existence of QTNs.

Different genetic effects, such as the additive ($\mathbf{a}_1$ and $\mathbf{a}_2$), dominance ($\mathbf{d}_1$ and $\mathbf{d}_2$) and additive $\times$ additive ($\mathbf{i}_{aa}$), additive $\times$ dominance ($\mathbf{i}_{ad}$), dominance $\times$ additive ($\mathbf{i}_{da}$), and dominance $\times$ dominance effects ($\mathbf{i}_{dd}$) between QTN $\mathbf{A}$ and $\mathbf{B}$ can also be tested individually using equations (12.3), (12.4), (12.5), (12.6), (12.7), (12.8), (12.9), and (12.10), respectively. The critical thresholds for these individual effects can be determined on the basis of simulation studies.

One alternative test for the genetic control of overall drug responses can be performed on the area under curve (AUC). The AUC can be calculated by taking integral for each composite diplotype, expressed as

$$\text{AUC}_{j_1 j_2} = \int_1^C \left[\frac{\text{E}_{\max j_1 j_2} c^{\text{H}_{j_1 j_2}}}{\text{EC}_{50 j_1 j_2}^{\text{H}_{j_1 j_2}} + c^{\text{H}_{j_1 j_2}}} \right] dc.$$

The null hypothesis based on the AUC can be formulated as $\text{AUC}_{j_1 j_2} = \text{AUC}$. The permutation tests can be used to determine the critical value for AUC-related hypothesis tests.

Example 12.2

Revisit Example 12.2.6. Lin et al. (2007) used the dynamic model of epistasis to analyze this example for drug response. Dobutamine was injected into a group of 155 men and women ages 32 to 86 years old to investigate their response in heart rate to this drug. The subjects received increasing doses of dobutamine, until they achieved target heart rate response or predetermined maximum dose. The dose levels used were 0 (baseline), 5, 10, 20, 30, and 40 mcg/min, at each of which heart rate was measured. The time interval of 3 minutes is allowed between two successive doses for subjects to reach a plateau in response to that dose. Only those (98) in whom there were heart rate data at all the six dose levels were included for data analysis.

TABLE 12.7

Likelihood ratios for 16 possible combinations of assumed risk haplotypes with one from candidate gene β1AR and the second from candidate gene β2AR.

β2AR	β1AR			
	AC	AG	GC	GG
AC	38.16	36.79	31.19	15.84
AG	24.87	19.00	21.50	8.93
GC	19.67	15.16	22.42	4.62
GG	40.52	37.57	**56.32**	27.21

The maximum likelihood ratio value is detected when [GG] at β2AR and [GC] at β1AR are used as the risk haplotypes.

TABLE 12.8

Maximum likelihood estimates (MLEs) of total genetic, additive, dominance, and interaction effects for AUC and their significance tests under the optimal haplotype model [(GG)(GC)].

Genetic Effects	MLE	(p-value)
a_1	2.68	39.96 (<0.05)
a_2	-0.18	39.78 (<0.05)
d_1	-2.18	41.32 (<0.05)
d_2	-1.49	46.83 (<0.05)
i_{aa}	1.81	44.36 (<0.05)
i_{ad}	-3.62	40.91 (<0.05)
i_{da}	2.68	40.78 (<0.05)
i_{dd}	-4.02	40.35 (<0.05)

The dose-specific heart rate data were normalized to remove the heterogeneity in baseline heart rates among subjects (Fig. 12.1).

At the β1AR gene, two SNPs at codons 49 (with two alleles A and G) and 389 (with two alleles C and G), and at the β2AR gene, two SNPs at codon 16 (with two alleles A and G)and 27 (with two alleles C and G) were typed. The estimates of population genetic parameters of these SNPs were given in Table 12.6. These SNPs were associated with drug response data to determine whether sequence variants at the two polymorphisms from one gene interact with those from the other gene to affect patients' responsiveness to dobutamine.

☐

The model detected the risk haplotypes that display significant main and interaction effects on drug response. Using haplotypes [AC], [AG], [GC], and [GG] as a risk haplotype at the β1AR gene, respectively, in conjunction with a risk haplotype selected from [AC], [AG], [GC], and [GG] at the β2AR gene, the corresponding

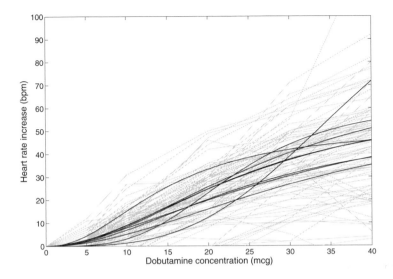

FIGURE 12.1
Profiles of heart rate in response to different dosages of dobutamine for nine across-QTN composite diplotypes (foreground) identified for SNPs within two genes. The profiles of 98 studied subjects are also shown (background).

log-likelihood-ratio (LR) test statistics was calculated (Table 12.7) based on the hypotheses test (12.31). Because of unobserved genotypes at a candidate gene, some composite diplotypes do not exist for several risk-haplotype or risk and non-risk haplotype combinations. The maximal LR value (56.32) with all the nine composite genotypes was obtained when GC as the risk haplotype of the $\beta 1$AR gene is combined with [GG] as the risk haplotype of the $\beta 2$AR gene. The LR values for the other combinations range from 4.62 to 40.52. The optimal risk-haplotype combination for drug response is statistically tested with 1000 permutation tests which obtained the critical threshold value of 54.08 at the 5% significance level. This test suggests that there exist significant haplotype effects at these two candidate genes on heart rate curves. Figure 12.1 displays the profile of heart rate to increasing concentration levels of dobutamine for nine across-QTN composite diplotypes on the basis of the MLEs of curve parameters. The longitudinal patterns of additive, dominance, and epistatic effects between two QTNs on heart rate response are shown in Fig. 12.2.

Statistical tests for individual genetic effects for DNA sequence variants based on equations (12.3), (12.4), (12.5), (12.6), (12.7), (12.8), (12.9), and (12.10) suggest that the additive and dominance effects at each of the candidate genes are highly significant (Table 12.8). Four kinds of epistasis between the two genes are all significant at the 5% significance level. Considerable over-crossing among different curves, as shown by Fig. 12.1, suggests sequence-sequence interactions within $\beta 1$AR and $\beta 2$AR. Figure 12.2 illustrates the longitudinal changes of different genetic effects

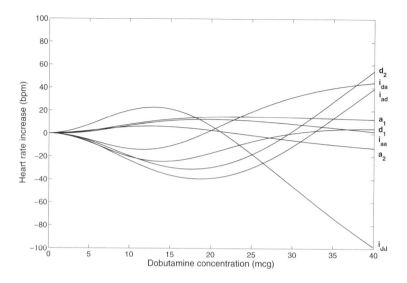

FIGURE 12.2
Profiles of different genetic components (including additive, dominance, and epistatic effects) for heart rate in response to different dosages of dobutamine.

over drug dose. It appears that the additive and dominance effects of SNP at codon 16, the additive effect of SNP at codon 27, and additive $\times$ additive epistatic effect between the two SNPs are more consistent over dose, as compared with the dominance effect of SNP at codon 27, additive $\times$ dominance, dominance $\times$ additive, and dominance $\times$ dominance epistatic effects. The dominance $\times$ dominance effect displays a dramatic dose-dependent change, with the pattern different from the other effects.

Example 12.3
Mapping Genetic Interactions for Growth Curves. A second example is used to show how an epistatic model can explain the temporal change of genetic actions and interactions in a time course. This example is derived from quantitative trait locus (QTL) mapping in the mouse. In the 1980s, a high growth gene, called *hg*, was detected (Bradford and Famula, 1984; Horvat and Medrano, 1995). This gene is a spontaneous mutation that results in a 30–50% increase in postnatal body growth in mice. Earlier physiological studies suggest that the increase of growth efficiency by the *hg* locus stems from increased energy metabolism without altering overall body composition (Calvert et al., 1985; Horvat and Medrano, 2001). Using genetic mapping, Horvat and Medrano (1995) localized the *hg* locus near Dl0Mit41 on the distal half of mouse chromosome 10 in both female and male F₂ populations. Wu et al. (2005) used functional mapping to test how the *hg* gene interacts with QTLs on

other chromosomes to affect the growth process.

The *hg* mutation was introgressed into the C57BL/6J (C57) background through nine backcrosses to create congenic line C57BL/6J-*hg/hg* (HG). A mapping population was founded by mating smaller CAST/EiJ (CAST) males to HG females, which produced a total of 75 F_1 and 1132 F_2 mice (Corva et al., 2001). To test the segregation of *hg* in the mapping population, these F_2 mice were genotyped by using D10Mit41 on chromosome 10 detected to be linked with *hg* and D10Mit69, a marker that maps within the *hg* deletion (Horvat and Medrano, 1995). Mice homozygous for HG alleles at D10Mit41 and without a PCR amplification product for D10Mit69 (indicating homozygosity for the *hg* deletion) were thought to be homozygous for the mutant allele (expressed as *hg/hg*). On the other hand, mice homozygous for CAST alleles at D10Mit41 and amplifying for D10Mit69 were regarded as being homozygous for the wide type allele (expressed as +/+). It was found that there were 274 +/+ mice, 596 +/*hg* mice, and 262 *hg/hg* mice in the F_2 cross, which conforms to Mendelian segregation ratios (Corva et al., 2001). However, in the data analyzed by Wu et al. (2005), only two subpopulations *hg/hg* (1) and +/+ (2) were available.

Both the *hg/hg* and +/+ F_2 subpopulations were typed for markers on chromosomes 1, 2, 4, 9, and X (Corva et al., 2001) to construct a linkage map that covers part of the mouse genome. Mice in the *hg/hg* and +/+ subpopulations were measured for their body weights on a weekly basis from 2 to 9 weeks of age. However, about one third of the mice from each subpopulation were measured only at weeks 3, 6, and 9. Data for body weights at different ages were corrected for the effects of dam, litter, sex, and parity. Let $\mathbf{y}_{i|1} = (y_{i|1}(t_{i1|1}), ..., y_{i|1}(t_{iT_i|1}))$ and $\mathbf{y}_{i|2} = (y_{i|2}(t_{i1|2}), ..., y_{i|1}(t_{iT_i|2}))$ be the observation vector for body mass in an unequally-spaced time course for mouse i from subpopulations 1 and 2, respectively, and $\mathbf{M}$ be the marker genotypes for an interval that harbors a QTL with two alleles Q and q, leading to three QTL genotypes QQ (2), Qq (1), and qq (1).

A joint mixture-based likelihood of unknown parameters Θ for integrating the *hg/hg* and +/+ F_2 subpopulations was constructed as

$$\log L(\Theta|\mathbf{y}_1, \mathbf{y}_2, \mathbf{M}) = \sum_{i=1}^{n_1} \log \left[\omega_{2|i} f_2(\mathbf{y}_{i|1}) + \omega_{1|i} f_1(\mathbf{y}_{i|1}) + \omega_{0|i} f_0(\mathbf{y}_{i|1}) \right],$$

$$= \sum_{i=1}^{n_2} \log \left[\omega_{2|i} f_2(\mathbf{y}_{i|2}) + \omega_{1|i} f_1(\mathbf{y}_{i|2}) + \omega_{0|i} f_0(\mathbf{y}_{i|2}) \right],$$

where $\omega_{j|i}$ is the conditional probabilities of QTL genotype j ($j = 2, 1, 0$) given the marker genotype of progeny i, which is the same for the two subpopulations, and $f_j(\mathbf{y}_{i|1}$ and $f_j(\mathbf{y}_{i|2}$ are two multivariate normal distributions with QTL genotype-specific means $\mathbf{u}_{j|1} = (u_{i|1}(t_{i1|1}), ..., u_{i|1}(t_{iT_i|1}))$ and $\mathbf{u}_{j|2} = (u_{i|2}(t_{i1|2}), ..., u_{i|2}(t_{iT_i|2}))$, and covariance matrix for two subpopulations, respectively.

Wu et al. (2005) used a logistic curve, $g(t) = a/(1 + be^{-rt})$, to model the genotype-specific means, and AR(1) to model the covariance structure. Thus, Θ contains two sets of genotype-specific logistic parameters $(a_{j|1}, b_{j|1}, r_{j|1})$ and $(a_{j|2}, b_{j|2}, r_{j|2})$, as well as the covariance-structuring parameters. After the maximum likelihood estimates (MLEs) of the parameters are obtained, we can estimate the dynamic changes of differen types of genetic effects for the *hg* and QTL across ages by

$$a_1(t) = \frac{1}{4}[u_{2|1}(t) + u_{0|1}(t) - u_{2|2}(t) - u_{0|2}(t)],$$

for the additive genetic effect of the *hg* locus,

$$a_2(t) = \frac{1}{4}[u_{2|1}(t) + u_{2|2}(t) - u_{0|1}(t) - u_{0|2}(t)],$$

for the additive genetic effect of the QTL,

$$d_2(t) = \frac{1}{2}[u_{1|1}(t) + u_{1|2}(t)] - \frac{1}{4}[u_{2|1}(t) + u_{0|1}(t) + u_{2|2}(t) + u_{0|2}(t)],$$

for the dominance genetic effect of the QTL,

$$I(t) = \frac{1}{4}[u_{2|1}(t) + u_{0|2}(t) - u_{0|1}(t) - u_{2|2}(t)],$$

for the additive × additive genetic effect, and

$$J(t) = \frac{1}{2}[u_{1|1}(t) - u_{1|2}(t)] - \frac{1}{4}[u_{2|1}(t) + u_{0|1}(t) - u_{2|2}(t) - u_{0|2}(t)],$$

for the additive × dominance genetic effect between *hg* and QTL.

Genome-wide search for QTLs leads to the profile of the log-likelihood ratio (LR) test statistics for the two subpopulations. One QTL on chromosome 2 and two separate QTLs on chromosome X were detected for body mass growth. The three growth curves each determined by a genotype at each of these QTLs detected were drawn separately for the *hg/hg* and +/+ mice (Fig. 12.3) using the MLEs of curve parameters. As expected, the *hg* locus displays a striking (additive) effect on growth trajectories. The growth trajectories of the same QTL genotype are different between the two subpopulations, suggesting that the genetic expression of QTL is affected by genetic background. In general, the three detected QTL start to exert their effects on growth in both subpopulations when the mice are 3 weeks of age (Fig. 12.4). After this age, the QTL effects tend to increase with age.

Further tests were made on the genetic effects of the QTLs and their interactions with the *hg* locus to affect growth trajectories. The QTL detected on chromosome 2 has highly significant additive and dominance effects on growth trajectories, operating in a dominant gene action manner as shown by small differences between genotypes *Qq* and *QQ* in both subpopulations (see Fig. 12.3). This QTL also displays significant additive × additive and additive × dominant epistatic effects with the *hg* locus (Fig. 12.4).

Located on the same chromosome, the two QTL detected on chromosome X exhibit different modes of gene action for growth. The first QTL at 3 cM from the first marker has a nonsignificant additive effect but highly significant dominant effect ((Fig. 12.3**B**). When interacting with the *hg* gene, however, this dominant QTL

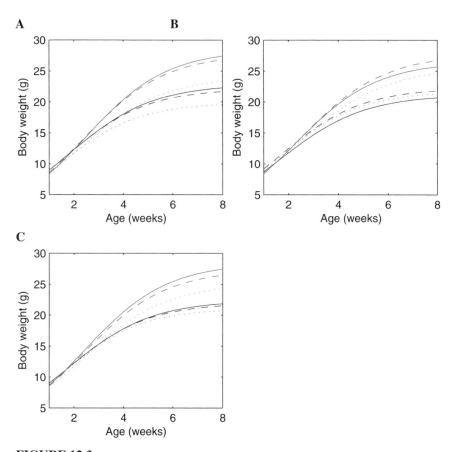

FIGURE 12.3
Three growth curves each presenting a group of genotype, *QQ* (solid curves), *Qq* (broken curves), and *qq* (dot curves), in the *hg/hg* (red) and +/+ (blue) mice at the QTL on chromosomes 2 (**A**) and X (**B** and **C**).

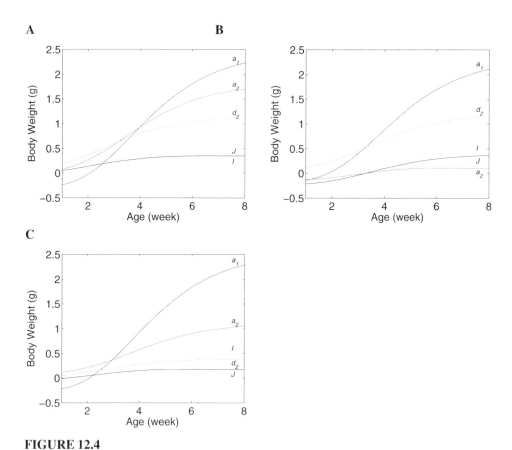

FIGURE 12.4

Dynamic changes of the genetic effects of the *hg* gene (a_1) and QTL (a_2 and d_2) as well as their epistatic interactions (*I* and *J*) for the QTL on chromosomes 2 (**A**) and X (**B** and **C**).

displays an inverse pattern, i.e., with a significant additive $\times$ additive but nonsignificant additive $\times$ dominant epistatic effects. The second QTL at 37 cM from the first marker seems to act in a partial dominant manner (Fig. 12.3**C**), with both types of epistatic effects being significant. Except for the first QTL on chromosome X with a nonsignificant additive effect, the favorable allele at the other QTL that contributes to greater growth originates from the HG parent.

All these age-dependent changes of genetic effects are illustrated in (Fig. 12.4). The additive effect (a_1) of the *hg* locus increases rapidly with age, so does the additive (a_2) and/or dominant effects (d_2) of the QTL, but with a lesser extent. The interaction effects (I and J) between the QTL and *hg* is quite stable over age, contributing to a significant part of the genetic variation throughout growth ontogeny.

[]

12.4 Prospects

The development of an effective strategy for the characterization of the genes responsible for drug response is one of the most pressing challenges for pharmacogenetic researchers. Statistical models described in this chapter present a conceptual framework for haplotyping genes and their interactions at the sequence level that control the longitudinal aspects of drug response. By comparing the curves of drug response among different genetic variants, we can test and identify the presence of sequenced QTN controlling the entire response process, and also reveal the genetic architecture of various developmental aspects responsible for the direction of pharmacodynamic reactions. The model provides a useful platform for identifying the genetic factors that regulate the process, rate, and shape of drug response. This model estimates the stochastic environmental components of drug response based on a non-stationary model, but it can be extended to specify the effects on pharmacodynamic response of interactions between QTNs and predictable environments, such as nutrients or physical exercise. In the ultimate, the approach described will provide the mechanistic principles to guide the development of efficient gene therapy strategies for diseases.

13

Mapping Genotype-Environment Interactions in Drug Response

"How can we select an optimal medicine for individual patients?" This emerging question has become a significant challenge since the successful completion of the sequencing of the human genome. The understanding of how genes regulate drug response is a first important step toward developing better drugs based on individual genetic make-up. But this is not enough because the expression of any gene must cope with, or be stimulated by, some particular biotic and abiotic environmental factors. For example, the same form of a gene may be expressed differently among people with different ancestral origins. The N-acetyltransferase 2 (NAT2) is an enzyme that is heavily involved in the detoxification of many carcinogens and the metabolism of many commonly used drugs (Burchard et al., 2003) The genetic expression of NAT2 results in two distinct phenotypic forms, slow and rapid acetylators. These two forms perform differently among races, with the frequency of the slow acetylator ranging from about 14% in East Asians, 34% in African Americans, and to 54% in European Americans.

There are other numerous examples about the impacts of race, sex, and life style on the expression of genes. For this reason, the study of genetic and environmental interactions that modulate an individual's responsiveness to medications has become one of the most vigorous areas in pharmacogenomic and toxicogenomic research (Borlak, 2005). Because of different genes they carry, and because of different environments in which these genes can be expressed, some people are more responsive to a drug than others. Thus, by identifying a network of interactions between the genes and enviroments, many fundamental questions in health sciences can be better addressed. These questions include how disease occurs and can be predicted, how potential environmental hazards and exposed individuals can be identified, and, in the ultimate, how disease can be prevented.

In this chapter, we will describe statistical models for detecting differential expressions of a gene across environmental changes in a pharmacogenomic study. The model allows the test for differences in allele frequencies and linkage disequilibria among individuals reared in different environments, and the characterization of specific genes that interact with environmental factors in a coordinated way to determine pharmacodynamic response. A detailed estimation and test step will be described with real-world example analyses, elucidating the usefulness and utilization of the models.

13.1 Haplotyping Genotype-Environment Interactions

13.1.1 Environmental Sensitivity and Genotype-Environment Interactions

The difference of a single genotype in its phenotypic value across a range of environments is called the environmental sensitivity or phenotypic plasticity. This phenomenon, pervading the kingdom of biology, was recognized by many early biologists (Waddington, 1942; Schmalhausen, 1949) and has now received a resurgence of interest and a renaissance in its fundamental role in shaping evolutionary adaptation and consequences (Via et al., 1995; Scheiner, 1993; Schlichting and Smith, 2002; West-Eberhard, 2003, 2005; Wu et al., 2004a; de Jong, 2005). The biochemical cause of environmental sensitivity can be explained in Fig. 13.1 (Beavis and Keim, 1996). Consider a gene (with alleles A and a) that encodes enzymes involved in a single step of metabolic pathway such as nutrient transport or DNA binding and the subsequent phenotype. The activity of such enzymes is affected by environment, such as substrate and temperature. Allele A may be favorably expressed in a high substrate and high temperature, whereas allele a prefers a low substrate and low temperature. Thus, under four environmental conditions, low substrate and low temperature (I), low substrate and high temperature (II), high substrate and low temperature (III), and high substrate and high temperature (IV), we can see different performances between two genotypes AA and aa, and in particular, the changes in the order of phenotypic expression between the two genotypes. The analysis with one locus can be extended to consider two genes at each of which the encoded enzymes are sensitive to different environments (Fig. 13.2). Relative to the one-locus model, the two-locus model can explain more complicated changes of the phenotypes across different environmental conditions (from I to IV) among genotypes.

Variation in environmental sensitivity among different genotypes is called genotype by environment interaction. Three hypotheses have been proposed to explain the genetic causes of genotype by environment interaction. The overdominance hypothesis proposes that a heterozygote at relevant genes shows higher stability, or lower plasticity, than a homozygote and the degree of stability is proportional to heterozygosity for these loci (homeostasis (Gillespie and Turelli, 1989)). The pleiotropic hypothesis states that differential expression of the same loci across environments causes phenotypic plasticity (allelic sensitivity (Via and Lande, 1985)). The epistatic hypothesis suggests that there exist specific plasticity genes that interact epistatically with the loci for the mean value of the trait to regulate environmental sensitivity (gene regulation (Scheiner and Lyman, 1989)). These hypotheses have been tested with results from quantitative trait locus (QTL) mapping in different species from *Arabidopsis* (Kliebenstein et al., 2002; Ungerer et al., 2003) to *Populus* (Wu, 1998; Rae et al., 2007), *Drosophila* (Leips and Mackay, 2000; Geiger-Thornsberry and Mackay, 2002; Anholt and Mackay, 2004), and *Caenorhabditis* (Gutteling et al., 2007).

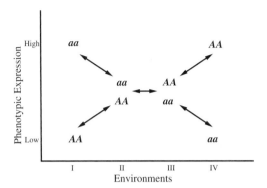

FIGURE 13.1

A one-locus model for explaining genotypic sensitivity to different environmental conditions. Adapted from Beavis and Keim (1996).

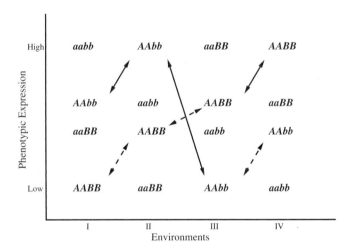

FIGURE 13.2

A two-locus model for explaining genotypic sensitivity to different environmental conditions. Adapted from Beavis and Keim (1996).

The best system for genotype by environment interaction studies is a plant or animal for which study materials can be well controlled under changing environments. For some particular systems, the same genotype can be repeated and grown or reared under different environmental conditions. This allows the test of how a given gene is expressed differently across different environments with no need of any background control.

The best system for developing comprehensive designs and methodologies for studies of genotype-environment interactions is plants and animals because these organisms can be manipulated in controlled conditions and also because genetically modified or chemical-induced mutant animals can be made available. For this reason, plant/animal models have offered unique opportunities to explore the role of genotype-environment interactions as the means of understanding the pathways to chemical-induced diseases (Andrews et al., 2002). However, there are many aspects of complexity in humans in which results from animals cannot be easily used to make inference about humans. Therefore, some informative populations in humans such as twins and multi-generation families and migrants have been developed to identify environment-related genes and their interrelationships with environmental factors. Another commonly used approach for studying genotype-environment interactions is to compare the genetic difference between two or more groups of subjects that are each from a different race (Asian, black, or white), sex (male or female), or geographic origin. Also, of two groups of subjects, one is exposed to a chemical (case) and a second is not (control).

The uniqueness of human genetic studies requires powerful genetic and statistical models that can not only estimate the genetic and environmental impacts on inter-individual variability in drug response, but also characterize the differences in the population properties of genes, such as heterozygosity, allele frequencies, and associations of alleles at different genes. The differences in these population genetic parameters will inform our understanding of how these different "environments" (sexes or races) have experienced the toxic effect of drugs and their susceptibility to environmental agents.

13.1.2   Genetic Design

To study genotype-environment interactions for a complex trait, such as drug response, in humans, we need to sample subjects from two or more natural populations, such as different sexes, races, or geographic origins, which are assumed to be at Hardy–Weinberg equilibrium. For different populations, the same set of single nucleotide polymorphisms (SNPs) are typed for sampled subjects from each population. Different from a genetic experiment with controlled materials in plants or animals, human genetic studies need to consider the differences in the population structure of genes that are due to various evolutionary forces. For simplicity, consider two SNPs, each with two alleles (designated as 1 and 0, respectively), which construct a diversity of haplotypes [11], [10], [01], and [00]. Let p and q be the 1-allele frequencies for the first and second SNP, respectively. Thus, the 0-allele frequencies at different SNPs will be $1 - p$ and $1 - q$. In a natural human population, the frequencies of four haplotypes can be expressed, in terms of allele frequencies and linkage disequilibrium (D) between the two SNPs, as

$$
\begin{aligned}
p_{11} &= pq + D \\
p_{10} &= p(1-q) - D \\
p_{01} &= (1-p)q - D \\
p_{00} &= (1-p)(1-q) + D
\end{aligned}
\tag{13.1}
$$

The parameters contained in equation (13.1) can be used to describe some important aspects of the genetic structure and diversity of a natural population. Thus, differences in genetic architecture between the two different populations can be characterized by these population-specific parameters. Because it is easy to derive the closed forms for estimating haplotype frequencies (Liu et al., 2004), we will estimate the linkage disequilibrium from the estimated haplotype frequencies.

Let $\Omega_{p|1} = (p_{11|1}, p_{10|1}, p_{01|1}, p_{00|1})$ and $\Omega_{p|2} = (p_{11|2}, p_{10|2}, p_{01|2}, p_{00|2})$ be the vectors of haplotype frequencies in two different populations 1 and 2, respectively. All the genotypes for the two SNPs are consistent with diplotypes, except for the double heterozygote, 10/10, that contains two different diplotypes [11][00] and [10][01] (Table 13.1). Because both populations are assumed to be at Hardy-Weinberg equilibrium, the frequency of a diplotype is expressed as the product of the frequencies of the two haplotypes that construct it. Table 13.1 characterizes the differences in diplotype frequencies between the two populations.

Consider a drug response trait that differs between the two populations studied. We assume a quantitative trait nucleotide (QTN) (composed of two SNPs) that are responsible for variation in the trait between and within populations. By assuming that one haplotype performs differently from the rest of the haplotypes, we can construct composite diplotypes for different populations. Without loss of generality, assume that such a distinct haplotype (i.e., risk haplotype) is l_1 ($l_1 = [11], [10], [01], [00]$) (denoted by A) for population 1, and l_2 ($l_2 = [11], [10], [01], [00]$) (denoted by B) for population 2, respectively. The non-risk haplotype is denoted as $\bar{A}$ for population 1 and $\bar{B}$ for population 2. Three composite diplotypes constituted by the risk and non-risk haplotypes are AA (**2**), $A\bar{A}$ (**1**), and $\bar{A}\bar{A}$ (**0**) for population 1 and BB (**2**), $B\bar{B}$ (**1**), and $\bar{B}\bar{B}$ (**0**) for population 2. The genotypic values of the three composite diplotypes may be arrayed in $(\mu_{2|1}, \mu_{1|1}, \mu_{0|0})$ for population 1 and $(\mu_{2|2}, \mu_{1|2}, \mu_{0|2})$ for population 2. The distribution of composite diplotypes among observed genotypes are listed in Table 13.1.

13.1.3 Likelihoods

Assume that a total of n_1 and n_2 subjects are sampled for populations 1 and 2, respectively. In each population, there are nine possible genotypes for the two SNPs that constitute a QTN, each genotype with an observed number generally expressed as $n_{r_1 r_1'/r_2 r_2'|1}$ for population 1 and $n_{r_1 r_1'/r_2 r_2'|2}$ for population 2 ($r_1 \geq r_1', r_2 \geq r_2' = 1, 0$).

The phenotypic value of a quantitative trait for subject i within populations 1 and

TABLE 13.1

Diplotypes and their frequencies for each of nine genotypes at two SNPs, and composite diplotypes for each of all possible risk haplotypes in two different populations.

	Diplotype			Risk Haplotype in Population 1				Risk Haplotype in Population 2			
Genotype	Configuration	Freq. in Pop. 1	Freq. in Pop. 2	[11]	[10]	[01]	[00]	[11]	[10]	[01]	[00]
11/11	[11][11]	$p_{11\|1}^2$	$p_{11\|2}^2$	AA	$A\bar{A}$	$A\bar{A}$	$\bar{A}\bar{A}$	BB	$B\bar{B}$	$B\bar{B}$	$\bar{B}\bar{B}$
11/10	[11][10]	$2p_{11\|1}p_{10\|1}$	$2p_{11\|2}p_{10\|2}$	$A\bar{A}$	$A\bar{A}$	$A\bar{A}$	$\bar{A}\bar{A}$	$B\bar{B}$	$B\bar{B}$	$B\bar{B}$	$\bar{B}\bar{B}$
11/00	[10][10]	$p_{10\|1}^2$	$p_{10\|2}^2$	$\bar{A}\bar{A}$	AA	$\bar{A}\bar{A}$	$\bar{A}\bar{A}$	$\bar{B}\bar{B}$	BB	$\bar{B}\bar{B}$	$\bar{B}\bar{B}$
10/11	[11][01]	$2p_{11\|1}p_{01\|1}$	$2p_{11\|2}p_{01\|2}$	$A\bar{A}$	$\bar{A}\bar{A}$	$A\bar{A}$	$\bar{A}\bar{A}$	$B\bar{B}$	$\bar{B}\bar{B}$	$B\bar{B}$	$\bar{B}\bar{B}$
10/10	{ [11][00] [10][01] }	{ $2p_{11\|1}p_{00\|1}$ $2p_{10\|1}p_{01\|1}$ }	{ $2p_{11\|2}p_{00\|2}$ $2p_{10\|2}p_{01\|2}$ }	{ $A\bar{A}$ $A\bar{A}$ }	{ $\bar{A}\bar{A}$ AA }	{ $\bar{A}\bar{A}$ AA }	{ AA $\bar{A}\bar{A}$ }	{ $B\bar{B}$ $B\bar{B}$ }	{ $\bar{B}\bar{B}$ BB }	{ $\bar{B}\bar{B}$ BB }	{ BB $\bar{B}\bar{B}$ }
10/00	[10][00]	$2p_{10\|1}p_{00\|1}$	$2p_{10\|2}p_{00\|2}$	$\bar{A}\bar{A}$	$A\bar{A}$	$\bar{A}\bar{A}$	$A\bar{A}$	$\bar{B}\bar{B}$	$B\bar{B}$	$\bar{B}\bar{B}$	$B\bar{B}$
00/11	[01][01]	$p_{01\|1}^2$	$p_{01\|2}^2$	$\bar{A}\bar{A}$	$\bar{A}\bar{A}$	AA	$A\bar{A}$	$\bar{B}\bar{B}$	$\bar{B}\bar{B}$	BB	$B\bar{B}$
00/10	[01][00]	$2p_{01\|1}p_{00\|1}$	$2p_{01\|2}p_{00\|2}$	$\bar{A}\bar{A}$	$\bar{A}\bar{A}$	$A\bar{A}$	$A\bar{A}$	$\bar{B}\bar{B}$	$\bar{B}\bar{B}$	$B\bar{B}$	$B\bar{B}$
00/00	[00][00]	$p_{00\|1}^2$	$p_{00\|2}^2$	$\bar{A}\bar{A}$	$\bar{A}\bar{A}$	$\bar{A}\bar{A}$	AA	$\bar{B}\bar{B}$	$\bar{B}\bar{B}$	$\bar{B}\bar{B}$	BB

Two alleles for each of the two SNPs are denoted as 1 and 0, respectively. Genotypes at different SNPs are separated by a slash. Diplotypes are the combination of two bracketed maternally and paternally derived haplotypes. By assuming different haplotypes as a risk haplotype (denoted as A for population 1 and B for population 2), composite diplotypes are accordingly defined and their genotypic values are given.

2 is expressed, in terms of the hypothesized QTN, as

$$y_{i|1} = \sum_{j=0}^{2} \xi_{i|1} \mu_{j|1} + e_{i|1},$$

$$(13.2)$$

$$y_{i|2} = \sum_{j=0}^{2} \xi_{i|2} \mu_{j|2} + e_{i|2},$$

where $\xi_{i|1}$ (or $\xi_{i|2}$) is the indicator variable defined as 1 if in population 1 (or 2) subject i has a composite diplotype j and 0 otherwise, and $e_{i|1}$ (or $e_{i|2}$) is the residual error, normally distributed as $N(0, \sigma_1^2)$ (or $N(0, \sigma_2^2)$). The genotypic values of composite diplotypes and variance are arrayed by a quantitative genetic parameter vector $\Omega_{q|1}$ for population 1 and $\Omega_{q|2}$ for population 2, respectively.

The log-likelihood of haplotype frequencies, genotypic values of the QTN and residual variances given sex-specific phenotypic (y_1, y_2) and SNP data (S_1, S_2) is factorized into two parts, expressed as

$$\log L(\Omega_{p|1}, \Omega_{q|1}; \Omega_{p|2}, \Omega_{q|2} | y_1, S_1, y_2, S_2)$$
$$= \log L(\Omega_{p|1}, \Omega_{p|2} | S_1, S_2) + \log L(\Omega_{q|1}, \Omega_{q|2} | y_1, S_1, \Omega_{p|1}; y_2, S_2, \Omega_{p|2}) \qquad (13.3)$$
$$= \log L(\Omega_{p|1} | S_1) + \log L(\Omega_{p|2} | S_2) + \log L(\Omega_{q|1} | y_1, S_1, \Omega_{p|1}) + \log L(\Omega_{q|2} | y_2, S_2, \Omega_{p|2})$$
$$(13.4)$$

where equation (13.4) is derived from equation (13.3) because the two populations are assumed to be independent of each other, and

$$\log L(\Omega_{p|k} | S_k) = \text{constant}$$
$$+ 2n_{11/11|k} \log p_{11|k}$$
$$+ n_{11/10|k} \log(2p_{11|k} p_{10|k})$$
$$+ 2n_{11/00|k} \log p_{10|k}$$
$$+ n_{10/11|k} \log(2p_{11|k} p_{01|k})$$
$$+ n_{10/10|k} \log(2p_{11|k} p_{00|k} + 2p_{10|k} p_{01|k})$$
$$+ n_{10/00|k} \log(2p_{10|k} p_{00|k})$$
$$+ 2n_{00/11|k} \log p_{01|k}$$
$$+ n_{00/10|k} \log(2p_{01|k} p_{00|k})$$
$$+ 2n_{00/00|k} \log p_{00|k}$$

$$\log L(\Omega_{q|k} | y_k, S_k, \hat{\Omega}_{p|k}) =$$
$$\sum_{i=1}^{n_{11/11|k}} \log f_{2|k}(y_{i|k})$$
$$+ \sum_{i=1}^{n_{11/10|k}} \log f_{1|k}(y_{i|k})$$
$$+ \sum_{i=1}^{n_{11/00|k}} \log f_{0|k}(y_{i|k})$$
$$+ \sum_{i=1}^{n_{10/11|k}} \log f_{1|k}(y_{i|k})$$
$$+ \sum_{i=1}^{n_{10/10|k}} \log[\phi_k f_{1|k}(y_{i|k}) + (1 - \phi_k) f_{0|k}(y_{i|k})]$$
$$+ \sum_{i=1}^{n_{10/00|k}} \log f_{0|k}(y_{i|k})$$
$$+ \sum_{i=1}^{n_{00/11|k}} \log f_{0|k}(y_{i|k})$$
$$+ \sum_{i=1}^{n_{00/10|k}} \log f_{0|k}(y_{i|k})$$
$$+ \sum_{i=1}^{n_{00/00|k}} \log f_{0|k}(y_{i|k})$$

$$(13.5)$$

where the risk haplotype for both populations is assumed to be [11], $f_{j|k}(y_{i|k})$ is a normal distribution density function of composite diplotype $j (= 2, 1, 0)$ for population k, and

$$\phi_k = \frac{p_{11|k} p_{00|k}}{p_{11|k} p_{00|k} + p_{10|k} p_{01|k}} \qquad (13.6)$$

is the relative proportion of diplotype [11][00] within the double heterozygote for population k.

It can be seen from equation (13.3) or (13.3) that maximizing $L(\mathbf{\Omega}_{p|1}, \mathbf{\Omega}_{q|1}; \mathbf{\Omega}_{p|2}, \mathbf{\Omega}_{q|2} | y_1, \mathbf{S}_1, y_2, \mathbf{S}_2)$ is equivalent to maximizing $\log L(\mathbf{\Omega}_{p|k} | \mathbf{S}_k)$ and $\log L(\mathbf{\Omega}_{q|k} | y_k, \mathbf{S}_k, \mathbf{\Omega}_{p|k})$ individually in equation (13.5).

13.1.4   The EM Algorithm

A closed-form solution for the EM algorithm has been derived to estimate the unknown parameters that maximize the two sex-specific likelihoods of (13.5) (Liu et al., 2004). The estimates of population-specific haplotype frequencies are based on the log-likelihood function $L(\mathbf{\Omega}_{p|k} | \mathbf{S}_k)$, whereas the estimates of population-specific genotypic values of composite diplotypes and the residual variance are based on the log-likelihood function $L(\mathbf{\Omega}_{p|k}, \mathbf{\Omega}_{q|k} | y_k, \mathbf{S}_k)$. These two different types of parameters can be estimated using a two-stage hierarchical EM algorithm (see (Liu et al., 2004) for a detailed implementation).

13.1.5   Model Selection

According to equation (13.5), the summed likelihood across the populations, $L(\mathbf{\Omega}_{p|1}, \mathbf{\Omega}_{q|1} | y_1, \mathbf{S}_1) + L(\mathbf{\Omega}_{p|2}, \mathbf{\Omega}_{q|2} | y_2, \mathbf{S}_2)$, is formulated by assuming that haplotype [11] is a risk haplotype. However, a real risk haplotype is unknown from raw data $(y_k, \mathbf{S}_k)$. An additional step for the choice of the most likely risk haplotype should be implemented. The simplest way to do so is to calculate the likelihood values by assuming that any one of the four haplotypes can be a risk haplotype (Table 13.1). Thus, we obtain four possible likelihood values as follows:

No.	Risk Haplotype	Likelihood
1	[11]	$L_1(\hat{\mathbf{\Omega}}_{p|1}, \hat{\mathbf{\Omega}}_{q_1|1} | y_1, \mathbf{S}_1) + L_1(\hat{\mathbf{\Omega}}_{p|2}, \hat{\mathbf{\Omega}}_{q_1|2} | y_2, \mathbf{S}_2)$
2	[10]	$L_2(\hat{\mathbf{\Omega}}_{p|1}, \hat{\mathbf{\Omega}}_{q_2|1} | y_1, \mathbf{S}_1) + L_2(\hat{\mathbf{\Omega}}_{p|2}, \hat{\mathbf{\Omega}}_{q_2|2} | y_2, \mathbf{S}_2)$
3	[01]	$L_3(\hat{\mathbf{\Omega}}_{p|1}, \hat{\mathbf{\Omega}}_{q_3|1} | y_1, \mathbf{S}_1) + L_3(\hat{\mathbf{\Omega}}_{p|2}, \hat{\mathbf{\Omega}}_{q_3|2} | y_2, \mathbf{S}_2)$
4	[00]	$L_4(\hat{\mathbf{\Omega}}_{p|1}, \hat{\mathbf{\Omega}}_{q_4|1} | y_1, \mathbf{S}_1) + L_4(\hat{\mathbf{\Omega}}_{p|2}, \hat{\mathbf{\Omega}}_{q_4|2} | y_2, \mathbf{S}_2)$

The largest likelihood value calculated is thought to correspond to the most likely risk haplotype. Under an optimal combination of risk haplotypes between two populations, we estimate population-specific quantitative genetic parameters $\hat{\mathbf{\Omega}}_{q_{l_1}|1}$ and $\hat{\mathbf{\Omega}}_{q_{l_2}|2}$ $(l_1, l_2 = 1, \cdots, 4)$.

13.1.6   Hypothesis Tests

The genetic architecture of a quantitative trait is characterized by population (including haplotype frequencies, allele frequencies, and linkage disequilibria) and quantitative genetic parameters (including haplotype effects and mode of inheritance for haplotypes). The model proposed provides a meaningful way for estimating the genetic architecture of a trait and further testing population-specific differences in genetic control.

After haplotype frequencies are estimated, allele frequencies and linkage disequilibrium between the two SNPs within each population k can be calculated as

Parameter	Population k	
Allele frequency for SNP 1	$p_k = p_{11|k} + p_{10|k}$	(13.7)
Allele frequency for SNP 2	$q_k = p_{11|k} + p_{01|k}$	(13.8)
Linkage disequilibrium	$D_k = p_{11|k}p_{00|k} - p_{10|k}p_{01|k}$	(13.9)

The estimated genotypic values for the three composite diplotypes can be used to estimate the overall mean, additive and dominant genetic effects of haplotypes, and the mode of inheritance for each sex by using the following formulas:

Parameter	Population k	
Overall mean	$\mu_k = (\mu_{2|k} + \mu_{0|k})/2$	(13.10)
Additive effect	$a_k = (\mu_{2|k} - \mu_{0|k})/2$	(13.11)
Dominance effect	$d_k = \mu_{1|k} - (\mu_{2|k} + \mu_{0|k})/2$	(13.12)
Inheritance mode	$\rho_k = d_k/a_k$	(13.13)

Each of the genetic parameters described in equations (13.7), (13.8), (13.9), (13.10), (13.11), (13.12), and (13.13) can be tested when appropriate hypotheses are formulated.

13.1.6.1 Overall Genetic Control

Haplotype effects on the trait, i.e., the existence of functional haplotypes, in both populations studied can be tested using the following hypotheses expressed as

$$
\begin{cases}
H_0: & \mu_{j|1} \equiv \mu_1 \text{ and } \mu_{j|2} \equiv \mu_2 \text{ for } j = 2,1,0 \\
H_1: & \text{At least one equality in } H_0 \text{ does not hold}
\end{cases}
\tag{13.14}
$$

The log-likelihood ratio test statistic (LR) under these two hypotheses can be calculated,

$$
\text{LR} = -2[\log L_0(\tilde{\mu}_1|y_1; \tilde{\mu}_2|y_2) - \log L_1(\hat{\Omega}_{q|1}|y_1, S_1, \hat{\Omega}_{p|1}; \hat{\Omega}_{q|2}|y_2, S_2, \hat{\Omega}_{p|2})],
\tag{13.15}
$$

where the L_0 and L_1 are the plug-in likelihood values under the null and alternative hypotheses of (13.14), respectively. Although the critical threshold for determining the existence of a functional haplotype can be based on empirical permutation tests, the LR may asymptotically follow a χ^2 distribution with four degrees of freedom, so that the threshold can be obtained from the χ^2 distribution table.

13.1.6.2   Population-Specific Population Genetic Architecture

The two populations may be different in terms of population genetic parameters. Such population-specific differences can be tested by formulating the following hypotheses

$$\begin{cases} H_0: & p_1 = p_2 \\ H_1: & p_1 \neq p_2 \end{cases} \tag{13.16}$$

for allele frequency at SNP 1,

$$\begin{cases} H_0: & q_1 = q_2 \\ H_1: & q_1 \neq q_2 \end{cases} \tag{13.17}$$

for allele frequency at SNP 2, and

$$\begin{cases} H_0: & D_1 = D_2 \\ H_1: & D_1 \neq D_2 \end{cases} \tag{13.18}$$

for the linkage disequilibrium between the two SNPs.

For each of the hypotheses (13.16)–(13.18), the LR values are calculated, which are each thought to asymptotically follow a χ^2-distribution with one degree of freedom. Population-specific differences in overall population genetic architecture can be tested with the null hypothesis H_0: $p_1 = p_2$, $q_1 = q_2$, and $D_1 = D_2$, with the corresponding LR value to be χ^2-distributed with three degrees of freedom.

13.1.6.3   Population-Specific Quantitative Genetic Architecture

Population-specific differences in overall quantitative genetic architecture can be tested by formulating the hypotheses

$$\begin{cases} H_0: & a_1 = a_2 \text{ and } d_1 = d_2 \\ H_1: & \text{At least one equality in } H_0 \text{ does not hold} \end{cases} \tag{13.19}$$

The LR value calculated under the null and alternative hypotheses is suggested to follow a χ^2-distribution with two degrees of freedom. The rejection of the null hypothesis implies that the effects of the same haplotype are different between the two populations. If there exists a population-specific difference, the next step is to test whether this difference is due to the additive or dominance genetic effects, or both.

13.1.6.4   Population-Specific Risk Haplotypes

It is possible that the two populations have different risk haplotypes. Let $\mu_{j_1|1}$ ($j_1 = \mathbf{2,1,0}$) and $\mu_{j_2|2}$ ($j_2 = \mathbf{2,1,0}$) be the genotypic values of composite diplotypes for populations 1 and 2 constructed by a population-specific risk haplotype. By formulating the likelihood $\log L(\boldsymbol{\Omega}_{q|k}|y_k, \mathbf{S}_k, \hat{\boldsymbol{\Omega}}_{p|k})$ of equation (13.5) based on population-specific composite diplotypes, these genotypic values can be estimated with the EM algorithm. A best combination of risk haplotypes between the two populations can be determined from the AIC values.

13.1.7 Haplotyping with Multiple SNPs

13.1.7.1 Three-SNP Model

Consider a QTN that is composed of three SNPs, S_1–S_3, each with two alleles denoted by 1 and 0. Let p, q, and r, and D_{12}, D_{13}, D_{23}, and D_{123} be the 1-allele frequencies for the three SNPs, and the linkage disequilibria between SNPs 1 and 2, SNPs 1 and 3, SNPs 2 and 3 and among the three SNPs, respectively. Eight haplotypes, [111], [110], [101], [100], [011], [010], [001], and [000], formed by these three SNPs, have population-specific frequencies arrayed in $\Omega_{p|k} = (p_{111|k}, p_{110|k}, p_{101|k},$ $p_{100|k}, p_{011|k}, p_{010|k}, p_{001|k}, p_{000|k})$ for population k. Each of these haplotype frequencies is constructed by allele frequencies at different SNPs and their linkage disequilibria of different orders, expressed as

$$p_{111|k} = p_k q_k r_k + p_k D_{23|k} + q_k D_{13|k} + r_k D_{12|k} + D_{123|k}$$

$$p_{110|k} = p_k q_k (1 - r_k) - p_k D_{23|k} - q_k D_{13|k} + (1 - r_k) D_{12|k} - D_{123|k}$$

$$p_{101|k} = p_k (1 - q_k) r_k - p_k D_{23|k} + (1 - q_k) D_{13|k} - r_k D_{12|k} - D_{123|k}$$

$$p_{100|k} = p_1 (1 - q_k)(1 - r_k) + p_k D_{23|k} - (1 - q_k) D_{13|k} - (1 - r_k) D_{12|k}$$
$$+ D_{123|k}$$

$$p_{011|k} = (1 - p_k) q_k r_k + (1 - p_k) D_{23|k} - q_k D_{13|k} - r_k D_{12|k} - D_{123|k}$$

$$p_{010|k} = (1 - p_k) q_k (1 - r_k) - (1 - p_k) D_{23|k} + q_k D_{13|k} - (1 - r_k) D_{12|k}$$
$$+ D_{123|k}$$

$$p_{001|k} = (1 - p_k)(1 - q_k) r_k - (1 - p_k) D_{23|k} - (1 - q_k) D_{13|k} + r_k D_{12|k}$$
$$+ D_{123|}$$

$$p_{000|k} = (1 - p_k)(1 - q_k)(1 - r_k) + (1 - p_k) D_{23|k} + (1 - q_k) D_{13|k} + (1 - r_k) D_{12|k}$$
$$- D_{123|k},$$

$$(13.20)$$

for population k.

Population-specific population genetic architecture can be tested by comparing the differences in allele frequencies (p_k, q_k, r_k) and linkage disequilibria of different orders ($D_{12|k}, D_{13|k}, D_{23|k}, D_{123|k}$) between the two populations studied.

In a natural population, there are 27 genotypes for the three SNPs. The frequency of each genotype is expressed in terms of haplotype frequencies. Some genotypes are consistent with diplotypes, whereas the others that are heterozygous at two or more SNPs are not. Each double heterozygote contains two different diplotypes. One triple heterozygote, i.e., 10/10/10, contains four different diplotypes, [111][000] (in a probability of $2p_{111|k}p_{000|k}$), [110][001] (in a probability of $2p_{110|k}p_{001|k}$), [101][010] (in a probability of $2p_{101|k}p_{010|k}$), and [100][011] (in a probability of $2p_{100}p_{011}$). The relative frequencies of different diplotypes for this double or triple heterozygote are a function of haplotype frequencies. The integrative EM algorithm can be employed to estimate the MLEs of haplotype frequencies. A general formula for estimating haplotype frequencies can be derived.

By assuming l_1 ($l_1 = [111], [110], [101], [100], [011], [010], [001], [000]$) as the risk haplotype (labeled by A) and all the others as the non-risk haplotype (labeled by $\bar{A}$) for population 1, the formulation of genotypic values for three composite diplotypes, μ_2 for AA, μ_1 for $A\bar{A}$, and μ_0 for $\bar{A}\bar{A}$ can be derived. For population 2, genotypic values of three composite diplotypes are μ_2 for BB, μ_1 for $B\bar{B}$, and μ_0 for $\bar{B}\bar{B}$. Similar procedures described for the two-SNP model can be obtained to estimate and test population-specific additive and dominance genetic effects when a haplotype contains three SNPs.

13.1.7.2 L-SNP Model

It is possible that the two- and three-SNP models are too simple to characterize genetic variants for quantitative variation. We can develop a model that includes an arbitrary number of SNPs whose sequences are associated with the phenotypic variation. A key issue for the multi-SNP sequencing model is how to distinguish among $2^{\ell-1}$ different diplotypes for the same genotype heterozygous at ℓ loci. The relative frequencies of these diplotypes can be expressed in terms of haplotype frequencies.

Consider a a functional haplotype that contains L SNPs among which there exist linkage disequilibria of different orders. The two alleles, 1 and 0, at each of these SNPs are symbolized by $r_1, \ldots, r_L$, respectively. Let $p_{r_1}^k, \ldots, p_{r_L}^k$ be the allele frequencies for these different SNPs within population k. A haplotype frequency, denoted as $p_{r_1 r_2 \cdots r_L}$, is decomposed into the following components:

$$p_{r_1 r_2 \ldots r_L}^k$$

$$= p_{r_1}^k p_{r_2}^k \cdots p_{r_L}^k \qquad\qquad \text{No LD}$$

$$+ (-1)^{r_{L-1}+r_L} p_{r_1}^k \cdots p_{r_{L-2}}^k D_{(L-1)L|k} + \ldots + (-1)^{r_1+r_2} p_{r_3}^k \cdots p_{r_L}^k D_{12|k}$$

$$\qquad\qquad \text{Digenic LD}$$

$$+ (-1)^{r_{L-2}+r_{L-1}+r_L} p_{r_1}^k \cdots p_{r_{L-3}}^k D_{(L-2)(L-1)L|k} + \ldots$$

$$+ (-1)^{r_1+r_2+r_3} p_{r_4}^k \cdots p_{r_L}^k D_{123|k} \qquad\qquad \text{Trigenic LD}$$

$$+ \ldots$$

$$+ (-1)^L (-1)^{r_1 + \ldots + r_L} D_{1 \ldots L|k} \qquad\qquad \text{L-genic LD}$$

where D_k's are the linkage disequilibria of different orders among particular SNPs for population k.

Population-specific difference in terms of allele frequencies and linkage disequilibria between different SNPs as well as haplotype additive and dominance effects can be tested by formulating the corresponding hypotheses.

Example 13.1

Wang et al. (2008) used the genotype-environment interaction model to detect differences in the genetic architecture of pain sensitivity between men and women. Differences in males and females (sexual dimorphism) is ubiquitous in many biological aspects (Mackay, 2001; Anholt and Mackay, 2004). In humans, sexually dimorphic traits include those from morphological shapes to brain development to disease susceptibility (Weiss et al., 2005, 2006). Substantial differences are also observed in sensitivity to pain, sensitivity to pain-killing drugs, and susceptibility to developing chronic pain between men and women (Fillingim, 2003a,b; Craft et al., 2004). All these sex-specific discrepancies are due to varying expression of genes on the X chromosome and autosomes, thought to result from differences in cellular and hormonal environments between the two sexes (Rinn and Snyder, 2005). A growing body of research has been conducted to elucidate the genetic control of sexual dimorphism in various complex phenotypes by gene mapping approaches (Mogil et al., 1997; Zhao et al., 2004a; Weiss et al., 2005, 2006). Despite these efforts, however, little is known about the underlying genetic architecture of sex-specific differences.

Genetic and phenotypic data were from a pain genetics project in which 237 subjects (including 143 men and 94 women) from five different races were sampled for six SNPs at three candidate genes. As a demonstration of the utilization of the model, we will focus on two SNPs, OPRDT80G (with two alleles T and G) and OPRDT921C (with two alleles T and C), at the delta opioid receptor. Pain testing procedures followed Fillingim et al. (2005). The phenotypic values of traits were adjusted to remove the effect due to races. Results from this example are shown in Table 13.2 and Figs. 13.3 and 13.3.

These two SNPs construct four haplotypes, [TC], [TT], [GC], and [GT], which yield nine genotypes, TT/CC, TT/CT, TT/TT, TG/CC, *TG*/CT, TG/TT, GG/CC, GG/CT, and GG/TT, with the 10 corresponding diplotypes, [TC][TC], [TC][TT], [TT][TT], [TC][GC], [TC][GT] or [TT][GC], [TT][GT], [GC][GC], [GC][GT], and [GT][GT]. Based on the observed numbers of each genotype in the male and female populations, we estimated sex-specific haplotype frequencies (Table 13.2). The pattern of haplotype distribution is consistent between the two sexes, with haplotypes [TC] and [TT] jointly occupying a majority proportion in the populations. Haplotype [GT] is very rare, with the frequency close to zero. SNP OPRDT80G has a low heterozygosity because the frequency of its commoner allele (T) is closer to 0.90, whereas there is a high heterozygosity for SNP OPRDT921C in terms of its averaged allele frequencies. The two SNPs are highly significantly associated ($p < 0.001$), with a normalized linkage disequilibrium of $D' = 1.00$, because alleles T from OPRDT80G and T from OPRDT921C as well as alleles G from OPRDT80G and C from OPRDT921C tend to form the same haplotypes more frequently than at random. There is no sex-specific difference in allele frequencies at the two SNPs and their linkage disequilibrium.

By assuming that one of the haplotypes is a risk haplotype, we estimated the effects of each haplotype on a pain trait in the pooled male and female population. The likelihoods of haplotype [TC], [TT], and [GC] as a risk haplotype are -594.8, -594.5, and -595.0, and thus the most likely risk haplotype is [TT]. The genotypic values of composite diplotypes constructed by this risk haplotype and its non-risk haplotype counterpart were estimated and compared between different sexes. In both males and female, the three composite diplotypes do not display significant genetic

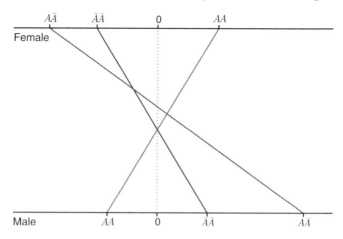

FIGURE 13.3
Different genotypic values of a pain sensitivity trait for composite diplotypes constructed by risk haplotype [TT] (denoted as A) and non-risk haplotype (denoted as
$\bar{A}$) in males and females.

differences in the pain trait studied, but the directions of the additive and dominance
effects are different between the two sexes (Table 13.2). In males, the non-risk haplotype tends to increase pain sensitivity, and it is overdominant to the risk haplotype,
leading to increased pain sensitivity at a marginal significance level ($p = 0.058$) (Fig.
13.3).

By contrast, in females, the non-risk haplotype tends to reduce pain sensitivity,
and it also tends to be overdominant to the risk haplotype by reducing pain sensitivity. These discrepancies in both effect size and direction (Fig. 13.4) make the
overall quantitative genetic architecture of the pain trait significantly different between the two sexes ($p = 1.49 \times 10^{-7}$) (Table 13.2). Although the additive genetic
effect displays a gene by sex interaction at the $p = 0.03$ significance level, a gene by
sex interaction for the dominance effect is highly significant at $p = 5.16 \times 10^{-7}$. No
significant difference was observed in inheritance mode between males and females.

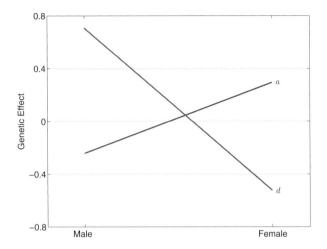

FIGURE 13.4
Different additive (*a*) and dominant genetic effects (*d*) of haplotypes on a pain sensitivity trait in males and females.

13.2 Haplotyping Genotype-Environment Interactions for Pharmacological Processes

13.2.1 Introduction

When drug response rises as a curve, the estimation and test of genotype-environment interactions that modulate the curve will be complicated. Statistical models that can handle genotype-environment interactions for a dynamic process have been proposed (Zhao et al., 2004a,b; Hou et al., 2006). Through the implementation of functional mapping, these models can examine the impact of genotype-environment interactions on the developmental change of a trait in time course and reveal the underlying biological control mechanisms for such a developmental change. By comparing differences in a set of curve parameters among the same genotype grown in different environments, functional mapping can quantify genotype-environment interactions of a biological process. These dynamic interaction models have been used to analyze human growth data (Hou et al., 2006), plant height growth (Zhao et al., 2004b), and animal body mass growth (Zhao et al., 2004b).

The modeling of genotype-environment interactions for a dynamic process help to construct an evolutionary and ecological developmental framework at the interplay among different biological disciplines (Rice, 1997; Klingenberg and Zimmermann, 1992). More specifically, it can test when a environment-dependent gene starts to trigger its effect on a biological process, how long this effect takes during the time

TABLE 13.2

The estimates and tests of population genetic structure for two SNPs, OPRDT80G (with alleles T and G) and OPRDT921C (with two alleles C and T), and quantitative genetic effects on a pain sensitivity trait of haplotypes constructed by these two SNPs in males and females.

Genetic Parameter	Male		Female		Sex-specific	Sex-specific
	MLE	p-value	MLE	p-value	LR	p-value
Haplotype Frequency						
$\hat{p}_{TC}$	0.364		0.396			
$\hat{p}_{TT}$	0.505		0.463			
$\hat{p}_{GC}$	0.131		0.142			
$\hat{p}_{GT}$	0.000		0.000			
Allele Frequency and Linkage Disequilibrium						
$\hat{p}_T$ (OPRDT80G)	0.869		0.858		0.079	0.778
$\hat{q}_C$ (OPRDT921C)	0.495		0.537		0.578	0.447
$\hat{D}$	-0.066	3.41×10^{-5}	-0.066	9.63×10^{-5}	0.001	0.973
Additive and Dominant Effects and Inheritance Mode for Risk Haplotype [TT]						
$\hat{a}$ and $\hat{d}$		0.129		0.548	31.44	1.49×10^{-7}
$\hat{a}$	-0.2	0.431	0.3	0.481	4.72	0.030
$\hat{d}$	0.7	0.058	-0.5	0.350	24.86	6.16×10^{-7}
$\hat{\rho}$	2.9		1.8		0.63	0.427

course of trait progression and in which gene action mode it can alter developmental trajectories. These questions, once incorporated into a pharmacogenetic and pharmacogenomic study, will help to better elucidate the genetic architecture of pharmacological reactions.

13.2.2 Dynamic Model

13.2.2.1 Experimental Setting

Suppose we use the same genetic design as described in Section 13.1. Population genetic parameters can be estimated, separately for different populations, from SNPs based on a multinomial likelihood. A mixture-based likelihood is used to model multiple measures of a trait across a range of doses or during a course time for the detection of haplotype effects. Let $\mathbf{y}_{i|k} = (y_{i|k}(t_{i1|k}), \cdots, y_{i|k}(t_{iT_{i|k}|k}))$ be the vector of $T_{i|k}$ measurements on subject i in population k and $\mathbf{t}_{i|k} = (t_{i1|k}, \cdots, t_{iT_{i|k}|k})$ be the corresponding vector of measurement times.

At a particular time $t_{i\tau|k}$ for subject i from population k, the relationship between the observation and genotypic value of a composite diplotype j_k can be described by a linear regression model,

$$y_{i|k}(t_{i\tau|k}) = \sum_{j_k=0}^{2} \xi_{i|k} u_{j|k}(t_{i\tau|k}) + e_{i|k}(t_{i\tau|k}),$$

where $\xi_{i|k}$ is the indicator variable denoted as 1 if composite diplotype j_k is considered for subject i and 0 otherwise, and $e_{i|k}(t_{i\tau|k})$ is the residual error that is iid normal with mean zero and variance $\sigma_{i|k}^2(t_{i\tau|k})$. The errors for subject i from population k at two different time points, $t_{i\tau_1|k}$ and $t_{i\tau_2|k}$, are correlated with covariance $\sigma_k(t_{i\tau_1|k}, t_{i\tau_2|k})$. These covariances comprise the $(T_{i|k} \times T_{i|k})$ matrix $\boldsymbol{\Sigma}_{i|k}$.

13.2.2.2 Mean-Covariance Modeling

Assume that the trait process considered is time-dependent, described by $g(t) = h(t : \Omega)$, where Ω is a set of parameters that are used to fit the curve. For a specific composite diplotype j_k ($j_k = 2, 1, 0$) in population k, the mean vector is then modeled by

$$\mathbf{u}_{j_k|k} = [u_{j_k|k}(t_{i1|k}), ..., u_{j_k|k}(t_{i\tau|k}), ..., u_{j_k|k}(t_{iT_{i|k}|k})],$$
$$= [h_k(t_{i1|k} : \boldsymbol{\Omega}_{j_k|k}), ..., h_k(t_{i\tau|k} : \boldsymbol{\Omega}_{j_k|k}), h_k(t_{iT_{i|k}|k} : \boldsymbol{\Omega}_{j_k|k})], \quad (13.21)$$

where $\boldsymbol{\Omega}_{j_k}$ is a unknown vector that contains curve parameters of interest.

The covariance matrix among longitudinal measures can be structured with an appropriate model. Statistical analysis of longitudinal data has established a number of structural models that capture most of the information contained in the matrix. Núñez-Antón and Woodworth (1994) proposed a power transformation model for fitting the correlations as a function of time. By estimating the power parameter, this nonstationary model is regarded as more general compared to the stationary AR(1) model.

Núñez-Antón and Woodworth (1994) and Núñez-Antón and Zimmerman (2000) proposed a general form of the covariance matrix (Σ_i) for subject i with $T_{i|k}$ observations at times $0 < t_{i1} < t_{i2} < \cdots < t_{iT_i}$, expressed as

$$\sigma(t_{i\tau_1}, t_{i\tau_2}) = \begin{cases} \sigma^2 \rho^{(t_{i\tau_2}^\lambda - t_{i\tau_1}^\lambda)/\lambda}, & \lambda \neq 0 \\ \sigma^2 \rho^{\ln(t_{i\tau_2}/t_{i\tau_1})}, & \lambda = 0 \end{cases} \quad 0 \leq \tau_1 \leq \tau_2 \leq T_i, 0 < \rho < 1, \quad (13.22)$$

where σ^2, ρ, and λ are the variance, correlation between a unit time interval, and power, respectively. These parameters are population-specific. For population k, they are expressed as $\Omega_{v|k} = (\sigma_k^2, \rho_k, \lambda_k)$ for the modeling of matrix $\Sigma_{i|k}$.

Based on the Barrett (1979) theorem that leads to the well-known tridiagonal or Jacobi matrix, Núñez-Antón and Woodworth (1994) derived the inverse of the covariance matrix expressed in equation (13.22). They also derived a closed form of the determinant by applying the Cholesky factorization to the covariance matrix. We can incorporate the closed forms of the inverse and determinant of the matrix into a mixture-based gene detecting model.

The EM algorithm can be derived to estimate curve parameters $\Omega_{j_k|k}$ and covariance function parameters $\Omega_{v|k}$. In general, to facilitate computation, the EM algorithm is integrated with the simplex algorithm or Newton-Raphson approach.

13.2.3   Hypothesis Testing

Functional mapping for longitudinal traits allows for the tests of a number of biologically meaningful hypotheses about genotype-environment interactions. These hypotheses tests can be made at the global level for testing the existence of a significant gene, the local level for testing the genetic effect on growth at a particular time point, the regional level for testing the overall effect of a gene on a particular period of biological process, or for a differential development test about the change of genetic expression across time. These tests at different levels can be formulated to test the effects of genotype-environment interaction on the shape of growth.

13.2.3.1   Global Test

Testing whether a specific QTN exists to affect the rate and shape of curves is a first step toward the understanding of the genetic architecture of growth and development. The genetic control over entire processes can be tested by formulating the following hypotheses:

$$\begin{cases} H_0 : \Omega_{j_k|k} \equiv \Omega_k, \text{irrespective of } j_k \ (k = 1, ..., K) \\ H_1 : \text{Not all these equalities above hold} \end{cases} \quad (13.23)$$

The H_0 states that there are no QTN affecting curves and different diplotypic curves in each population overlap (the reduced model), whereas the H_1 proposes that such a QTN does exist (the full model). The test statistic for testing the hypotheses is calculated as the log-likelihood ratio of the reduced to the full model. An empirical approach for determining the critical threshold is based on permutations tests (Churchill and Doerge, 1994).

13.2.3.2 Local Test

The local test can examine the significance of the main effects of a gene and its interaction with exposure on longitudinal traits measured at a time point (t^*) of interest. For example, the hypothesis for testing the effect of gene on growth at a given time t^* can be formulated as

$$\begin{cases} H_0 : u_{j_k|k}(t^*) = u_k(t^*), \ j_k = 2,1,0; k = 1,...,K \\ H_1 : \text{Not all the equalities hold} \end{cases} \tag{13.24}$$

which is equivalent to testing the difference of the full model with no restriction and the reduced model with a restriction as set in the null hypothesis.

13.2.3.3 Regional Test

We may be interested in testing the difference of trait processes in a time interval rather than simply at a time point. The question of how a gene exerts its effects on a period of time course $[t_1, t_2]$ can be tested using a regional test approach based on the areas,

$$A_{j_k|k} = \int_{t_1}^{t_2} \mathscr{G}(t : \mathbf{\Omega}_{j_k|k}) dt.$$

covered by longitudinal curves. The hypothesis test for the genetic effect on a period of trait process is equivalent to testing the difference between the full model with no restriction and the reduced model with a restriction.

13.2.3.4 Differential Development Test

The effects of gene may change with time, which suggests the occurrence of genotype $\times$ time interaction effects on longitudinal curves. The differentiation of the trait process with respect to time t represents the rate of trait change. If the change rates at a particular time point t^* are different between the curves of different genotypes, this means that significant genotype $\times$ time interaction occurs between this time point and the next.

13.2.3.5 Genotype-Environment Interaction Test

How detected gene(s) affect longitudinal curves differently between two populations can be tested. Such genotype-environment interactions can be caused by either environment-specific genes (i.e., those that trigger an effect only in a particular environment) or environment-sensitive genes (i.e., those that have effects in different environments but with varying degrees). The test of environment-specific genes is based on the null hypothesis

$$\Omega_{j_{k_1}|k_1} \equiv \Omega_{k_1}, \tag{13.25}$$

for population k_1, and

$$\Omega_{j_{k_2}|k_2} \equiv \Omega_{k_2}, \tag{13.26}$$

for population k_2. The rejection of only one of the two null hypotheses (13.25) and (13.26) detected the effect of environment-specific genes. If both the null hypotheses are rejected, this means that the detected gene(s) simultaneously affect developmental curves in two different environments.

After we detect a significant risk haplotype for longitudinal curves, we can test whether this risk haplotype contributes to genotype-environment interactions. Let **2**, **1**, and **0** be the composite diplotypes formed from this risk haplotype. This test can be done by formulating the null hypothesis

$$\Omega_{2|k_1} - \Omega_{2|k_2} \equiv \Omega_{1|k_1} - \Omega_{1|k_2} \equiv \Omega_{0|k_1} - \Omega_{0|k_2} \tag{13.27}$$

which states that the difference in genotypic longitudinal curves between two different groups is identical across different genotypes. The rejection of this hypothesis suggests that, although the same gene(s) affect developmental trajectories in different environments, there is a significant genotype-environment interaction to shape longitudinal curves.

Example 13.2
Mapping QTL-Environment Interactions for Growth Curves. Zhao et al. (2004a) used dynamic approaches to map genotype-environment interactions for quantitative trait loci (QTLs) that control growth trajectories in the mouse. Different from QTN haplotyping, QTL mapping makes use of molecular markers to infer the positions and effects of the underlying QTLs. Vaughn et al. (1999) constructed a $\sim$1780 cM long linkage map (in Haldane's units) with an average interval length of $\sim$23 cM using 96 microsatellite markers for 259 male and 243 female mice in an F_2 population derived from two strains, the Large (LG/J) and Small (SM/J). The analysis by Zhao et al. (2004a) was based on the data of Vaughn et al. (1999) to detect genotype-sex interactions for body mass growth curves.

The F_2 progeny was measured for their body mass at 10 weekly intervals starting at age 7 days. The raw weights were corrected for the effects of each covariate due to dam, litter size at birth, and parity. To study the genetic architecture of sexual differences in body mass growth, the raw weights were not corrected for the effect due to sex. Figure 13.5 illustrates growth curves of body weights separately for the male (**A**) and female F_2 mice (**B**). On average, males display different growth trajectories than females, with the former being heavier at all time points than the latter. Substantial variation in growth curve among different animals in each sex suggests that specific QTL may be involved in shaping developmental trajectories.

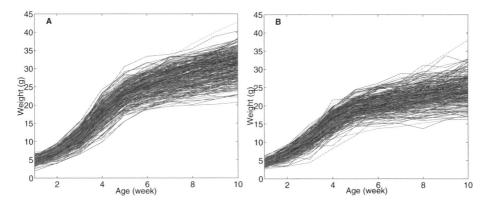

FIGURE 13.5

Plots of body mass against ages for 259 male (**A**) and 243 female (**B**) mice in an F_2 progeny.

Suppose there is a segregating QTL with alleles Q and q that affects growth curves of both the male and female mice in this F_2 population, but to different extents between the two sexes. For this QTL, there are three genotypes QQ (2), Qq (1), and qq (0) whose probabilities can be inferred from the observed genotypes of two flanking markers that bracket the QTL. The log-likelihood function of unknown parameters Θ_k given growth ($\mathbf{y}_k$) and marker data ($\mathbf{M}_k$) for sex k of size n_k can be constructed as

$$\log L_k(\Theta_k|\mathbf{y}_k,\mathbf{M}) = \sum_{i=1}^{n_k} \log \sum_{j=0}^{2} \left[\omega_{j|i} f_j(\mathbf{y}_{i|k})\right],$$

where $\omega_{j|i}$ is the conditional probability of QTL genotype j given the marker genotype of progeny i, which is the same for the two sexes. By letting the derivatives of differentiation,

$$\frac{\partial}{\partial \Theta_k} \log L_k(\Theta_k|\mathbf{y}_k,\mathbf{M}) = \sum_{i=1}^{n_k} \frac{\omega_{j|i} \frac{\partial}{\partial \Omega_k} f_j(\mathbf{y}_{i|k})}{\sum_{j'=0}^{2} \left[\omega_{j'|i} f_{j'}(\mathbf{y}_{i|k})\right]}$$

$$= \sum_{i=1}^{n_k} \frac{\omega_{j|i} f_j(\mathbf{y}_{i|k})}{\sum_{j'=0}^{2} \left[\omega_{j'|i} f_{j'}(\mathbf{y}_{i|k})\right]} \frac{\partial \log f_j(\mathbf{y}_{i|k})}{\partial \Omega_k}$$

$$= \sum_{i=1}^{n_k} \sum_{j=0}^{2} \Psi_{j|i} \frac{\partial \log f_j(\mathbf{y}_{i|k})}{\partial \Omega_k} = 0,$$

we derive a series of log-likelihood equations of obtaining the maximum likelihood estimates (MLEs) of Θ_k.

Zhao et al. (2004a) used a logistic curve, $g(t) = a/(1 + be^{-rt})$, to describe growth trajectories in the mouse. Thus, Θ_k contains three sets of logistic parameters (a_j, b_j, r_j) $(j = 2, 1, 0)$, as well as the AR(1) parameters that model the covariance structure.

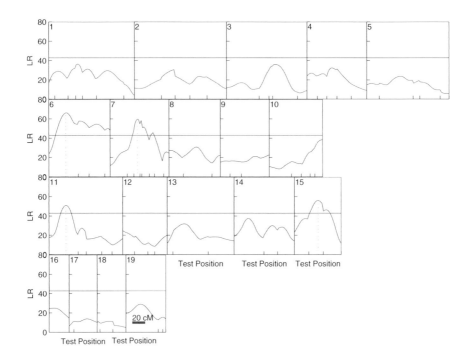

FIGURE 13.6

The profile of the LR values between the full and reduced (no QTL) model for body mass growth trajectories across the entire genome using the linkage map constructed from microsatellite markers. The genomic positions corresponding to the peak of the curve, as indicated by vertical dot lines, are the MLEs of the QTL positions. The genome-wide threshold value for claiming the existence of QTL is given as the horizonal line. Tick marks on the *x*-axis represent the positions of markers on the linkage group.

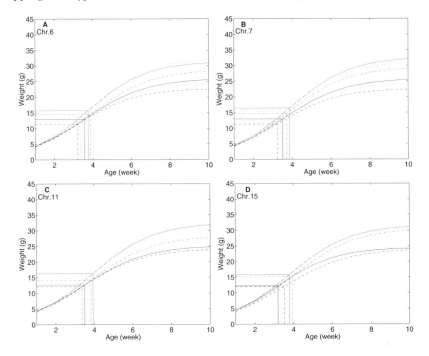

FIGURE 13.7

Three growth curves each presenting three groups of genotypes, QQ (solid curves), Qq (dot curves), and qq (broken curves), in the male (red) and female (blue) mice at the QTLs on chromosomes 6, 7, 11, and 15. The times at the inflection point ($t_{I_{j|k}}$) are indicated by the vertical lines each corresponding to a QTL genotype in each sex.

In order to use the AR(1) model for the covariance structure, Zhao et al. (2004a) transformed the growth data and then used TBS-based functional mapping of Wu et al. (2004b) to map dynamic QTLs. They scanned the entire linkage map at every 2 cM to test the existence of any QTL by calculating the log-likelihood ratio (LR) and comparing it against the critical value from permutation tests. Four peaks of the LR profile beyond the critical threshold were found on chromosomes 6, 7, 11, and 15 (Fig. 13.6). This thus suggests the existence of significant QTLs for growth curves at the corresponding LR peaks on these chromosomes.

The three growth curves each determined by a genotype at each of these significant QTL are drawn separately for males and females (Fig. 13.6) using the MLEs of curve parameters $(\hat{a}_j, \hat{b}_j, \hat{r}_j)$. The growth trajectories of the same QTL genotype are different between the two sexes, suggesting that the genetic expression of QTL is affected by sex-related background. In general, these four QTL start to exert their effects on growth when the mice age 3 or 4 weeks. These ages are just the timing at which maximal growth rate occurs (inflection point, calculated by $t_{I_{j|k}} = \ln b_{j|k}/r_{j|k}$; Fig. 13.6). After the inflection point, the QTL effects tend to increase with age.

We further tested how the QTL interact with sex to affect growth trajectories. We calculate the LR values for QTL × sex interaction effects for all the four QTL. Significant interactions were detected for QTL on chromosomes 6 and 11. The QTL on chromosome 6 is significant for both males and females (Fig. 13.7**A**), but the modes of gene action are different between the two sexes. In males, this QTL appears to be overdominant because the heterozygote (Qq) outgrows the better homozygote (QQ). But in females, this QTL operates in a partial dominant fashion since the heterozygote is between the two homozygotes. Given these analyses, the QTL on chromosome 6 triggers significant interaction effects with sex through a so-called "allelic sensitivity" mechanism in which phenotypic changes result from differential expression of the same QTL. The second interacting QTL on chromosome 11 has a different mechanism. It exerts an effect on growth only in one sex (Fig. 13.7**C**). Although this QTL affects growth trajectories in a dominant fashion in males, it displays nonsignificant effect in females. Thus, this QTL may use a "regulatory mechanism" to affect differentiation in growth curves in that phenotypic changes rely upon the formation of novel genes.

Males reach the inflection point 3–5 days earlier than females. As demonstrated in Fig. 13.7, the QTL with significant effects on overall growth curves also affect the timing of the occurrence of maximal growth rate. Moreover, these QTL interact significantly with sexes to affect the timing of the inflection point. With different growth curves, each corresponding to a QTL genotype, we can investigate possible pleiotropic effects of each of these growth QTL on many other developmental events, such as the timing of sexual maturity and reproductive fitness, or biomedically important traits, such as metabolic rate and fatness. We can therefore integrate growth and development, which are historically regarded as two different biological problems, into a comprehensive framework under which their common or unique underlying genetic machineries are identified.   ☐

13.3   Genetic Considerations

Three hypotheses have been proposed to explain the genetic basis of phenotypic plasticity:

(1) The overdominance hypothesis, in which the ability of an individual to respond to a changing environment is associated with its heterozygosity. The more heterozygous, the more stable, leading to a theory called homeostasis (Gillespie and Turelli, 1989);

(2) The pleiotropic hypothesis, in which differential expression of a gene across environments causes phenotypic plasticity. This hypothesis is illustrated in Fig. 13.3 for one gene and Fig. 13.4 for two genes. That is, the expression of a gene varies when the environment changes. The extreme condition is that

a gene changes its direction of expression from one environment to the next. The production of phenotypic plasticity is through allelic sensitivity (Via and Lande, 1985);

(3) The epistatic hypothesis that suggests the existence of specific plasticity genes that interact epistatically with the genes for the mean value of a trait. Thus, there are two sets of genes, performance genes and plasticity genes. Phenotypic plasticity is produced through gene regulation (Scheiner and Lyman, 1989).

The models we described above can well be used to test the allelic sensitivity hypothesis. It is possible that the models can be modified to test the homeostasis and gene regulation hypotheses. To test the gene regulation hypothesis, we need to assume different but epistatically interacting QTLNs or QTLs for plastic responses and means of a trait across different environments. Characterizing how the genetic architecture of a complex trait differs across environments is of paramount importance toward elucidating the mechanistic basis of genotype by environment interactions and ultimately predicting final phenotypes in a given environment.

We are now in a good position to identify genes underlying quantitative variation in plastic response of various complex traits (including dynamic traits) to changes in environments and to determine how these genes act singly or interact epistatically to affect different phenotypes across ecologically relevant environments. Statistical models will provide a powerful tool to gain insights into how genomes and environments interact to determine phenotypes and further alter the microevolutionary process of the phenotypes.

14

Nonparametric Functional Mapping of Drug Response

Original models for functional mapping to study dynamic changes of genetic expression over time or other independent variables were established on the belief that biological processes can be described by mathematical functions (Ma et al., 2002; Wu et al., 2004b). One of the most significant examples for this is the use of S-shaped logistic curves to model growth trajectories. West et al. (2001) indicated from fundamental principles of biophysical processes that logistic forms of growth are biologically crucial for the maintenance of optimal metabolic level and, thereby, the best use of available resources for an organism from birth to adulthood. The advantage of using a parametric function to map dynamic genetic control lies in the estimation and test of biologically meaningful parameters that define curve rates and shapes and the computational prediction of biological processes and events not observed in the study. Such parametric functional mapping provides a quantitative test framework within which a number of biological hypotheses can be asked and tested at the interplay between genetic actions and development.

However, parametric functional mapping is often limited in some practical uses. First, computation is expensive especially when the underlying function is complicated. This mainly arises from the fact that the derivation of a closed form for the estimation of curve parameters is not possible. Second, more importantly, there are many situations in which no appropriate mathematical function can be used to describe a biological process. To model an arbitrary shape of curves in a flexible way, a different statistical model based on nonparametric theory should be formulated. Legendre polynomial analyses that can be specified by varying orders were incorporated into functional mapping to map dynamic quantitative trait loci (QTLs) independently by Lin and Wu (2006b) and Yang et al. (2006). A nonparametric approach based on B-spline fitting was formulated for QTL mapping in a recent study by Yang et al. (2008). These approaches were detected to provide consistent results with those from parametric functional mapping, but offer a new advantage when functional data can be simply modeled by a mathematical function.

In this chapter, we will introduce the basic principle for nonparametric functional mapping of longitudinal growth traits. We will describe different treatments of nonparametric modeling for genetic mapping and haplotyping. Several examples will be shown for the utilization of nonparametric approaches to detecting dynamic genes for developmental patterns of a trait. We will provide discussion on the advantages of nonparametric functional mapping in pharmacogenetic studies of drug response and pinpoint the areas in which the model can be further refined.

14.1 Nonparametric Modeling with Legendre Polynomial

14.1.1 Legendre Orthogonal Polynomials

Kirkpatrick and Heckman (1989) were among the first who modeled the covariance function to study evolutionary trajectories by introducing Legendre orthogonal polynomials (LOP). K. Meyer and L. R. Schaeffer provided a series of pioneering work on the incorporation of LOP to into a random regression model (RRM) for the longitudinal pattern of covariances for genetic evaluation of milk production based on test day records in dairy (Schaeffer and Dekkers, 1994; Meyer, 1998, 2001). Now, RRM has become one of the most popular approaches for animal genetic and breeding studies (Van Der Werf et al., 1998; Norberg et al., 2006; Togashi and Lin, 2007). By fitting curves with arbitrary shapes by varying orders, the LOP has been shown to have great power for functional mapping of QTLs (Yang et al., 2006; Wu and Lin, 2006). There are several favorable properties for Legendre polynomials to be utilized in curve fitting, i.e., (1) the functions are orthogonal, (2) it is flexible to fit sparse data, (3) higher orders are estimable for high levels of curve complexity, and (4) computation is fast because of good convergence. As compared to the B-spline approach that constructs curves from pieces of lower degree polynomials smoothed at selected pointed (knots), Legendre polynomials are simpler in which only fewer regression coefficients are needed to model the curve.

The Legendre polynomials are solutions to a very important differential equation, the Legendre equation,

$$(1-x^2)\frac{d^2z}{dx^2} - 2x\frac{dz}{dx} + r(r+1)z = 0.$$

The polynomials may be denoted by $P_r(x)$, called the Legendre polynomial of order r. The polynomials are either even or odd functions of x for even or odd orders r.

The general form of a Legendre polynomial of order k is given by the sum,

$$P_r(x) = \sum_{k=0}^{K}(-1)^k \frac{(2r-2k)!}{2^r k!(r-k)!(r-2k)!}x^{r-2k}, \qquad (14.1)$$

where $K = r/2$ or $(r-1)/2$ whichever is an integer. This polynomial is defined over the interval [-1, 1]. From equation (14.1), we show the first few polynomials as

$$P_0(x) = 1$$
$$P_1(x) = x$$
$$P_2(x) = \frac{1}{2}(3x^2 - 1)$$

$$P_3(x) = \frac{1}{2}(5x^3 - 3x)$$

$$P_4(x) = \frac{1}{8}(35x^4 - 30x^2 + 3)$$

$$P_5(x) = \frac{1}{8}(63x^5 - 70x^3 + 15x)$$

$$P_6(x) = \frac{1}{16}(231x^6 - 315x^4 + 105x^2 - 5).$$

In this modeling, independent variable x is expressed as time t, which is adjusted, to rescale the measurement times to the range of the orthogonal function [-1, 1], by

$$t^* = -1 + \frac{2(t - t_{min})}{t_{max} - t_{min}},$$

where t_{min} and t_{max} are respectively the first and last time points. Assume that there are 11 equally-spaced time points. The coefficients of the Legendre polynomials for the first seven orders are calculated with the adjusted times (Table 14.1). By choosing different orders of orthogonal polynomials, the Legendre function has potential to approximate the functional relationships between trait values and times to any specified degree of precision.

14.1.2 Genetic Design

Consider a random sample of n subjects from a natural population at Hardy–Weinberg equilibrium. To understand the non-random pattern of human genetic variation and its link to drug response, these subjects are typed for single nucleotide polymorphisms (SNPs) genome-wide or from candidate chromosomal regions that may harbor important sequence variants. Although the model has been available to haplotype a complex trait with any number of SNPs at the same time, we focus our description on two associated SNPs at which there is an association between haplotype diversity and trait phenotype. Let 1 and 0 be two alleles at each SNP, thus forming four different haloptypes denoted as [11], [10], [01], and [00]. The haplotypes derived from the maternal and paternal parents are combined to yield 10 diplotypes and nine observed genotypes, symbolized as 11/11, 11/10, 11/00, 10/11, 10/10, 10/00, 00/11, 00/10, and 00/00, where the numbers over and under the slashes denote genotypes at the first and second SNP, respectively. The observations of two-SNP genotypes are denoted as n subscripted by the genotype symbols.

A drug response trait is repeatedly measured for each subject at multiple points during a time course or across a range of dose levels. Unlike the situation considered in the preceding chapters, the measured data here cannot be fit explicitly by a mathematical function, although dynamic modeling of genetic control patterns of this trait is still useful in clinical pharmacological trials.

In this and following sections, we will provide nonparametric models for functional mapping of genes or DNA sequences that are responsible for any aspect of drug response. We assume that a risk haplotype (A) is associated with drug response,

TABLE 14.1
Coefficients of the first seven Legendre polynomials for adjusted time points (t^*)

t	0	1	2	3	4	5	6	7	8	9	10
t^*	-1	$-4/5$	$-3/5$	$-2/5$	$-1/5$	0	1/5	2/5	3/5	4/5	1
$P_0(t^*)$	1	1	1	1	1	1	1	1	1	1	1
$P_1(t^*)$	-1	$-4/5$	$-3/5$	$-2/5$	$-1/5$	0	1/5	2/5	3/5	4/5	1
$P_2(t^*)$	1	23/50	1/25	$-13/50$	$-11/25$	$-1/2$	$-11/25$	$-13/50$	1/25	23/50	1
$P_3(t^*)$	-1	$-2/25$	9/25	11/25	7/25	0	$-7/25$	$-11/25$	$-9/25$	2/25	1
$P_4(t^*)$	1	$-24/103$	$-51/125$	$-93/823$	29/125	3/8	29/125	$-93/823$	$-51/125$	$-24/103$	1
$P_5(t^*)$	-1	167/418	107/701	$-59/218$	$-274/891$	0	274/891	59/218	$-107/701$	$-167/418$	1
$P_6(t^*)$	1	$-172/439$	132/767	163/557	$-56/695$	$-5/16$	$-56/695$	163/557	132/767	$-172/439$	1
$P_7(t^*)$	-1	110/459	$-10/31$	13/891	231/787	0	$-231/787$	$-13/891$	10/31	$-110/459$	1

which, with its non-risk haplotype counterpart ($\bar{A}$), forms three composite diplotypes, AA (**2**), $A\bar{A}$ (**1**), and $\bar{A}\bar{A}$ (**0**). We will need to discern the differences among these three composite diplotypes.

14.1.3 Likelihoods

Let $\mathbf{S}$ denote the SNP data and $\mathbf{y} = (y(1), \ldots, y(T))$ denote the observation vector from time 1 to T for each subject. Let Ω_p be the unknown population genetic parameters (e.g., haplotype frequencies, p_{11}, p_{10}, p_{01}, and p_{00}) that describe the segregation, distribution, and diversity of the SNPs studied, and Ω_q be the unknown quantitative genetic parameters that specify the genetic effects of haplotypes on drug response and the covariance structure. A joint likelihood for the longitudinal data ($\mathbf{y}$) and marker data ($\mathbf{S}$) is then formulated as

$$\log L(\Omega_p, \Omega_q | \mathbf{y}, \mathbf{S}) = \log L(\Omega_p | \mathbf{S}) + \log L(\Omega_q | \mathbf{y}, \mathbf{S}, \Omega_p), \qquad (14.2)$$

where

$$
\begin{aligned}
\log L(\Omega_p | \mathbf{S}) &= \text{constant} & \log L(\Omega_q | \mathbf{y}, \mathbf{S}, \Omega_p) = \\
&+ 2n_{11/11} \log p_{11} & \sum_{i=1}^{n_{11/11}} \log f_2(\mathbf{y}_i) \\
&+ n_{11/10} \log(2p_{11}p_{10}) & + \sum_{i=1}^{n_{11/10}} \log f_1(\mathbf{y}_i) \\
&+ 2n_{11/00} \log p_{10} & + \sum_{i=1}^{n_{11/00}} \log f_0(\mathbf{y}_i) \\
&+ n_{10/11} \log(2p_{11}p_{01}) & + \sum_{i=1}^{n_{10/11}} \log f_1(\mathbf{y}_i) \\
&+ n_{10/10} \log(2p_{11}p_{00} + 2p_{10}p_{01}) & + \sum_{i=1}^{n_{10/10}} \log[\phi f_1(\mathbf{y}_i) + (1 - \phi)f_0(\mathbf{y}_i)] \\
&+ n_{10/00} \log(2p_{10}p_{00}) & + \sum_{i=1}^{n_{10/00}} \log f_0(\mathbf{y}_i) \\
&+ 2n_{00/11} \log p_{01} & + \sum_{i=1}^{n_{00/11}} \log f_0(\mathbf{y}_i) \\
&+ n_{00/10} \log(2p_{01}p_{00}) & + \sum_{i=1}^{n_{00/10}} \log f_0(\mathbf{y}_i) \\
&+ 2n_{00/00} \log p_{00} & + \sum_{i=1}^{n_{00/00}} \log f_0(\mathbf{y}_i)
\end{aligned}
$$

$$(14.3)$$

assuming that [11] is a risk haplotype, where $f_j(\mathbf{y}_i)$ is a multivariate normal distribution density function of composite diplotype j ($j = \mathbf{2}, \mathbf{1}, \mathbf{0}$).

The previous chapters have witnessed the estimation of haplotype frequencies based on the multinomial likelihood $\log L(\Omega_p | \mathbf{S})$ with the EM algorithm. The key issue is to estimate quantitative genetic parameters with a mixture-based likelihood $\log L(\Omega_q | \mathbf{y}, \mathbf{S}, \hat{\Omega}_p)$ constructed on the estimation of $\log L(\Omega_p | \mathbf{S})$.

14.1.3.1 Modeling the Mean Vector with LOP

Our goal is to model the time-dependent genotypic values of different composite diplotypes using the orthogonal Lengedre polynomial with a particular order r. A family of such polynomials is denoted by

$$\mathbf{P}_r(t^*) = [P_0(t^*), P_1(t^*), \cdots, P_r(t^*)]$$

and a vector of basis genotypic values, which is time-independent, denoted by

$$\mathbf{v}_j^r = (v_{j0}, v_{j1}, \cdots, v_{jr})'. \tag{14.4}$$

Thus, the time-dependent genotypic values $u_j(t)$ in $f_j(\mathbf{y}_i)$ can be described as a linear combination of $\vec{v}_j$ weighted by the family of the polynomials, i.e.,

$$u_j(t) = \mathbf{P}_r(t^*)\mathbf{v}_j^r. \tag{14.5}$$

Now, the mean vector of composite diplotype j is expressed as

$$\begin{aligned}
\mathbf{u}_j &= [u_j(1),...,u_j(T)] \\
&= [P_0(1)v_{j0} + P_1(1)v_{j1} + ... + P_r(1)v_{jr}, ..., P_0(T)v_{j0} + P_1(T)v_{j1} + ... + P_r(T)v_{jr}].
\end{aligned}$$

Modeling the mean vector is now equivalent to estimating the basis genotypic values for different composite diplotypes $\mathbf{v}_j^r$. The maximum likelihood estimates (MLEs) of basis vector $\mathbf{v}_j^r$ will be much more easily obtained, compared with that of curve parameters as in parametric functional mapping, because a closed form of estimating $\mathbf{v}_j^r$ can be derived in the EM framework.

14.1.3.2   Modeling the Covariance Structure

Many approaches can be used to model the structure of longitudinal covariance matrix. A simplest model, AR(1), will be used, assuming a constant variance and stationary correlation between two different time points, t_1 and t_2, by $\mathrm{corr}(t_1, t_2) = \rho^{|t_1 - t_2|}$, with $0 < \rho < 1$. In practice, a TBS-based model is recommended because of its removal of heteroscedasity of variance. The covariance-structuring parameters are arrayed in $\mathbf{\Omega}_v$.

14.1.4   Model Selection

For genetic haplotyping with LOP-based nonparametric approaches, we need to perform two types of model selection. Possibly, the third hypothesis is needed, depending on the data nature.

14.1.4.1   Optimal Risk Haplotype

The choice of an optimal risk haplotype is a key for gene detection. By assuming all possible risk haplotypes and then constructing the corresponding composite diplotypes in the likelihood (14.10), we estimate the parameters and plug in these estimates into the likelihood. The largest likelihood is consistent with an optimal risk haplotype.

14.1.4.2   Optimal LOP Order

The determination of the best order for the Legendre polynomial is also important to best fit the data. For the selection of the best LOP order, we need to use the AIC information criterion or the Bayesian Information Criterion (BIC) (Schwarz, 1978) because the number of parameters is different among different orders. The AIC value at a particular order r is calculated by

$$\text{AIC} = -2\ln L(\hat{\mathbf{\Omega}}_p, \{\hat{\mathbf{v}}_j^r\}_{j=0}^2, \hat{\mathbf{\Omega}}_v | r) + 2\,\text{dimension}(\mathbf{\Omega}_p, \{\mathbf{v}_j^r\}_{j=0}^2, \mathbf{\Omega}_v | r), \quad (14.6)$$

where $(\hat{\mathbf{\Omega}}_p, \{\hat{\mathbf{v}}_j^r\}_{j=0}^2, \hat{\mathbf{\Omega}}_v | r)$ is the the MLE of parameters for the Legendre polynomial of order r and dimension$(\mathbf{\Omega}_p, \{\mathbf{v}_j^r\}_{j=0}^2, \mathbf{\Omega}_v | r)$ represents the number of independent parameters under order r.

The BIC value for determining the optimal order of the Legendre function is calculated by

$$\text{BIC} = -2\ln L(\hat{\mathbf{\Omega}}_p, \{\hat{\mathbf{v}}_j^r\}_{j=0}^2, \hat{\mathbf{\Omega}}_v | r) + 2\,\text{dimension}(\mathbf{\Omega}_p, \{\mathbf{v}_j^r\}_{j=0}^2, \mathbf{\Omega}_v | r)\ln[nT] \quad (14.7)$$

As compared to AIC, BIC adjusts the effects of sample size and the number of time points measured. The minimum AIC or BIC value corresponds to the best LOP order for curve fitting.

14.1.4.3 Diplotype-Dependent Optimal LOP Orders

If there is strong evidence that shows difference shapes of curves among different composite diplotypes, we may fit these diplotype-specific curves by using different LOP orders. This type of selection should be based on AIC or BIC. If the curves of three composite diplotypes AA, $A\bar{A}$, and $\bar{A}\bar{A}$ are fit by the LOP of order r_2, r_1, and r_0, respectively, the AIC is calculated in this case as

$$\text{AIC} = -2\ln L[\hat{\mathbf{\Omega}}_p, (\hat{\mathbf{v}}_2^{r_2}, \hat{\mathbf{v}}_1^{r_1}, \hat{\mathbf{v}}_0^{r_0}), \hat{\mathbf{\Omega}}_v | r] + 2\,\text{dimension}[\mathbf{\Omega}_p, (\mathbf{v}_2^{r_2}, \mathbf{v}_1^{r_1}, \mathbf{v}_0^{r_0}), \mathbf{\Omega}_v | r].$$

Similarly, the BIC is calculated accordingly.

14.1.5 Hypothesis Tests

The genetic control of entire curves can be tested by formulating the following hypotheses,

$$H_0 : \mathbf{v}_j^r \equiv \mathbf{v}^r \text{ vs. } H_1 : \mathbf{v}_j^r \neq \mathbf{v}^r, \quad j = 2, 1, 0. \quad (14.8)$$

The H_0 states that there is no difference in basis genotypic values among different composite diplotypes (reduced model), whereas the H_1 proposes that such differences do exist (full model). The test statistic for testing the hypotheses is calculated as the log-likelihood ratio of the reduced to the full model:

$$\text{LR} = -2[\ln L_0(\hat{\mathbf{\Omega}}_p, \tilde{\mathbf{v}}^r, \tilde{\mathbf{\Omega}}_v) - \ln L_1(\hat{\mathbf{\Omega}}_p, \hat{\mathbf{v}}_j^r, \hat{\mathbf{\Omega}}_v)], \quad (14.9)$$

where the tildes and hats denote the MLEs of the unknown parameters under the H_0 and H_1, respectively. The LR is asymptotically χ^2-distributed. An empirical approach for determining the critical threshold is based on permutation tests (Churchill and Doerge, 1994).

If diplotype-specific LOP orders are detected, the test of haplotype effects cannot be based on hypotheses (14.8) because $\mathbf{v}_j^r$ has different dimensions among three composite diplotypes. In this case, we can calculate the areas under curves for each composite diplotype by

$$A_j = \int_1^T u_j(t)dt,$$

where $u_j(t)$ is calculated by equation (14.5). By formulating the null hypothesis,

$$H_0 : A_j \equiv A, \quad j = 2,1,0,$$

we can perform the test of whether there are differences in genetic control of haplotypes among composite diplotypes.

14.2 Nonparametric Modeling of Event Processes with Legendre Polynomial

14.2.1 Introduction

Nonparametric modeling of functional mapping is flexible to model the genetic control of a longitudinal trait and its associated time-to-events. In clinical trials, toxic responses of a drug may be related with an event trait, such as mortality. It is interesting in practice to test whether these two processes share the common genetic basis. For example, the identification of specific genetic variants responsible for an HIV patient's time-dependent CD4 count and for the time to onset of AIDS symptoms can help to design individualized drugs to control this patient's progression to AIDS. Similarly, in studies of prostate cancer, a shared genetic basis between prostate specific antigen, repeatedly measured for patients following treatment for prostate cancer, and the time to disease recurrence can be used to make optimal treatment schedules for patients.

14.2.2 Model and Estimation

Assume that we use the same experimental setting as described in Section 14.1. Let z be the time-to-event data measured for all subjects. The haplotype frequencies (Ω_p) are estimated with the EM algorithm. Quantitative genetic parameters associated with drug response and an event should be estimated by constructing a joint model. Assume that these two types of traits have different risk haplotypes, [11] (denoted as A) for drug response ($\mathbf{y}$), and [10] (denoted as B) for the event (z). Let j_1 and j_2 be an arbitrary composite diplotype associated with drug response and the event, $j_1 = AA$ (**2**), $A\bar{A}$ (**1**), and $\bar{A}\bar{A}$ (**0**), and $j_2 = BB$ (**2**), $B\bar{B}$ (**1**), and $\bar{B}\bar{B}$ (**0**). Thus, the likelihood of a joint model is constructed as

$$
\begin{aligned}
\log L(\boldsymbol{\Omega}_q|\mathbf{y},z,\mathbf{S},\hat{\boldsymbol{\Omega}}_p) = & \\
\sum_{i=1}^{n_{11/11}} & \log f_{\mathbf{20}}(\mathbf{y}_i,z_i) \\
+ \sum_{i=1}^{n_{11/10}} & \log f_{\mathbf{11}}(\mathbf{y}_i,z_i) \\
+ \sum_{i=1}^{n_{11/00}} & \log f_{\mathbf{02}}(\mathbf{y}_i,z_i) \\
+ \sum_{i=1}^{n_{10/11}} & \log f_{\mathbf{10}}(\mathbf{y}_i,z_i) \\
+ \sum_{i=1}^{n_{10/10}} & \log[\phi f_{\mathbf{10}}(\mathbf{y}_i,z_i) + (1-\phi) f_{\mathbf{01}}(\mathbf{y}_i,z_i)] \\
+ \sum_{i=1}^{n_{10/00}} & \log f_{\mathbf{01}}(\mathbf{y}_i,z_i) \\
+ \sum_{i=1}^{n_{00/11}+n_{00/10}+n_{00/00}} & \log f_{\mathbf{00}}(\mathbf{y}_i,z_i),
\end{aligned}
\tag{14.10}
$$

where $f_{j_1 j_2}(\mathbf{y}_i,z_i)$ is modeled by a multivariate normal distribution with mean vector $\mathbf{u}_{j_1 j_2}$ and covariance $\boldsymbol{\Sigma}$.

Diplotype-specific mean vectors for drug response are modeled by the LOP of order r ($\mathbf{P}_r(t*)$) and basis genotypic values ($\mathbf{v}_j^r$), as shown by equation 14.5. We thus have

$$
\begin{aligned}
\mathbf{u}_{j_1 j_2} &= [u_{j_1}(1),...,u_{j_1}(T),u_{j_2}^z] \\
&= [\mathbf{P}_r(1)\mathbf{v}_{j_1}^r,...,\mathbf{P}_r(T)\mathbf{v}_{j_1}^r,u_{j_2}^z],
\end{aligned}
$$

in which the unknown parameters include $(\{\mathbf{v}_{j_1}^2\}_{j_1=0}^2, \{u_{j_2}^z\}_{j_2=0}^2)$.

The covariance matrix has a structure like

$$
\boldsymbol{\Sigma} = \begin{pmatrix} \boldsymbol{\Sigma}_y & \boldsymbol{\Sigma}_{yz} \\ \boldsymbol{\Sigma}_{zy} & \sigma_z^2 \end{pmatrix},
$$

where $\boldsymbol{\Sigma}_y$ and σ_z^2 are the covariance matrix and variance for drug response and the event, respectively, and $\boldsymbol{\Sigma}_{yz} = \boldsymbol{\Sigma}_{zy}'$ is the covariance matrix between these two processes. The structures of $\boldsymbol{\Sigma}_y$ and $\boldsymbol{\Sigma}_{yz}$ can be empirically modeled on the basis of prior knowledge or results. Several approaches for parametric modeling of the covariance matrix, reviewed in Zimmerman and Núñez-Antón (2001), can be utilized. The correlation with drug response and the event can be fitted by the autoregressive model, expressed as

$$
\text{corr}[y_i(t),z(\tau_i)] = \lambda^{|\tau_i-t|}, \quad -1 \le \lambda \le 1,
\tag{14.11}
$$

where τ_i is the time at which the event is measured for subject i. The variance of the event is denoted by σ_z^2. If the AR(1) is used to model Σ_y, we will have covariance-structuring parameters $(\sigma_y^2, \rho, \sigma_z^2, \lambda)$.

As in Section 14.1.4, it is necessary to select optimal risk haplotypes for drug response and event among 16 possible combinations (assuming two SNPs). Also, we need to determine an optimal LOP order that best fit the data of drug response and event.

14.2.3 Hypothesis Testing

We can now test whether there are significant haplotype effects on drug response and event trait. This can be done by formulating the following hypotheses:

$$H_0: \ \mathbf{v}_{j_1}^r \equiv \mathbf{v}^r, \ u_{j_2}^z \equiv u^z, \ j_1, j_2 = 2, 1, 0$$
$$H_1: \ \text{At least one of the equalities in } H_0 \text{ is not true,}$$
(14.12)

from which the log-likelihood ratio is calculated. Whether the same haplotype [11] controls the two traits can be tested by formulating the null hypotheses as follows:

$$H_0: \mathbf{v}_{j_1}^r \equiv \mathbf{v}^r, \ j_1 = 2, 1, 0,$$
(14.13)

and

$$H_0: u_{j_1}^z \equiv u^z, \ j_1 = 2, 1, 0.$$
(14.14)

If the null hypotheses of (14.13) and (14.14) are both rejected, this means that the haplotype [11] pleiotropically determines drug response and the event. A similar test for the pleiotropic effect of haplotype [10] can be done with the null hypotheses as follows:

$$H_0: \mathbf{v}_{j_2}^r \equiv \mathbf{v}^r, \ j_2 = 2, 1, 0,$$
(14.15)

and

$$H_0: u_{j_2}^z \equiv u^z, \ j_2 = 2, 1, 0.$$
(14.16)

Example 14.1

We used a quantitative trait locus (QTL) mapping result by Lin and Wu (2006b) to show the usefulness of nonparametric functional mapping based on Legendre polynomials. The study material used was described in Wu et al. (2002a), derived from the interspecific hybridization of *Populus* (poplar), *P. deltoides* and *P. euramericana*. Genetic maps were constructed with test markers that are heterozygous in one parent but homozygous in the other. Planted in the field, this hybrid population was measured for the total stem heights and diameters at the end of each of the first 11 growing seasons. Ma et al. (2002) showed that a logistic curve can well be used to fit growth trajectories in both height and diameter growth. But for stem volume indices,

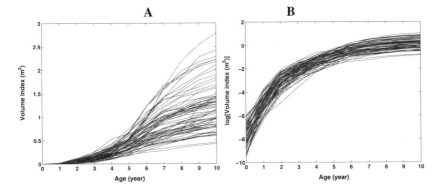

FIGURE 14.1

Plots of stem volume index growth vs. ages for poplar hybrids. The relationships between growth and age are displayed for untransformed (**A**) and log-transformed data (**B**).

y, (expressed as the product of heights and squared diameters) as a more important stemwood production trait, no such a logistic equation exists to fit mainly because stem volume has not reached its asymptotic growth during this measurement period (Fig. 14.1). Also, an event trait, the ages to first flower (z), was predicted with using prior knowledge.

The Legendre function (with the coefficients derived from the adjusted ages; Table 14.1) was used to model the mean vector for volume indices, and TBS-based AR(1) and equation (14.11) used to model the covariance structure. When scanning the existence of a QTL throughout the entire linkage map, three significant QTLs on linkage groups 2, 5, and 12 were detected at the 5% chromosome-wide testing level. The estimates of the Legendre parameters were used to draw growth curves for two segregating genotypes at each QTL detected. Figure 14.2 provides such an example for the QTL detected on linkage group 12. In general. these QTL are switched on to affect the overall stem growth process after age 4-5 years at which strong inter-tree competition sets in the stand due to canopy closure. There are genotypic differences in the age to first flower at each of the growth QTL. But as tested, only QTL on linkage group 12 has a significant impact on the age to first flower (Fig. 14.2). At this QTL, the slower-growing genotype flowers about 0.7 year earlier than the faster-growing genotype. Through this QTL, the fast-growing attribute and the capacity to efficiently occupy growth resources can be transmitted to the next generation.

□

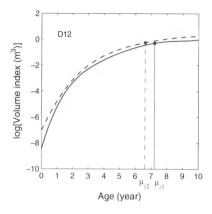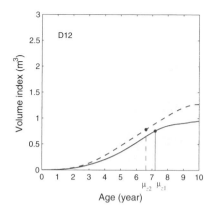

FIGURE 14.2

Two growth curves each presenting two groups of genotypes at the QTL detected linkage groups 12 in the ***Populus deltoides***-specific map based on log-transformed data (at the left panel). The estimated two genotypic growth curves were ante-transformed back to recover original scales shown at the right panel. Growth trajectories for all the individuals studied are indicated in yellow background.

14.3 Nonparametric Functional Mapping with B-Spline

Nonparametric regression methods for modeling the mean structure of longitudinal data have been based on more commonly used B-spline basis functions (Rice and Wu, 2001). Brown et al. (2005) extended the B-spline basis to model multiple longitudinal variables. Statistical models with B-spline provide a simple, flexible, and computationally efficient approach for functional data analyses. It is therefore worthwhile implementing more flexible B-spline basis functions into the nonparametric functional mapping model.

14.3.1 Basics of B-Splines

The term B-spline, short for basis spline, was coined by Isaac Jacob Schoenberg (de Boor, 2001). B-spline aims to produce a spline curve with observed functional data, parametrized by spline functions that are expressed as linear combinations of basis splines. Over a series of time points, define $m+1$ knots in a sequence $\tau_0 \leq \tau_1 \leq ... \leq \tau_m$. A B-spline of degree r is a parametric curve in time interval $[\tau_0, \tau_m]$ which is composed of basis B-splines of degree r, i.e.,

$$u(t) = \sum_{l=0}^{m-r-1} v_l B_{l,r}(t), \quad t \in [t_r, t_{m-r}], \tag{14.17}$$

where v_l are called control points or de Boor points.

The $m - r$ basis B-splines of degree r can be defined using the Cox-de Boor recursion formula

$$B_{l,0}(t) := \begin{cases} 1, & \text{if } \tau_l \leq t \leq \tau_{l+1} \\ 0 & \text{otherwise} \end{cases} \tag{14.18}$$

$$B_{l,r}(t) := \frac{t - \tau_l}{\tau_{l+r} - \tau_l} B_{l,r}(t) + \frac{\tau_{l+r+1} - t}{\tau_{l+r+1} - \tau_{l+1}} B_{l+1,r-1}(t). \tag{14.19}$$

The B-spline is said to be uniform if the knots are equidistant. Otherwise, the B-spline is non-uniform. For a uniform B-spline, the basis B-splines for a given degree r are just shifted copies of each other. An alternative non-recursive definition for the $m - r$ basis B-splines is

$$B_{l,r}(t) = B_r(t - \tau_l), \qquad l = 0, ..., m - r - 1,$$

with

$$B_r(t) = \frac{r+1}{r} \sum_{l=0}^{r+1} \psi_{l,r}(t - \tau_l)_+^r$$

and

$$\psi_{l,r} := \prod_{l'=0, l' \neq l}^{r+1} \frac{1}{\tau_{l'} - \tau_l},$$

where $(t - \tau_l)_+^r$ is the truncated power function.

There are many types of B-splines, which include:

14.3.1.1 Constant B-spline

The constant B-spline is the most simple spline. It is defined on only one knot span and is not even continuous on the knots. It is a just indicator function for the different knot spans.

$$B_{l,0}(t) = 1_{[\tau_l, \tau_{l+1}]} = \begin{cases} 1, & \text{if } \tau_l \leq t \leq \tau_{l+1} \\ 0 & \text{otherwise.} \end{cases}$$

14.3.1.2 Linear B-spline

The linear B-spline is defined on two consecutive knot spans and is continuous on the knots, but not differentiable.

$$B_{l,1}(t) = \begin{cases} \frac{t - \tau_l}{\tau_{l+1} - \tau_l}, & \text{if } \tau_l \leq t \leq \tau_{l+1} \\ \frac{\tau_{l+2} - t}{\tau_{l+2} - \tau_{l+1}}, & \text{if } \tau_{l+1} \leq t \leq \tau_{l+2} \\ 0 & \text{otherwise} \end{cases}$$

14.3.1.3   Uniform Quadratic B-spline

Quadratic B-splines with uniform knot-vector is a commonly used form of B-spline. The blending function can easily be precalculated, and is equal for each segment in this case.

$$B_{l,2} = \begin{cases} \frac{1}{2}t^2, \\ -t^2 + t + \frac{1}{2} \\ \frac{1}{2}(1-t)^2 \end{cases}$$

Written in matrix form, we have

$$u_l(t) = [t^2\ t\ 1]\frac{1}{2} \begin{bmatrix} 1 & -2 & 1 \\ -2 & 2 & 0 \\ 1 & 1 & 0 \end{bmatrix} \begin{bmatrix} v_{l-1} \\ v_l \\ v_{l+1} \end{bmatrix} \quad \text{for } t \in [0,1]$$

14.3.1.4   Cubic B-spline

A B-spline formulation for a single segment can be written as

$$u_l(t) = \sum_{l'=0}^{3} v_{l-3+l'} B_{l-3+l',3}(t), \quad \text{for } t \in [0,1],$$

where u_l is the lth B-spline segment and v is the set of control points, segment l and l' is the local control point index. A set of control points would be $v_l^w = (w_l x_l, w_l y_l, w_l z_l, w_l)$ where the w_l is weight, pulling the curve towards control point v_l as it increases or moving the curve away as it decreases.

An entire set of segments, $m-2$ curves $(u_3, u_4, ..., u_m)$ defined by $m+1$ control points $(v_0, v_1, ..., v_m, m > 3)$, as one B-spline in t would be defined as

$$u(t) = \sum_{l=0}^{m} v_l B_{l,3}(t),$$

where l is the control point number, and t is a global parameter giving knot values. This formulation expresses a B-spline curve as a linear combination of B-spline basis functions.

14.3.1.5   Uniform Cubic B-spline

Cubic B-splines with uniform knot-vector is the most commonly used form of B-spline. The blending function can easily be precalculated, and is equal for each segment in this case. Put in matrix-form, it is

$$u_l(t) = [t^3\ t^2\ t\ 1]\frac{1}{6} \begin{bmatrix} -1 & 3 & -3 & 1 \\ 3 & -6 & 3 & 0 \\ -3 & 0 & 3 & 0 \\ 1 & 4 & 1 & 0 \end{bmatrix} \begin{bmatrix} v_{l-1} \\ v_l \\ v_{l+1} \\ v_{l+2} \end{bmatrix} \quad \text{for } t \in [0,1].$$

14.3.2   Haplotyping Model for DNA Sequence Variants

Following Section 14.1.2, we assume that there is a natural population from which n subjects are sampled at random for genotyping with SNPs and phenotyping with a functional drug response trait. The multivariate normal distribution of drug response data contain composite diplotype-specific mean vectors and covariance matrix. We used a similar procedure to model the mean-covariance structure, except for the mean vector modeled by one of the B-splines described above. All the hypotheses tests and their procedures described in Section 14.1.5 can be directly applied to S-spline approaches for genetic haplotyping.

B-spline-implemented functional mapping is attractive in its computational advantage. However, there is a big issue about the choices of the spline basis and of the number of basis functions. All these will in turn involve choices of the number and location of knots. Rice and Wu (2001) suggested using a cross-validation procedure or information based criteria such AIC and BIC for the choice of knots, but this is still an open question in which there is a lot of room for improvement and exploration.

For some complex curves with many local features such as peaks, a large number of basis functions is needed, in which case the B-spline basis approach may have difficulties in fitting and smoothing because of inadequate degrees of freedom. The solutions into this problem include the use of a local basis, such as local polynomials, which requires a low degree polynomial, and the reduced rank procedure that involves the use of principal components.

14.4   Nonparametric Functional Mapping of Pharmacokinetics and Pharmacodynamics

In pharmacogenetic research, we are often faced with a high-dimensional analysis of different aspects that define a system of pharmacological reaction processes. A univariate B-spline model may not be sufficient to capture information contained in these processes. The extension of B-splines into a multivariate case has been possible in the past two decades (Mueller, 1986; Johnson and Marsh, 1999; Brown et al., 2005). This newly developed computational and statistical approach intends to model not only a functional curve, but also a surface with geometric shapes, and will open up excellent opportunities for statistical pharmacogeneticists to study a network of genetic regulation and structure that is related to pharmacokinetic and pharmacodynamic processes.

14.5   Nonparametric Modeling of the Covariance Structure

Functional mapping relies upon the mathematical description of biological and developmental processes and constructs hypothesis tests within the framework for the estimation of parameters that define the process. However, many biological processes may not be readily described by a mathematical function. To map the QTL that control those biological processes, nonparametric approaches that do not need any form of mathematical functions should be developed.

Nonparametric analogues of the parametric modeling approach have also been developed recently in the statistical literature. Nonparametric regression methods using kernel estimators have been considered for the mean structure of growth curve data by Hart and Wehrly (1986) and Ferreira et al. (1997). All of these nonparametric approaches have in common that the unknown mean response curve over time is estimated by smoothing the raw data, and time is the only explanatory variable. Muller (1988) applied nonparametric regression methods to longitudinal data but without considering a serial correlation structure.

Relative to nonparametric modeling of the mean structure, nonparametric covariance modeling has received little attention. Most authors have considered only the stationary case (Hall et al., 1994). However, two recent papers have considered the possibility of estimating the nonstationary case. Diggle and Verbyla (1998) used kernel-weighted local linear regression smoothing of sample variograms ordinates and of squared residuals to provide a nonparametric estimator for the covariance structure without assuming stationarity. In addition, they used the value of the estimator as a diagnostic tool but did not study the use of the estimator in more formal statistical inference concerning the mean profiles. Wang (2003) used kernel estimators to estimate covariance functions in a nonparametric way. His only assumption was to have a fully unstructured smooth covariance structure, together with a fixed effects model. The proposed kernel estimator was consistent with complete but irregularly spaced follow-ups, or when the missing mechanism is strongly ignorable missing at random (Rosenbaum and Rubin, 1983).

Zeger and Diggle (1994) studied a semiparametric model for longitudinal data in which the covariates entered parametrically and only the time effect entered nonparametrically. To fit the model, they extended to longitudinal data the backfitting algorithm of Hastie and Tibshirani (1990) for semiparametric regression.

15

Semiparametric Functional Mapping of Drug Response

Parametric and nonparametric functional mapping can be viewed as alternative to each other in terms of their statistical applications, biological relevance and computational efficiency. While the parametric approach is biologically sensible, the nonparametric approach displays tremendous flexibility and computational advantage. For this reason, both approaches should be able to find their applications in practical data analysis, depending on the biological and statistical features of a particular data set. For some biological processes, there are multiple different phases of development in each of which a variety of genetic and environmental factors play a role through their interactions, leading to changes in the dynamic pattern of development and response. For these processes, a single parametric or nonparametric model may not be adequate. Very often, a combined model that takes advantage of each approach is capable of discerning the differences between developmental phases, and therefore provides a powerful way for gene identification and haplotyping. Such a combined approach is called the semiparametric modeling of functional mapping.

Semiparametric analysis of longitudinal or functional data has become a vital area and received considerable attention in recent statistical research because of increasing popularity of this type of data related to genetics, bioinformatics, and bioimaging (Liang et al., 2003; Fan and Li, 2004; Sun and Wu, 2005; Wu and Zhang, 2006; Xue and Zhu, 2007). A handful of semiparametric time-varying coefficient regression models has been established, where the influences of some covariates vary nonparametrically with time while the effects of the remaining covariates follow certain parametric functions of time (Sun and Wu, 2005). By integrating counting process techniques into the semiparametric analysis of longitudinal data, Martinussen and Scheike (1999, 2000, 2001) and Lin and Ying (2001) provided a new bridge between survival analysis, recurrent events, and time-dependent observations. All these recent developments in statistics can be embedded within the framework of functional mapping, thus opening up novel opportunities to estimate the genetic architecture of complex longitudinal or dynamic biological problems.

In this chapter, we will provide a motivation to integrate semiparametric longitudinal analysis into genetic studies, and then describe a semiparametric procedure for genetic haplotyping of dynamic traits. We will provide some key steps for deriving semiparametric functional mapping and discuss its statistical behavior through simulation studies of HIV dynamics and analyzing a real data set in plants. We will also pinpoint some areas in which further work is likely to lead to a success in pharmacogenetic studies.

15.1   Problems

In this section, we introduce two biological examples in which semiparametric analysis is essential in statistical modeling of functional mapping. In pharmacokinetic and pharmacodynamic areas, such examples are enormous, showing the commonality of a semiparametric longitudinal problem in general biomedical research.

15.1.1   Long-Term HIV Dynamics

HIV dynamics by modeling viral load trajectory after initiation of potent antiviral therapy has been instrumental for predicting the pathogenetic progress of HIV infection and providing scientific guidance about the control and prevention of HIV/AIDS. In their seminar article, H. Wu and colleague derived a bi-exponential function based on a compartmental analysis (Wu and Ding, 1999) and modeled this function to real HIV dynamic data from an AIDS clinical study, AIDS Clinical Trials Group (ACTG) Protocol 315, leading to a nice inference and prediction of HIV viral trajectory. This bi-exponential equation is expressed as

$$V(t) = P_1 e^{-\lambda_1 t} + P_2 e^{-\lambda_2 t}, \tag{15.1}$$

where $V(t)$ is the plasma HIV load at time t, λ_1 and λ_2 are two different viral decay rates in the first and second phase, representing the minimum turnover rates of two compartments, i.e., productively infected cells and latently/long-lived infected cells, respectively, and P_1 and P_2 are related to the baseline viral loads for the two above compartments when the treatment is initiated. By estimating both viral decay rates λ_1 and λ_2, the shape of the viral load trajectory can be described and, furthermore, the antiviral effect and antiviral treatment can be quantified and assessed in clinical studies.

   However, this equation can only be used for earlier stages of HIV viral load trajectory and will fail to model the change of HIV loads in late stages. As shown in Fig. 15.1, all patients in ACTG Protocol 315 follow a consistent exponential reduction in HIV load in earlier stages (<12 weeks) after initiation of highly active anti-retroviral therapy (HAART), but many of them change the form of HIV dynamics after a certain time point. While some keep their HIV load stable, others rebound in later stages. The curve of HIV load trajectory is changed in the second phase mainly because of the impact of drug resistance, noncompliance and other clinical factors. As a result of this, functional mapping based on equation (15.1) to study the genetic basis of HIV progression will be limited if our interest is to detect how genes control the dynamic change of viral loads during a long-term period. Indeed, such

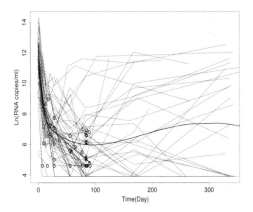

FIGURE 15.1
HIV dynamics for different patients from ACTG Protocol 315 data observed on days 0, 2, 7, 10, 14, 21, and 28 and weeks 8, 12, 24, and 48 after initiation of HAART. The circles are the points before which HIV load trajectories can be fit by the bi-exponential equation (15.1). After these points many patients change the form of HIV dynamics that do not follow the bi-exponential equation. The solid curve indicates the mean curve.

information is helpful in practice to design personalized medications for the control and eradication of HIV/AIDS infection.

As the decay rate of HIV load in the second phase, the change of λ_2 will influence the form of HIV dynamics. Based on this idea, Wu and Zhang (2002) extended the bi-exponential equation (15.1) to allow λ_2 to change over time. By modeling λ_2 with a flexible smooth function, one can capture the pattern of time-dependent change of HIV load in the second phase. The extended equation is expressed as

$$V(t) = P_1 e^{-\lambda_1 t} + P_2 e^{-\lambda_2(t)t}, \qquad (15.2)$$

where $\lambda_2(t)$ is fit as a function of time. The advantage of equation (15.2) lies in its preservation of the interpretability of the original parametric equation (15.1) and flexibility to accommodate the long-term change of HIV loads by a nonparametric model. Thus, equation (15.2), called the semiparametric model, should be more applicable to study HIV dynamics and AIDS progression, compared with the parametric model (15.1).

This practical problem encourages Wu et al. (2007c) to incorporate the semiparametric model (15.2) into functional mapping to precisely elucidate the genetic architecture of HIV dynamics during a long-term period. The new model allows the test of differences in the genetic control of short- and long-term HIV dynamics and the characterization of the effects of viral-host genome interaction. The model will provide an important tool for genetic and genomic studies of human complex diseases like HIV/AIDS and their pathological progression.

15.1.2   Different Phases of Programmed Cell Death

The question of how an organism develops into a fully functioning adult from a mass of undifferentiated cells has always attracted top researchers in diverse areas of developmental biology (Horvitz, 2003). Fundamentally, to produce a functioning adult form, a living organism should coordinate various complementary and sometimes antagonistic processes, which include cell proliferation and programmed cell death (PCD) or apoptosis, during its development (Cashio et al., 2005). In general, the whole process of development can be described by five reasonably distinct phases (Fig. 15.2) (Fogg, 1987): lag, exponential, declining growth rate, stationary, and death. Each of the phases is defined below:

(1) **Lag phase:** This is the initial growth phase, during which cell number remains relatively constant prior to rapid growth. During this phase, the organism prepares to grow, and unseen biochemical changes, cell division, and differentiation of tissues occur during this time.

(2) **Exponential phase:** During this phase the tissues are growing and dividing rapidly to take advantage of abundant nutrients. Growth rate, as a measure of the increase in biomass over time, is determined from the exponential phase. Growth rate is one important way of expressing the relative success of an organism in adapting to the biotic or abiotic environment imposed upon it. The duration of exponential phase depends upon the growth rate and the abundance of nutrients to support tissue growth. If the growth phase is plotted (time on x-axis and biomass on logarthmic y-axis), the exponential phase will be straightened out.

(3) **Declining growth:** Declining growth normally occurs when either a specific requirement for cell division is limiting or something else is inhibiting reproduction. During this phase growth slows or the death rate increases. As a result, the initiation of new tissues and the senescence and death of old ones start to come into equilibrium. This phase typically occurs as nutrients become limiting for growth.

(4) **Stationary phase:** Tissues enter stationary phase when net growth is zero, and within a matter of times cells may undergo dramatic biochemical changes. The nature of the changes depends upon the growth limiting factor. The shutdown of many biochemical pathways as stationary phase proceeds means that the longer the cells are held in this condition the longer the lag phase will be when cells are returned to good growth conditions.

(5) **Death phase:** When cell metabolism can no longer be maintained the death phase of a tissue is generally very rapid. The steepness of the decline is often more marked than that represented in the accompanying growth figure.

The duration and extent of each phase will depend on the organism and environmental conditions. For example, if tissues from the stationary phase are supplied

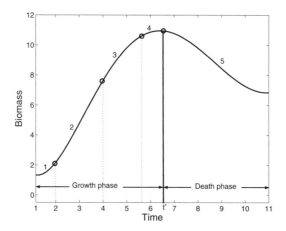

FIGURE 15.2

A typical example of development that includes five different stages, (1) lag, (2) exponential, (3) declining, (4) stationary, and (5) death.

with fresh nutrients, the lag phase will be longer than for the case of tissues from the declining phase. For growing tissues from the exponential phase, organisms supplied with fresh nutrients will likely skip the lag phase. If the growth nutrient is rich, organisms will remain in the exponential growth phase for a longer period and produce a greater biomass. Furthermore, their rate of growth in the exponential phase may also be greater.

Growth curves must be drawn from a series of growth measurements at different times during the growth curve. Mathematical equations have been derived to model the growth from the lag to stationary phase (West et al., 2001), but there is no specific mathematical equation for the death phase. A semiparametric model that describes the growth phase parametrically and the death phase with nonparametric modeling will be needed to equip functional mapping for the detection of genes that control an entire process of development.

15.2 Semiparametric Modeling of Functional Mapping: HIV Dynamics

15.2.1 Genetic Design

We start with the description of a genetic mapping population. Suppose this population is composed of n subjects randomly sampled from a natural population at Hardy–Weinberg equilibrium. The sampled subjects are typed for single nucleotide polymorphisms (SNPs) from candidate chromosomal regions or through the entire genome. The objective of this project is to detect the association between haplo-type or diplotypes constructed by a set of SNPs and longitudinal responses over a time course composed of parametric and nonparametric phases. For simplicity, we consider two SNPs, each with two alleles 1 and 0, which form four haplo-types [11], [10], [01], and [00], 10 diplotypes [11][11], [11][10], [11][01], [11][00], [10][10], [10][01], [10][00], [01][01], [01][00], and [00][00], and nine genotypes 11/11, 11/10, 11/10, 10/11, 10/10, 10/00, 00/11, 00/10, and 00/00. The genotypes are observable. Let $n_{r_1 r'_1 / r_2 r'_2}$ $(r_1 \geq r'_1, r_2 \geq r'_2 = 1, 0)$ be the sizes of each genotype. Let p_{11}, p_{10}, p_{01}, and p_{00} be the haplotype frequencies. The expected frequencies of genotypes are expressed in terms of haplotype frequencies (see Table 2.1). A multi-nomial likelihood (13.5) is then constructed to estimate the haplotype frequencies expressed as $\hat{\Omega}_p$ (see Section 2.2).

All subjects are measured for HIV loads repeatedly at different time points over a time course spanning from early to late stages. For subject i, viral loads are measured at a finite set (T_i) of time points. We recognize that measurement times can be irregularly spaced, with different measurement schedules for all subjects. Let $\mathbf{y}_i = \{y(t_{i1}), \cdots, y(t_{iT_i})\}$ be the vector of T_i measurements on subject i and let $\mathbf{t}_i = (t_{i1}, \cdots, t_{iT_i})$ be the corresponding vector of measurement times. The data struc-ture of longitudinal HIV response is similar to that shown in Fig. 15.1. Thus, the parametric bi-exponential equation (15.1) cannot well describe the dynamic change of HIV load during a long-time period. Instead, the semiparametric equation (15.2) is used for functional mapping, in which parameters P_0, P_1, and λ_1 are modeled directly whereas $\lambda_2(t)$ is modeled with a nonparametric approach.

15.2.2   Model Structure

We assume that there are distinct risk haplotypes that are associated with HIV dy-namics. Looking at the profiles of HIV load changes in a long term (Fig. 15.1), HIV dynamics seems to vary more substantially in early than late stages. This natu-rally leads us to think whether there is the same genetic basis for these two different but related processes. For general modeling, we assume that there are different risk haplotypes associated with the forms of early and late stages of HIV dynamics. Let [11] (symbolized as A) and [10] (symbolized as B) be the risk haplotypes for the short- and long-term dynamics, respectively. Let j_1 be any one of the three compos-ite diplotypes for the early stage, $j_1 = AA$ (2), $A\bar{A}$ (1), $\bar{A}\bar{A}$ (0), and j_2 be any one of the three composite diplotypes for the late stage, $j_2 = BB$ (2), $B\bar{B}$ (1), $\bar{B}\bar{B}$ (0).

Based on the structure of Table 2.1, we can formulate a mixture-based likelihood for unknown parameters related to haplotype effects (Ω_u) and residual variances and covariances (Ω_v) based on longitudinal ($\mathbf{y}$) and SNP data ($\mathbf{S}$) and estimated haplotype frequencies $(\boldsymbol{\Omega}_p)$, i.e.,

$$
\begin{aligned}
\log L(\boldsymbol{\Omega}_u, \boldsymbol{\Omega}_v | \mathbf{y}, \mathbf{S}, \hat{\boldsymbol{\Omega}}_p) = \\
\sum_{i=1}^{n_{11/11}} \log f_{20}(\mathbf{y}_i) \\
+ \sum_{i=1}^{n_{11/10}} \log f_{11}(\mathbf{y}_i) \\
+ \sum_{i=1}^{n_{11/00}} \log f_{02}(\mathbf{y}_i) \\
+ \sum_{i=1}^{n_{10/11}} \log f_{10}(\mathbf{y}_i) \\
+ \sum_{i=1}^{n_{10/10}} \log[\phi f_{10}(\mathbf{y}_i) + (1 - \phi) f_{01}(\mathbf{y}_i)] \\
+ \sum_{i=1}^{n_{10/00}} \log f_{01}(\mathbf{y}_i) \\
+ \sum_{i=1}^{n_{00/11} + n_{00/10} + n_{00/00}} \log f_{00}(\mathbf{y}_i),
\end{aligned}
\tag{15.3}
$$

where $f_{j_1 j_2}(\mathbf{y}_i)$ is modeled by a multivariate normal distribution with diplotype-specific mean vector $\mathbf{u}_{j_1 j_2 | i}$ and covariance $\boldsymbol{\Sigma}_i$.

15.2.2.1 Modeling the Mean Vector

The key of functional mapping is to model the mean and covariance structures. For a long-term HIV dynamics, we use P_0, P_1, and λ_1 to model its early stage and $\lambda_2(t)$ to model its late stage. These two stages are modeled by different haplotypes. Thus, for subject i, its expected mean vector determined by a combination of risk haplotypes for early and late stages can be expressed as

$$
\begin{aligned}
\mathbf{u}_{j_1 j_2 | i} &= [u_{j_1 j_2 | i}(t_{i1}), \dots, u_{j_1 j_2 | i}(t_{iT_i})] \\
&= [P_{1 j_1} e^{-\lambda_{1 j_1} t_{i1}} + P_{2 j_1} e^{-\lambda_{2 j_2}(t_{i1}) t_{i1}}, \dots, P_{1 j_1} e^{-\lambda_{1 j_1} t_{iT_i}} + P_{2 j_1} e^{-\lambda_{2 j_2}(t_{iT_i}) t_{iT_i}}].
\end{aligned}
\tag{15.4}
$$

In equation (15.4), $\lambda_{2 j_2}$ is an important parameter for describing the late stage of HIV dynamics for composite diplotype j_2. This parameter can be modeled nonparametrically with approaches described in Chapter 14. Wu et al. (2007c) used natural cubic splines (Knott, 2000) to model $\lambda_{2 j_2}$ as a function of time. Natural cubic splines display many favorable properties. For example, they can be constructed easily, have a good flexibility to fit the underlying curves of various shapes, and are the best twice-continuously differentiable interpolate for a twice-continuously differentiable function.

A cubic spline is a piecewise third-order polynomial function that passes through a set of control points. Let $\tau = (\tau_0, \tau_1, ..., \tau_m)$ be a nondecreasing sequence of control points, where τ_0 and τ_m are the endpoints and $\tau_1, ..., \tau_{m-1}$ are the interior points. Let (τ_l, x_l) $(l = 0, ..., m)$ be the coordinates of the control points. For a point $t \in [\tau_l, \tau_{l+1})$, we define $t^* = (t - \tau_l)/(\tau_{l+1} - \tau_l)$ with $t \in [0, 1)$. The l-th piece of the spline is then represented as

$$x_l(t^*) = a_l + b_l t^* + c_l t^{*2} + d_l t^{*3}, \qquad (15.5)$$

for $l = 0, 1, ..., m - 1$. The spline function is a summation of these piecewise functions, i.e.,

$$x(t) = \sum_{l=0}^{m-1} x_l(t^*),$$

where

$$t^* = \frac{t - \tau_l}{\tau_{l+1} - \tau_l} \quad \text{and } \bar{t} \in [\tau_l, \tau_{l+1}]$$

To ensure the second-order smoothness through all the interior points, the following conditions should be satisfied:

$$x_l(1) = x_{l+1}(0) = x_l; x_0(0) = x_0; x_{m-1}(1) = x_m$$
$$x_l'(1) = x_{l+1}'(0)$$
$$x_l''(1) = x_{l+1}''(0)$$

for $l = 0, 1, ..., m - 2$. The above constraints yield $(4m - 2)$ equations but there are $4m$ unknown parameters. To obtain two more equations, usually some boundary conditions need to be specified; for example, set the second derivative of each polynomial to zero at the endpoints, i.e.,

$$x_0''(0) = x_{m-1}''(1) = 0.$$

These particular boundary conditions produce the so-called *natural cubic spline* and lead to a simple tridiagonal system that can be solved easily to give the coefficients of the polynomials.

The natural cubic spline $x(t)$ can also be constructed via B-spline bases of degree 3 (Knott, 2000). Terminologically, the control points above are now called knots, with boundary knots of τ_0 and τ_m, and interior knots of $\tau_1, ..., \tau_{m-1}$. In general, for a B-spline of degree r with $(m - 1)$ interior knots, $(m + r)$ basis functions are needed to span the linear space formed by all the B-spline functions. Since a natural cubic spline has two additional constraints at the endpoints, the dimension of the linear space formed by natural cubic B-spline functions is $(m + r - 2) = m + 1$. Additionally, because a B-spline of degree r globally has $(r - 1)$-continuous derivatives except at the endpoints, the smoothness conditions specified on the interior points for a cubic spine are automatically satisfied by a B-spline representation. To describe a curve, another important issue for B-spline is the location of the knots, since different locations of knots yield different shapes of the spline functions. Usually, the knots are placed uniformly across the interval, which would generate a simple function representation. An alternative way is to place the knots unevenly, which could fit the curve more flexibly but with a more complex expression. In our study, the knots are placed unevenly according to the sample percentiles of the design time, i.e. more knots are placed where more design points are available. In this way, the knot placement is design-oriented and, thus, is more practical. For computational convenience, Wu et al. (2007c) simply used R to generate the B-spline basis matrix for natural cubic spline.

The parameters that model diplotype-specific mean vectors of HIV dynamics are arrayed in $\Omega_u = (\{P_{1j_1}, P_{2j_1}, \lambda_{1j_1}\}_{j_1=0}^2, \{a_{lj_2}, b_{lj_2}, c_{lj_2}, d_{lj_2}\}_{j_2=0,l=0}^{2,m})$. The first part of this unknown vector Ω_u is related to parametric parameters, whereas the second part is related to diplotype-specific genotypic values modeled by a natural cubic spline (15.5) with an optimal number of control points (m).

15.2.2.2 Irregularly Spaced Structured Antedependence Model

The ($T_i \times T_i$) within-subject covariance matrix, Σ_i, are all subject-specific, which contains variances and covariances among residual effects of subject i $\mathbf{e}_i = \{e(t_{i1}), \cdots, e(t_{iT_i})\}'$ distributed as $MVN(\mathbf{0}, \Sigma_i)$. Here, the structure of Σ_i is approximated by the SAD(1) model to accommodate non-stationary variances and correlations. Since time points are unequally spaced, an *irregularly spaced structured antedependence* (ISSAD) model, modified from the SAD(1) model, is used.

For the structured antedependence model, it is assumed that the residual error at a particular time t depends on the previous ones, with the degree of dependence (defined by antedependence coefficients) decaying with time lag, and the one that is formed at the current time point. The error formed at the current time point, denoted by $\epsilon_i = (\varepsilon(t_{i1}), \cdots, \varepsilon(t_{iT}))'$, is called the *innovation error* with $\varepsilon(t_{i\tau}) \sim (0, v^2(t_{i\tau}))$. If an error at time t is independent of all errors before $t - r$, this antedependence model is said to be of order r expressed by SAD(r). For a simple SAD(1) model, the errors at a particular time point can be expressed as $e(t_{i\tau}) = \phi e(t_{i(t-1)}) + \varepsilon(t_{i\tau})$. Thus, the relationships among the residual errors, antedependence parameters, and innovation errors occurring at irregularly spaced time points for subject i in an HIV/AIDS study can be expressed in matrix form as

$$A_i \epsilon_i = \mathbf{e}_i$$

where

$$A_i = \begin{pmatrix} 1 & 0 & \cdots & \cdots & 0 \\ -\phi^{t_{i2}-t_{i1}} & 1 & \cdots & \cdots & 0 \\ 0 & -\phi^{t_{i3}-t_{i2}} & \cdots & \cdots & 0 \\ \vdots & \vdots & \ddots & \ddots & \vdots \\ 0 & 0 & \cdots & -\phi^{t_{iT_i}-t_{i(T_i-1)}} & 1 \end{pmatrix}$$

and

$$\mathbf{e}_i = \begin{pmatrix} \varepsilon(t_{i1}) \\ \varepsilon(t_{i2}) + \phi^{t_{i2}-t_{i1}}\varepsilon(t_{i1}) \\ \varepsilon(t_{i3}) + \phi^{t_{i3}-t_{i2}}\varepsilon(t_{i2}) + \phi^{t_{i3}-t_{i1}}\varepsilon(t_{i1}) \\ \vdots \\ \varepsilon(t_{iT_i}) + \phi^{t_{iT}-t_{i(T-1)}}\varepsilon(t_{i(T_i-1)}) + \cdots + \phi^{t_{iT_i}-t_{i1}}\varepsilon(t_{i1}) \end{pmatrix}.$$

If the innovative variance is assumed as a constant v^2, the residual variance-covariance matrix of the longitudinal trait is then expressed as

$$\Sigma_i = \text{var}(\mathbf{e}_i) = \mathbf{A}_i^{-1}\mathbf{V}_i(\mathbf{A}_i')^{-1},$$

where $\mathbf{V}_i$ is the innovation variance-covariance matrix expressed as

$$\mathbf{V}_i = \begin{pmatrix} \sigma^2(t_{i1}) & 0 & \cdots & 0 \\ 0 & \sigma^2(t_{i2}) & \cdots & 0 \\ \vdots & \vdots & \ddots & \vdots \\ 0 & 0 & \cdots & \sigma^2(t_{iT_i}) \end{pmatrix}$$

$$= \begin{pmatrix} v^2 & 0 & \cdots & 0 \\ 0 & v^2 & \cdots & 0 \\ \vdots & \vdots & \ddots & \vdots \\ 0 & 0 & \cdots & v^2 \end{pmatrix}.$$

Closed forms for the inverse and determinant of matrix Σ_i can be obtained, which facilitates the estimation of the parameters, $\Omega_v = (\phi, v^2)$, that model the matrix

$$\begin{aligned} \Sigma_i^{-1} &= \mathbf{A}_i'\mathbf{V}_i\mathbf{A}_i \\ |\Sigma_i| &= |\mathbf{V}_i| = \prod_{t=1}^{T_i}\sigma^2(t_{i\tau}) = T_iv^2. \end{aligned} \qquad (15.6)$$

Notice that even if innovation variances are assumed to be constant over time points, the residual variance can still change with time. Furthermore, the correlation function in ISSAD(1) is design-dependent because the correlation between the residuals of two measurements does not depend on the order of the measurements, but rather depends on the actual time interval between them.

15.2.3 Model Estimation

Although unknown parameters (Ω_u, Ω_v) can be theoretically estimated with the EM algorithm, it is practically very difficult to derive the expressions of their estimators because of the nonlinearity of semiparametric equations. An EM-simplex algorithm has been shown to be very efficient for the parameter estimation. Computational efficiency can be increased if the closed forms (15.6) are incorporated into the algorithm.

An optimal combination between risk haplotypes for the early and late stages of HIV dynamics can be determined on the basis of likelihoods by assuming all possible $4^2 = 16$ risk haplotype combinations. The largest likelihood corresponds to an optimal risk haplotype combination between the two stages.

15.2.4 Hypothesis Testing

As shown in Wang and Wu (2004), functional mapping for HIV dynamics can make a number of biologically and clinically meaningful hypothesis tests about the genetic control of HIV progression as shown below:

15.2.4.1 Global Test

It tests the existence of significant risk haplotypes for the whole period of HIV dynamics. The hypothesis is formulated as

$$H_{01} : \quad (P_{1j_1}, P_{2j_1}, \lambda_{1j_1}, a_{lj_2}, b_{lj_2}, c_{lj_2}, d_{lj_2}) \equiv (P_1, P_2, \lambda_1, a_l, b_l, c_l, d_l)$$
$$j_1, j_2 = 2, 1, 0; l = 0, ..., m \qquad (15.7)$$

$$H_1 : \quad \text{Not all these equalities above hold}$$

The H_{01} states that there is a risk haplotype that is associated with HIV dynamics (the reduced model), whereas H_1 proposes that such a risk haplotype does exist (the full model). The test statistics for testing the hypotheses is the log-likelihood ratio (LR) of the reduced model to the full model. The LR value asymptotically follows a χ^2-distribution. The critical threshold can also be determined from permutation tests. The global effects of different genetic components, additive and dominant, on the shape of the entire HIV dynamics can also be tested.

15.2.4.2 Regional Test

Instead of testing the whole dynamic curve, we can additionally test how the detected risk haplotypes trigger their effects on the difference of viral load trajectories during a particular period $[t_1, t_2]$. This test can be based on the area under curve with the null hypothesis expressed as

$$H_{02} : \quad \int_{t_1}^{t_2} (P_{1j_1} e^{-\lambda_{1j_1} t} + P_{2j_1} e^{-\lambda_{2j_2}(t)t}) dt$$
$$= \int_{t_1}^{t_2} (P_{1j} e^{-\lambda_1 t} + P_{2j} e^{-\lambda_2(t)t}) dt. \qquad (15.8)$$

15.2.4.3 Local Test

It is also important to know how the detected risk haplotypes affect viral load at a given time point (t^*) of interest. The null hypothesis for this test is constructed as

$$H_{03} : P_{1j_1} e^{-\lambda_{1j_1} t^*} + P_{2j_1} e^{-\lambda_{2j_2}(t)t^*} = P_1 e^{-\lambda_1 t^*} + P_{2j} e^{-\lambda_2(t)t^*}. \qquad (15.9)$$

By moving the time point from 1 to T, this so-called local test allows for the characterization of the timing at which the risk haplotypes are switched on or off. The effects of different genetic components for a particular time point can be tested accordingly.

15.2.4.4 Interaction Test

The effects of risk haplotypes may change with time, which suggests the occurrence of haplotype × time interaction effects on the HIV dynamics. The differentiation of $\mu_{j_1 j_2}(t)$ with respect to time t represents a slope of the viral load curve (decay rate). If the slopes at a particular time point t^* are different between the curves of different composite diplotypes, this means that significant haplotype × time interaction occurs between this time point and next. The test for haplotype × time interaction can be formulated with the null hypothesis:

$$H_{04} : \frac{d}{dt}\mu_{j_1 j_2}(t^*) = \frac{d}{dt}\mu(t^*). \tag{15.10}$$

The effect of haplotype × time interaction on HIV pathogenesis can be examined during entire viral load trajectories.

15.2.4.5 Pleiotropic Test

It tests whether a risk haplotype pleiotropically affects the earlier and later stages of HIV dynamics. The effect of risk haplotype [11] on the earlier and later stage can be tested by formulating the following null hypotheses, respectively

$$H_{051} : (P_{1j_1}, P_{2j_1}, \lambda_{1j_1}) \equiv (P_1, P_2, \lambda_1), \tag{15.11}$$

and

$$H_{052} : (a_{lj_1}, b_{lj_1}, c_{lj_1}, d_{lj_1}) \equiv (a_l, b_l, c_l, d_l) \tag{15.12}$$

When both null hypotheses (15.11) and (15.12) are rejected, this means that risk haplotype [11] exerts a pleiotropic effect on the two stages of HIV dynamics.

Similarly, the pleiotropic effect of risk haplotype [10] on the early and late stages of HIV dynamics can be tested by formulating the following null hypotheses:

$$H_{051} : (P_{1j_2}, P_{2j_2}, \lambda_{1j_2}) \equiv (P_1, P_2, \lambda_1), \tag{15.13}$$

and

$$H_{052} : (a_{lj_2}, b_{lj_2}, c_{lj_2}, d_{lj_2}) \equiv (a_l, b_l, c_l, d_l) \tag{15.14}$$

The LR values for testing hypotheses (15.8)–(15.12) can be thought to be χ^2-distributed with the degrees of freedom equal to the difference in the number of unknown parameters being estimated between the null hypothesis and its alternative.

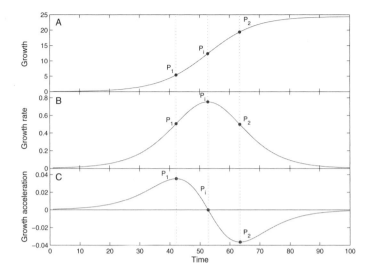

FIGURE 15.3
Diagrams for (**A**) growth curve (equation 15.15), (**B**) growth rate curve (equation 15.16), and (**C**) growth acceleration curve (equation 15.17). Three critical points P_1, P_i, and P_2 are shown on the curves.

15.3 Semiparametric Modeling of Functional Mapping: PCD

15.3.1 Phase Dissection of Growth

There are a number of mathematical models that describe the growth process of a trait (von Bertalanffy, 1957; West et al., 2001). Among others, Richards's (1959) model derived from the balance between cell birth and death rates presents a general growth equation (Fig. 15.3A), expressed as

$$g(t) = a(1 + be^{-rt})^{\frac{1}{1-k}}, \qquad (15.15)$$

where a is the asymptotic value of a trait, b is a parameter to position the curve on the time axis, r is the growth rate constant of the trait, and k is the shape parameter of the curve. Such a universal growth model (15.15) has been used to explain the exponential law for ontogenetic growth of any tissue, organ, and organism based on basic cellular mechanisms (West et al., 2001), thus providing fundamental principles for modeling organismic development at the mechanistic level.

As a general growth model, equation (15.15) not only describes an overall shape of growth trajectory (Fig. 15.3A) used for cancer diagnosis, but also provides a number of hypotheses that can be tested in a biologically meaningful manner. In the growth curve, there are three physiologically important points, denoted as P_1, P_i, and P_2, respectively (Gregorczyk, 1998); Fig. 15.3). The point P_i, known as the inflection point, is one at which growth rate reaches its maximum (Fig. 15.3**B**). Because of this point, the curve is divided into two phases, the exponential growth (from time $t = 0$ to P_i) and the asymptotic growth (from P_i to the infinite time). Thus, the determination of the timing of the inflection point can help us to better understand the shape and process of tumor growth.

The points P_1 and P_2 present the timing of maximum acceleration and maximum deceleration of growth (Fig. 15.3**C**), which are the first and second inflection points of growth rate curve, respectively. These two points partition the growth curve into three phases, the exponential growth (from time $t = 0$ to P_1), the linear growth (from P_1 to P_2) and the ageing (from P_2 to the infinite). Gregorczyk (1998) provided the coordinates of these three points by calculating the first, second, and third derivatives of equation (15.15) with respect to time given as

$$\frac{dg}{dt} = g' = \frac{abr}{k-1}e^{-rt}(1+be^{-rt})^{\frac{k}{1-k}}, \tag{15.16}$$

$$\frac{d^2g}{dt^2} = g'' = \frac{abr^2 e^{-rt}}{(k-1)^2(b+e^{rt})}(1+be^{-rt})^{\frac{k}{1-k}}[b+(1-k)e^{rt}], \tag{15.17}$$

$$\frac{d^3g}{dt^3} = g''' = \frac{r^2 g}{(k-1)^2}\left[1-\left(\frac{g}{a}\right)^{k-1}\right]$$
$$\left\{\frac{r}{k-1}\left[1-k\left(\frac{g}{a}\right)^{k-1}\right]^2 - \frac{r}{k-1}k(k-1)\left[1-\left(\frac{g}{a}\right)^{k-1}\right]\left(\frac{g}{a}\right)^{k-1}\right\}. \tag{15.18}$$

By letting these derivatives equal to zero, the timing (abscissa) of each of the three points (P_i, P_1 and P_2) is derived, respectively, as

$$t_i = \frac{1}{r}\ln\left[\frac{b}{k-1}\right], \tag{15.19}$$

$$t_1 = t_i + \frac{1}{r}\ln\left[\frac{(k-1)\eta_1^{k-1}}{1-\eta_1^{k-1}}\right], \tag{15.20}$$

$$t_2 = t_i + \frac{1}{r}\ln\left[\frac{(k-1)\eta_2^{k-1}}{1-\eta_2^{k-1}}\right], \tag{15.21}$$

where

$$\eta_{1,2} = \left[\frac{k(k+1)\mp(k-1)\sqrt{k(k+4)}}{2k(2k-1)}\right]^{\frac{1}{k-1}}.$$

Based on these times (15.19), (15.20), and (15.21), we can quantify different phases of growth as described in Section 15.1.2. They are

(1) The lag phase spanning $[0, t_1]$,

(2) The exponential phase spanning $[t_1, t_i]$,

(3) The declining growth rate spanning $[t_i, t_2]$,

(4) The stationary phase spanning $[t_2, t^*]$.

The transition time point, t^*, is a point which marks the end of growth phase and beginning of the death phase. The time duration for the linear growth phase (or grand period of growth) (Zelawski and Lech, 1980) is calculated as

$$\Delta t = t_2 - t_1 = \frac{1}{r} \ln \left[\frac{\eta_2^{k-1}(1 - \eta_1^{k-1})}{\eta_1^{k-1}(1 - \eta_2^{k-1})} \right]. \tag{15.22}$$

It is interesting to estimate the amounts of growth during each of these phases, which can be calculated by the area under the curve, expressed as

$$A_{[0,t_1]} = \int_0^{t_1} g(t)dt, \text{ for the lag phase,}$$

$$A_{[t_1,t_i]} = \int_{t_1}^{t_i} g(t)dt, \text{ for the exponential phase,}$$

$$A_{[t_i,t_2]} = \int_{t_i}^{t_2} g(t)dt, \text{ for the declining growth rate phase,}$$

$$A_{[t_2,t^*]} = \int_{t_2}^{t^*} g(t)dt, \text{ for the stationary phase.}$$

In practice, the amount of growth $A_{[t_1,t_2]}$ between time points t_1 and t_2 is also interesting and can be calculated in a similar way.

While equation (15.15) is the growth curve (Fig. 15.3A), the first (15.16) and second derivatives (15.17) are the growth rate curve (Fig. 15.3B) and the growth acceleration curve (Fig. 15.3C), respectively. The growth rate curve describes the growth change per unit time and the growth acceleration curve describes the growth rate change per unit time. It is easy to derive the ordinates of these three points, respectively, as

$$g_i = ak^{\frac{1}{1-k}}, \tag{15.23}$$

$$g_1 = a\eta_1, \tag{15.24}$$

$$g_2 = a\eta_2, \tag{15.25}$$

for the growth curve,

$$g_i' = ark^{\frac{k}{1-k}}, \tag{15.26}$$

$$g_1' = \frac{ar\eta_1}{k-1}(1 - \eta_1^{k-1}), \tag{15.27}$$

$$g_2' = \frac{ar\eta_2}{k-1}(1 - \eta_2^{k-1}), \tag{15.28}$$

for the growth rate curve, and

$$g_i'' = 0, \tag{15.29}$$

$$g_1'' = \frac{ar^2\eta_1}{(k-1)^2}(1 - \eta_1^{k-1})(1 - k\eta_1^{k-1}), \tag{15.30}$$

$$g_2'' = \frac{ar^2\eta_2}{(k-1)^2}(1 - \eta_2^{k-1})(1 - k\eta_2^{k-1}), \tag{15.31}$$

for the growth acceleration curve.

15.3.2 Haplotyping Model

15.3.2.1 Likelihood

We assume the same genetic design, as described in Section 15.2.1, used to haplotype PCD with two associated SNPs in a natural population. It is general to assume that the growth and death phases are associated with different risk haplotypes, denoted as A and B, respectively. Thus, different composite diplotypes should be assigned for these two phases with the diplotype information in Table 2.1. Let j_1 and j_2 denote a composite diplotype for the growth and death phases, respectively, $j_1 = AA$ (2), $A\bar{A}$ (1), $\bar{A}\bar{A}$ (0) and $j_1 = BB$ (2), $B\bar{B}$ (1), $\bar{B}\bar{B}$ (0). A trait for each subject i is measured at a series of time points, $(t_{i1}, ..., t_{iT_i})$. Thus, a mixture-based likelihood similar to (15.3) can be constructed, in which the mean vector $\mathbf{u}_{j_1 j_2 | i}$ and covariance $\mathbf{\Sigma}_i$ of the multivariate normal distribution for a composite diplotype combination between the growth and death phases can be modeled.

15.3.2.2 Semiparametric Modeling of the Mean Vector

Different from likelihood (15.3), the likelihood to be constructed here should consider the transition time for each subject. At the transition time t_i^*, the PCD process of subject i is divided into two phases, growth $[t_{i1}, ..., t_i^*]$ and death $[t_i^*, ..., t_{iT_i}]$. Let $\mathbf{y}_i = [y_i(t_{i1}), ..., y_i(t_i^*), y_i(t_i^*), ..., y_i(t_{iT_i})]$ be a new observation vector for subject i, where $y_i(t_i^*)$ is repeated twice. Now, the diplotype-specific mean vector of the trait for subject i is modeled as

$$
\begin{aligned}
\mathbf{u}_{j_1 j_2 | i} &= \left(\mathbf{u}_{G j_1 | i}; \mathbf{u}_{D j_2 | i} \right) \\
&= \left[u_{G j_1}(t_{i1}), ..., u_{G j_1}(t_{it_i^*}); u_{D j_1}(t_{it_i^*}), ..., u_{G j_1}(t_{iT_i}) \right] \\
&= \left[a_{j_1} \left(1 + b_{j_1} e^{-r_{j_1} t_{i1}} \right)^{\frac{1}{1-k_{j_1}}}, ..., a_{j_1} \left(1 + b_{j_1} e^{-r_{j_1} t_i^*} \right)^{\frac{1}{1-k_{j_1}}}; \right. \\
&\qquad \left. \mathbf{v}_{j_2}^{\mathrm{T}} \mathbf{P}(\tau_i^*), ..., \mathbf{v}_{j_2}^{\mathrm{T}} \mathbf{P}(\tau_{iT_i}) \right]
\end{aligned}
\qquad (15.32)
$$

where $\mathbf{u}_{G j_1 | i}$ is the growth vector from t_{i1} to t_i^*, modeled by a growth equation (15.15), where $\mathbf{u}_{D j_2 | i}$ is the death vector, modeled by a nonparametric approach, such as B-splines or Legendre polynomial. Note that $\mathbf{v}_{j_2}$ is the basis genotypic value vector for composite diplotype j_2, which needs to be estimated, and $\mathbf{P}(\tau_i^*)$, ..., $\mathbf{P}(\tau_{iT_i})$ are the basis spline functions or Legendre polynomials at normalized time τ_i^*, ..., τ_{iT_i}, respectively. All the parameters that model the mean vector (15.32) are arrayed in $\Omega_u = (\{a_{j_1}, b_{j_1}, r_{j_1}, k_{j_1}\}_{j_1=0}^2; \{\mathbf{v}_{j_2}\}_{j_2=0}^2)$.

This semiparametric approach has great interpretability and flexibility for modeling longitudinal data that cannot be obtained by a parametric or nonparametric model alone. By choosing an appropriate order, the nonparametric model can better capture the intrinsic pattern of developmental PCD. The number of parameters can be reduced if the order of the polynomial should be less than the number of time points.

15.3.2.3 Modeling the Covariance Structure

Because $y_i(t_i^*)$ is repeated twice in the semiparametric modeling of the mean vector (15.32), the covariance matrix Σ_i becomes $(T_i + 1)$-dimensional. This matrix is factorized into four parts as follows:

$$\Sigma_i = \begin{pmatrix} \Sigma_{G|i} & \Sigma_{GD|i} \\ \Sigma_{DG|i} & \Sigma_{D|i} \end{pmatrix}, \tag{15.33}$$

where $\Sigma_{G|i}$ and $\Sigma_{D|i}$ are the t_i^*- and $(T_i - t_i^*)$-dimensional covariance matrices for the growth and death phases, respectively, and $\Sigma_{GD|i} = \Sigma_{DG|i}$ is the covariance matrix between the two phases. If the matrix (15.33) is modeled by an AR(1) model, the following parameters will be needed and arrayed in $\Omega_v = (\rho_G, \sigma_G^2, \rho_D, \sigma_D^2, \rho_{GD})$, where we assume that the growth and death phases follow different AR(1) structures and that there is a correlation between the two phases.

15.3.3 Computation Algorithms

The EM and simplex algorithms have served as a standard approach for obtaining the maximum likelihood estimates (MLEs) of the parameters (Ω_u, Ω_v) in functional mapping. However, because of the complex nonlinear function being minimized by simplex algorithm, it cannot always guarantee the correct convergence of covariance parameters during the minimization process. This consequently results in negative infinity of the log-likelihood function and convergence will never be reached. Based on these concerns, the simplex algorithm can be used to estimate the mean parameters (Ω_u), namely the logistic curve and spline or Legendre polynomial parameters, and the EM algorithm used to estimate the parameters (Ω_v) that model the AR(1) structure of the covariance matrix (Cui et al., 2006).

Under the semiparametric modeling framework, two mean functions, growth and death, need to be connected. Two constraints are imposed to make the PCD curve continuous at the transition time point t_i^* for each subject. The first constraint is to make the growth mean equal to the death mean at time t_i^*. The second constraint is that the two functions have the same score at time t_i^* (Cui et al., 2006).

It is possible that the algorithm described above may generate local maxima for the likelihood surface. An empirical approach for reducing the possibility of local maxima is to use multiple sets of initial values of the parameters. The initial values are determined in the light of parameters estimates from the data by assuming no QTL involved. We will obtain the global maxima when no further increase of the likelihood is found in a space of parameters.

As usual, an optimal combination between risk haplotypes for the growth and death phases should be determined, which can be based on the likelihood. Also, for nonparametric modeling, an optimal polynomial order and an optimal number of knots (for B-splines) should be estimated with the AIC or BIC criteria.

15.3.4 Hypothesis Testing

15.3.4.1 Overall Haplotype Effects

Testing whether specific risk haplotypes exist to affect the PCD process is a first step toward the understanding of the detailed genetic architecture of complex phenotypes. The genetic control of the entire PCD process can be tested by formulating the following hypotheses:

$$\begin{cases} H_0 : (a_{j_1}, b_{j_1}, r_{j_1}, k_{j_1}; \mathbf{v}_{j_2}) \equiv (a, b, r, k; \mathbf{v}) \\ H_{1a} : \text{At least one of the equalities above does not hold} \end{cases} \tag{15.34}$$

The H_0 states that there is no risk haplotype affecting the dynamic PCD process (the reduced model), whereas the H_{1a} proposes that such a risk haplotype does exist (the full model). The test statistic for testing the hypotheses is calculated as the log-likelihood ratio of the reduced to the full model.

15.3.4.2 Phase-specific Haplotype Effects

Other hypotheses can be made to test if the detected risk haplotype A only controls the growth phase with the following null hypothesis:

$$H_0 : (a_{j_1}, b_{j_1}, r_{j_1}, k_{j_1}) \equiv (a, b, r, k) \tag{15.35}$$

or if it only controls the death phase with the following hypothesis:

$$H_0 : \mathbf{v}_{j_1} \equiv \mathbf{v} \tag{15.36}$$

If both the null hypotheses are rejected, this indicates that risk haplotype A pleiotropically affects the growth and death phases. Similarly, the pleiotropic effect of risk haplotype B on the two phases can be tested.

The model can be used to test the influence of risk haplotype A on different stages of growth, lag, exponential, declining growth rate, and stationary. These tests can be based on the area under curve during a time course of interest, with the null hypotheses being given below:

$$\begin{aligned} H_0 &: A_{[0,t_1]j_1} \equiv A_{[0,t_1]}, &&\text{for the lag phase,} \\ H_0 &: A_{[t_1,t_i]j_1} \equiv A_{[t_1,t_i]}, &&\text{for the exponential phase,} \\ H_0 &: A_{[t_i,t_2]j_1} \equiv A_{[t_i,t_2]}, &&\text{for the declining growth rate phase,} \\ H_) &: A_{[t_2,t^*]j_1} \equiv A_{[t_2,t^*]}, &&\text{for the stationary phase.} \end{aligned}$$

Example 15.1
Semiparametric Functional Mapping of QTLs: We used Cui et al.'s (2006) functional mapping of quantitative trait loci (QTLs) to show the utilization of semiparametric modeling in longitudinal data analysis. The analysis was performed for a a data set of rice genetics. Two inbred lines, semi-dwarf IR64 and tall Azucena, were

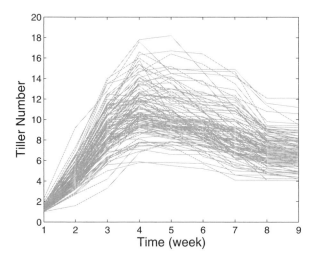

FIGURE 15.4

Dynamic changes of the number of tillers for 123 DH lines of rice as an example of PCD in plants. Adapted from Cui et al. (2006).

crossed to generate an F_1 progeny population. By doubling haploid chromosomes of the gametes derived from the heterozygous F_1, a doubled haploid (DH) population of 123 lines were founded (Huang et al., 1997). Such a DH population is equivalent to a backcross population because its marker segregation follows 1:1. A genetic linkage map of length 2005 cM with an average distance of 11.5 cM was then constructed, representing a good coverage of 12 rice chromosomes.

Starting from 10 days of transplanting, tiller numbers were measured every 10 days for each DH line. Figure 15.4 illustrates the dynamics of tiller numbers for all the DH lines measured at 9 time points. Cui et al. (2006) used the semiparametric model to map QTLs for the dynamic changes of tiller number during ontogeny. A growth equation was used to model the growth phase of tiller numbers, whereas a nonparametric approach based on the Legendre polynomial was used to model the death phase. A modified approach was used to model the covariance structure (Cui et al., 2006). According to the AIC and BIC information criteria, the death phase of tiller number can be best explained by the Legendre polynomial of order 3. Genomewide scanning for QTLs at every 2 cM within each marker interval leads to the LR profile of Fig. 15.5. Three QTLs were detected to trigger their effects on the overall PCD process of tiller number, which are located between markers RG146 and RG345 and between markers RZ730 and RZ801 on chromosome 1 and on marker RZ792 on chromosome 9.

To know more about the behavior of the detected QTLs, the MLEs of parameters for the growth and death phases were used to draw the developmental trajectories of tiller number for the two different QTL genotypes (Fig. 15.6). Each QTL shows

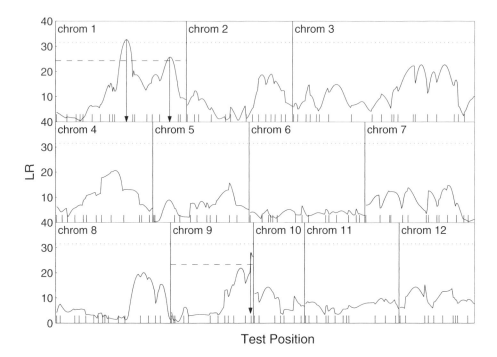

FIGURE 15.5

The profile of the log-likelihood ratios between the full and reduced (no QTL) model for tiller number trajectories across the 12 rice chromosomes. The genomic positions corresponding to the peak of the curve are the MLEs of the QTL localization (indicated by the arrows). The threshold value for claiming the existence of QTL is given as the horizonal solid line for the genome-wide level and broken line for the chromosome-wide level. The positions of markers on the linkage groups (Huang et al., 1997) are indicated at ticks. Adapted from Cui et al. (2006).

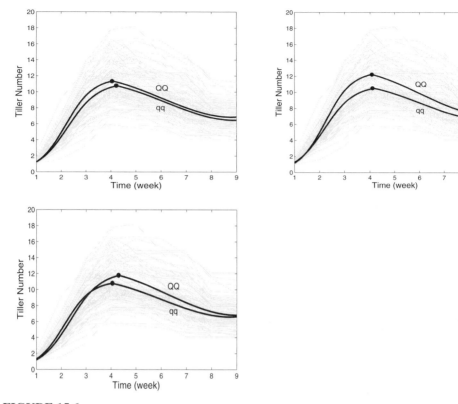

FIGURE 15.6

Two curves for the dynamic changes of tiller numbers each presenting two groups of genotypes, *QQ* and *qq*, at each of the three QTL, detected between markers RG146 and RG345 (**A**) and between markers RZ730 and RZ801 (**B**) on chromosome 1 and near marker RZ792 on chromosome 9 (**C**). Tiller number trajectories for all the individuals studied are indicated in grey background. Adapted from Cui et al. (2006).

a unique developmental pattern over time. For example, the dynamic process of genetic effects for the QTL located between markers RZ730 and RZ801 on chromosome 1 is different than those for the other two QTL. Statistical tests based on hypotheses 15 and 16 show that the QTL detected between markers RG146 and RG345 on chromosome 1 and on marker RZ792 on chromosome 9 merely control the growth phase, whereas the second QTL on chromosome 1 controls the entire developmental process ($P < 0.05$).

Tiller growth is thought to be an excellent example of PCD in plants (Greenberg, 1996) since it experiences several developmental stages when rices grow. At an early stage, tiller numbers increase dramatically corresponding to vegetative phase in rice. During reproductive phase, the increase of tiller numbers declines since the initiation of panicle, emergence of the flag leaf (the last leaf) and booting, heading, and flowering of the spikelets. Tillers that do not bear panicles are called ineffective tillers, and will be killed, leading to the death phase. The number of ineffective tillers is a closely examined trait in plant breeding since they are undesirable for commercial varieties. The ineffective tillers result in many unwanted problems in rice such as the overconsumption of nutrition and competition of space. The genetical control system will play an important role in reducing overproduced tillers and balancing the rice metabolism system for the optimal use efficiency of nutrients.   □

References

Allison, D.B. (1997). Transmission-disequilibrium tests for quantitative traits. *American Journal of Human Genetics* **60**, 676–90.

Altshuler, D., L.D. Brooks, A. Chakravarti, F.S. Collins, M.J. Daly, and P. Donnelly (2005). A haplotype map of the human genome. *Nature* **437**, 1299–1320.

Andresen, B., J.S. Shiner, and D.E. Uehlinger (2002). Allometric scaling and maximum efficiency in physiological eigen time. *Proceedings of the National Academy of Sciences of the United States of America* **99**, 5822–5824.

Andrews, K.L., J.L. Mudd, C. Li, and J.H. Miner (2002). Quantitative trait loci influence renal disease progression in a mouse model of alport syndrome. *American Journal of Pathology* **160**, 721–730.

Anholt, R.R.H. and T.F.C. Mackay (2004). Quantitative genetic analyses of complex behaviours in *Drosophila*. *Nature Reviews Genetics* **5**, 838–849.

Attinger, E.O., A. Anne, and D.A. McDonald (1966). Use of fourier series for the analysis of biological systems. *Biophysical Journal* **6**, 291–304.

Avraham, S., C.W. Tung, K. Ilic, P. Jaiswal, E.A. Kellogg, S. McCouch, A. Pujar, L. Reiser, S.Y. Rhee, M.M. Sachs, M. Schaeffer, L. Stein, P. Stevens, L. Vincent, F. Zapata, and D. Wareothers (2008). The plant ontology database: a community resource for plant structure and developmental stages controlled vocabulary and annotations. *Nucleic Acids Research* **36**, D449.

Bader, J.S. (2001). The relative power of snps and haplotype as genetic markers for ssociation tests. *Pharmacogenomics* **2**, 11–24.

Banavar, J.R., A. Maritan, and A. Rinaldo (1999). Size and form in efficient transportation networks. *Nature* **399**, 130–132.

Bard, J.B., S.Y. Rhee, and S.M. Ashburner (2005). An ontology for cell types. *Genome Biology* **6**, R21.

Barrett, W.W. (1979). A theorem on inverses of tridiagonal matrices. *Linear Algebra and its Applications* **27**, 211–217.

Bateson, W. (1909). *Mendel's Principles of Heredity*. Cambridge University Press, New York.

Beavis, W.D. and P. Keim (1996). Identification of QTL that are affected by environment. In: *Genotype by Environment Interactions*, edited by Kang, M.S. and Gauch, H. CRC Press, Boca Raton, FL, pp. 123–149.

Begum, E., M. Bonno, M. Obata, H. Yamamoto, M. Kawai, and Y. Komada (2006). Emergence of physiological rhythmicity in term and preterm neonates in a neonatal intensive care unit. *Journal of Circadian Rhythms* **4**, doi:10.1186/1740–3391–4–11.

Bennett, J.H. (1954). On the theory of random mating. *Annals of Eugenics* **18**, 311–317.

Biernacki, C., G. Celeux, and G. Govaert (1999). An improvement of the nec criterion for assessing the number of clusters in a mixture model. *Pattern Recognition Letters* **20**, 267–272.

Bookstein, F.L. (1991). *Morphometric Tools for Landmark Data: Geometry and Biology*. Cambridge University Press, New York.

Borlak, J. (2005). *Handbook of Toxicogenomics: Strategies and Applications*. John Wiley & Sons, New York.

Boxenbaum, H. (1982). Interspecies scaling, allometry, physiological time, and the ground plan of pharmacokinetics. *Journal of Pharmacokinetics and Pharmacodynamics* **10**, 201–227.

Bradford, G.E. and T.R. Famula (1984). Evidence for a major gene for rapid post-weaning growth in mice. *Genetical Research* **44**, 293–308.

Broman, K.W. (2005). The genomes of recombinant inbred lines. *Genetics* **169**, 1133–1146.

Brown, E.R., J.G. Ibrahim, and V. DeGruttola (2005). A flexible B-spline model for multiple longitudinal biomarkers and survival. *Biometrics* **61**, 64–73.

Brown, G.R., D.L. Bassoni, G.P. Gill, J.R. Fontana, N.C. Wheeler, R.A. Megraw, M.F. Davis, M.M. Sewell, G.A. Tuskan, and D.B. Neale (2003). Identification of quantitative trait loci influencing wood property raits in loblolly pine (*Pinus taeda* l.). III. QTL verification and candidate gene mapping. *Genetics* **164**, 1537–1546.

Brown, J.H. and G.B. West (2000). *Scaling in Biology*. Oxford University Press, New York.

Bruskiewich, R., E.H. Coe, P. Jaiswal, S. McCouch, M. Polacco, L. Stein, L. Vincent, and D. Ware (2002). The plant ontology consortium and plant ontologies. *Comparative and Functional Genomics* **3**, 137–142.

Burchard, E.G., E. Ziv, N. Coyle, S.L. Gomez, H. Tang, A.J. Karter, J.L. Mountain, E.J. Perez-Stable, D. Sheppard, and N. Risch (2003). The importance of race and ethnic background in biomedical research and clinical practice. *New England Journal of Medicine* **348**, 1170–1175.

Burnham, K.P. and D.R. Anderson (2002). *Model Selection and Multimodel Inference: A Practical Information-Theoretic Approach.* Springer, New York.

Burton, P.R., D.G. Clayton, L.R. Cardon, N. Craddock, P. Deloukas, A. Duncanson, D.P. Kwiatkowski, M.I. McCarthy, W.H. Ouwehand, N.J. Samani, J.A. Todd, P. Donnelly, J.C. Barrett, P.R. Burton, D. Davison, P. Donnelly, et al. (2007). Genome-wide association study of 14,000 cases of seven common diseases and 3,000 shared controls. *Nature* **447**, 661–678.

Calder, W.A. (1996). *Size, Function, and Life History.* Courier Dover Publications, New York.

Calvert, C.C., T.R. Famula, J.F. Bernier, and G.E. Bradford (1985). Serial composition during growth in mice with a major gene for rapid postweaning growth. *Growth* **49**, 246–57.

Camp, N.J. (1998). Genomewide transmission/disequilibrium testing: A consideration of the genotype relative risks at disease loci. *American Journal of Human Genetics* **61**, 1424–1430.

Cardon, L.R. (2006). Genetics. delivering new disease genes. *Science* **314**, 1403–1405.

Carroll, R.J. and D. Ruppert (1984). Power transformations when fitting theoretical models to data. *Journal of the American Statistical Association* **79**, 321–328.

Cashio, P., T.V. Lee, and A. Bergmann (2005). Genetic control of programmed cell death in drosophila melanogaster. *Seminars in Cell and Developmental Biology* **16**, 225–235.

Celeux, G. and G. Soromenho (1996). An entropy criterion for assessing the number of clusters in a mixture model. *Journal of Classification* **13**, 195–212.

Chagnon, Y.C., T. Rankinen, E.E. Snyder, S.J. Weisnagel, L. Perusse, and C. Bouchard (2003). The Human Obesity Gene Map: The 2002 Update. *Obesity Research* **11**, 313–367.

Chen, H., J. Chen, and J.D. Kalbfleisch (2004). Testing for a finite mixture model with two components. *Journal of the Royal Statistical Society Series B* **66**, 95–115.

Cheverud, J.M. (2006). Genetic architecture of quantitative variation. In: *Evolutionary Genetics: Concepts and Case Studies*, edited by Fox, C.W. and Wolf, J.B. Oxford University Press, Oxford, UK, pp. 288–309.

Cheverud, J.M., E.J. Routman, F.A.M. Duarte, B. van Swinderen, K. Cothran, and C. Perel (1996). Quantitative trait loci for murine growth. *Genetics* **142**, 1305–1319.

Churchill, G.A. and R.W. Doerge (1994). Empirical threshold values for quantitative trait mapping. *Genetics* **138**, 963–971.

Cordell, H.J. (2002). Epistasis: what it means, what it doesn't mean, and statistical methods to detect it in humans. *Human Molecular Genetics* **11**, 2463–2468.

Corva, P.M., S. Horvat, and J.F. Medrano (2001). Quantitative trait loci affecting growth in highgrowth (*hg*) mice. *Mammalian Genome* **12**, 284–290.

Craft, R.M., J.S. Mogil, and A. Maria Aloisi (2004). Sex differences in pain and analgesia: the role of gonadal hormones. *European Journal of Pain* **8**, 397–411.

Crosthwaite, S.K. (2004). Circadian clocks and natural antisense RNA. *FEBS Letters* **567**, 49–54.

Cui, Y. H., J. Zhu, and R.L. Wu (2006). Functional mapping for genetic control of programmed cell death. *Physiological Genomics* **25**, 458–469.

Cuppen, E. (2005). Haplotype-based genetics in mice and rats. *Trends in Genetics* **21**, 318–322.

Dale, J.K., M. Maroto, M.L. Dequeant, P. Malapert, M. McGrew, and O. Pourquie (2003). Periodic notch inhibition by lunatic fringe underlies the chick segmentation clock. *Nature* **421**, 275–278.

Daly, M.J., J.D. Rioux, S.F. Schaffner, T.J. Hudson, and E.S. Lander (2001). High-resolution haplotype structure in the human genome. *Nature Genetics* **29**, 229–232.

Darvasi, A. (1998). Experimental strategies for the genetic dissection of complex traits in animal models. *Nature Genetics* **18**, 19–24.

Dawson, E., G.R. Abecasis, S. Bumpstead, Y. Chen, S. Hunt, D.M. Beare, J. Pabial, T. Dibling, E. Tinsley, S. Kirby, et al. (2002). A first-generation linkage disequilibrium map of human chromosome 22. *Nature* **418**, 544–548.

de Boor, C. (2001). *A Practical Guide to Splines*. Springer, New York.

de Jong, G. (2005). Research reviews: evolution of phenotypic plasticity: patterns of plasticity and the emergence of ecotypes. *New Phytologist* **166**, 101–118.

Dempster, A.P., N.M. Laird, and D.B. Rubin (1977). Maximum likelihood from incomplete data via the em algorithm. *Journal of the Royal Statistical Society Series B* **39**, 1–38.

Derendorf, H. and B. Meibohm (1999). Modeling of pharmacokinetic/pharmacodynamic (PK/PD) relationships: concepts and perspectives. *Pharmaceutical Research* **16**, 176–185.

DeWan, A., M. Liu, S. Hartman, S.S.M. Zhang, D.T.L. Liu, C. Zhao, P.O.S. Tam, W.M. Chan, D.S.C. Lam, M. Snyder, et al. (2006). HTRA1 promoter polymorphism in wet age-related macular degeneration. *Science* **314**, 989–992.

Dewey, T.G. and M. Delle Donne (1998). Non-equilibrium thermodynamics of molecular evolution. *Journal of Theoretical Biology* **193**, 593–599.

Diggle, P.J., P. Heagerty, K. Y. Liang, and S. L. Zeger (2002). *Analysis of Longitudinal Data*. Oxford University Press, Oxford, UK.

Diggle, P.J. and A.P. Verbyla (1998). Nonparametric estimation of covariance structure in longitudinal data. *Biometrics* **54**, 401–415.

Du, F.X., P. Sorensen, G. Thaller, and I. Hoeschele (2002). Joint linkage disequilibrium and linkage mapping of quantitative trait loci. *Proceedings of the 7th World Congress Genetics Applied to Livestock Production* **32**, 661–668.

Duerr, R.H., K.D. Taylor, S.R. Brant, J.D. Rioux, M.S. Silverberg, M.J. Daly, A.H. Steinhart, C. Abraham, M. Regueiro, A. Griffiths, et al. (2006). A genome-wide association study identifies IL23R as an inflammatory bowel disease gene. *Science* **314**, 1461–1463.

Ebert, T.A. and M.P. Russell (1994). Allometry and model ii non-linear regression. *Journal of theoretical biology* **168**, 367–372.

Enquist, B.J., J.H. Brown, and G.B. West (1998). Allometric scaling of plant energetics and population density. *Nature* **395**, 163–165.

Enquist, B., G. West, E. Charnov, and J. Brown (1999). Allometric scaling of production and life-history variation in vascular plants. *Nature* **401**, 907–911.

Evans, W.E. and J.A. Johnson (2001). Pharmacogenomics: the inherited basis for interindividual differences in drug response. *Annual Review of Genomics and Human Genetics* **2**, 9–39.

Evans, W.E. and H.L. McLeod (2003). Pharmacogenomics–drug disposition, drug targets, and side effects. *New England Journal of Medicine* **348**, 538–549.

Evans, W.E. and M.V. Relling (1999). Pharmacogenomics: translating functional genomics into rational therapeutics. *Science* **286**, 487–491.

Eyheramendy, S., J. Marchini, G. McVean, S. Myers, and P. Donnelly (2007). A model-based approach to capture genetic variation for future association studies. *Genome Research* **17**, 88–95.

Falconer, D.S. and T.F.C. Mackay (1996). *Introduction to Quantitative Genetics* (4 ed.). Longman, New York.

Fan, J. and R. Li (2004). New estimation and model selection procedures for semiparametric modeling in longitudinal data analysis. *Journal of the American Statistical Association* **99**, 710–724.

Feng, M.R., X. Lou, R.R. Brown, and A. Hutchaleelaha (2000). Allometric pharmacokinetic scaling: towards the prediction of human oral pharmacokinetics. *Pharmaceutical Research* **17**, 410–418.

Ferraty, F. and P. Vieu (2006). *Nonparametric Functional Data Analysis: Theory and Practice*. Springer-Verlag, New York.

Ferreira, E., V. Nunez-AntonNúñez-Antón, and J. Rodriguez-Poo (1997). Kernel regression estimates of growth curves using nonstationary correlated errors. *Statistics and Probability Letters* **34**, 413–423.

Fillingim, R.B. (2003a). Sex, gender and pain: The biopsychosocial model in action XX vs. XY. *The International Journal of Sex Differences in the Study of Health, Disease and Aging* **1**, 98–101.

Fillingim, R.B. (2003b). Sex-related influences on pain: a review of mechanisms and clinical implications. *Rehabilitation Psychology* **48**, 165–174.

Fillingim, R.B., L. Kaplan, R. Staud, T.J. Ness, T.L. Glover, C.M. Campbell, J.S. Mogil, and M.R. Wallace (2005). The A118G single nucleotide polymorphism of the μ-opioid receptor gene (OPRM1) is associated with pressure pain sensitivity in humans. *Journal of Pain* **6**, 159–167.

Fisher, R.A. (1918). The correlation between relatives on the supposition of mendelian inheritance. *Transactions of the Royal Society of Edinburgh* **52**, 399–433.

Flint, J., W. Valdar, S. Shifman, R. Mott, et al. (2005). Strategies for mapping and cloning quantitative trait genes in rodents. *Nature Reviews Genetics* **6**, 271–286.

Fogg, G.E. (1987). *Algal Cultures and Phytoplankton Ecology*. University of Wisconsin Press, Madison, Wisconsin.

Foulley, J.L., Quaas R.L., and C. Thaon d'Arnoldi (1998). A link function approach to heterogeneous variance components. *Genetics Selection Evolution* **30**, 27–43.

Fraley, C. and A.E. Raftery (1998). How many clusters? which clustering method? answers via model-based cluster analysis. *The Computer Journal* **41**, 578–588.

Frank, O. (1926). Die theorie der pulswellen. *Z Biol* **85**, 91–130.

Frary, A., T.C. Nesbitt, A. Frary, S. Grandillo, E. Knaap, B. Cong, J. Liu, J. Meller, R. Elber, K.B. Alpert, et al. (2000). fw2. 2: A quantitative trait locus key to the evolution of tomato fruit size. *Science* **289**, 85–88.

Frayling, T.M., N.J. Timpson, M.N. Weedon, E. Zeggini, R.M. Freathy, C.M. Lindgren, J.R.B. Perry, K.S. Elliott, H. Lango, N.W. Rayner, et al. (2007). A common variant in the FTO gene is associated with body mass index and predisposes to childhood and adult obesity. *Science* **316**, 889–894.

Frazer, K.A., D.G. Ballinger, D.R. Cox, D.A. Hinds, L.L. Stuve, R.A. Gibbs, J.W. Belmont, A. Boudreau, P. Hardenbol, S.M. Leal, S. Pasternak, D.A. Wheeler, T.D. Willis, F. Yu, H. Yang, et al. (2007). A second generation human haplotype map of over 3.1 million SNPs. *Nature* **449**, 851–861.

Fulker, D.W. and L.R. Cardon (1994). A sib-pair approach to interval mapping of quantitative trait loci. *American Journal of Human Genetics* **54**, 1092–103.

Gabriel, K.R. (1962). Ante-dependence analysis of an ordered set of variables. *The Annals of Mathematical Statistics* **33**, 201–212.

Gabriel, S.B., S.F. Schaffner, H. Nguyen, J.M. Moore, J. Roy, B. Blumenstiel, J. Higgins, M. DeFelice, A. Lochner, M. Faggart, S.N. Liu-Cordero, C. Rotimi, A. Adeyemo, R. Cooper, R. Ward, E.S. Lander, M.J. Daly, and D. Altshuler (2002). The structure of haplotype blocks in the human genome. *Science* **296**, 2225–2229.

Geiger-Thornsberry, G.L. and T.F.C. Mackay (2002). Association of single-nucleotide polymorphisms at the delta locus with genotype by environment interaction for sensory bristle number in *Drosophila melanogaster*. *Genetical Research* **79**, 211–218.

Genet, A.J.H. (2002). A polymorphism in the $\beta 1$ adrenergic receptor is associated with resting heart rate. *American Journal of Human Genetics* **70**, 935–942.

Georges, M. (2007). Mapping, fine mapping, and molecular dissection of quantitative trait loci in domestic animals. *Annual Review of Genomics and Human Genetics* **8**, 131–162.

Gibbs, R.A., J.W. Belmont, A. Boudreau, S.M. Leal, P. Hardenbol, S.R. Pasternak, D.A. Wheeler, T.D. Willis, F.L. Yu, H.M. Yang, Z.Q. Zeng, G. Yang, H. R. Hu, W.T. Hu, C.H. Li, W. Lin, et al. (2005). A haplotype map of the human genome. *Nature* **437**, 1299–1320.

Gibbs, R.A., J.W. Belmont, P. Hardenbol, T.D. Willis, F. Yu, H. Yang, L.Y. Ch'ang, W. Huang, B. Liu, Y. Shen, P. K.-H. Tam, L.-C. Tsui, M.M.Y. Waye, J.T.-F. Wong, C.Q. Zeng, et al. (2003). The International HapMap Project. *Nature* **426**, 789–796.

Gillespie, J.H. and M. Turelli (1989). Genotype-environment interactions and the maintenance of polygenic variation. *Genetics* **121**, 129–138.

Giraldo, J. (2003). Empirical models and hill coefficients. *Trends in Pharmacological Sciences* **24**, 63–65.

Goldbeter, A. (1995). A model for circadian oscillations in the drosophila period protein (PER). *Proceedings: Biological Sciences* **261**, 319–324.

Goldbeter, A. (2002). Computational approaches to cellular rhythms. *Nature* **420**, 238–245.

Goldman, S. (1963). *Information Theory*. Prentice Hall, New York.

Gong, Y., Z. Wang, T. Liu, W. Zhao, Y. Zhu, J.A. Johnson, and R.L. Wu (2004). A statistical model for functional mapping of quantitative trait loci regulating drug response. *The Pharmacogenomics Journal* **4**, 315–321.

Green, S.A., J. Turki, I.P. Hall, and S.B. Liggett (1995). Implications of genetic variability of human $\beta 2$-adrenergic receptor structure. *Pulmonary Pharmacology* **8**,

1–10.

Greenberg, J.T. (1996). Programmed cell death: a way of life for plants. *Proceedings of the National Academy of Sciences of the United States of America* **93**, 12094–12097.

Gregorczyk, A. (1998). Richards plant growth model. *Journal of Agronomy and Crop Science* **181**, 243–247.

Gronert, G.A., D.L. Fung, J.H. Jones, S.L. Shafer, S.V. Hildebrand, and E.A. Disbrow (1995). Allometry of pharmacokinetics and pharmacodynamics of the muscle relaxant metocurine in mammals. *American Journal of Physiology–Regulatory, Integrative and Comparative Physiology* **268**, 85–91.

Guénet, J.L. and F. Bonhomme (2003). Wild mice: an ever-increasing contribution to a popular mammalian model. *Trends in Genetics* **19**, 24–31.

Guryev, V., B.M.G. Smits, J. van de Belt, M. Verheul, N. Hubner, and E. Cuppen (2006). Haplotype block structure is conserved across mammals. *PLoS Genetics* **2**, e121.

Gutteling, E.W., J.A.G. Riksen, J. Bakker, and J.E. Kammenga (2007). Mapping phenotypic plasticity and genotype–environment interactions affecting life-history traits in Caenorhabditis elegans. *Heredity* **98**, 28–37.

Haddad, J.N. (2004). On the closed form of the covariance matrix and its inverse of the causal ARMA process. *Journal of Time Series Analysis* **25**, 443–448.

Hall, P., N.I. Fisher, and B. Hoffmann (1994). On the nonparametric estimation of covariance functions. *Annals of Statistics* **22**, 2115–2134.

Hardy, G.H. (1908). Mendelian proportions in a mixed population. *Science* **28**(706), 49–50.

Hart, J.D. and T.E. Wehrly (1986). Kernel regression estimation using repeated measurements data. *Journal of the American Statistical Association* **81**, 1080–1088.

Hastie, T. and R. Tibshirani (1990). *Generalized Additive Models.* Chapman & Hall/CRC.

Ho, D.D., A.U. Neumann, A.S. Perelson, W. Chen, J.M. Leonard, M. Markowitz, et al. (1995). Rapid turnover of plasma virions and CD4 lymphocytes in HIV-1 infection. *Nature* **373**, 123–126.

Hochhaus, G. and H. Derendorf (1995). Dose optimization based on pharmacokinetic-pharmacodynamic modeling. In: *Handbook of Pharmacokinetic-Pharmacodynamic Correlations*, edited by Derendorf, H. and Hochhaus, G. CRC Press, Boca Raton, FL, pp. 79–120.

Horvat, S. and J.F. Medrano (1995). Interval mapping of high growth (*hg*), a major locus that increases weight gain in mice. *Genetics* **139**, 1737–1748.

Horvat, S. and J.F. Medrano (2001). Lack of socs2 expression causes the high-growth phenotype in mice. *Genomics* **72**, 209–212.

Horvitz, H.R. (2003). Worms, life, and death (Nobel Lecture). *ChemBioChem* **4**, 697–711.

Hou, W., C.W. Garvan, R.C. Littell, M. Behnke, F.D. Eyler, and R.L. Wu (2006). A framework to monitor environment-induced major genes for developmental trajectories: implication for a prenatal cocaine exposure study. *Statistics In Medicine* **25**, 4020–4035.

Hou, W., C.W. Garvan, W. Zhao, M. Behnke, F.D. Eyler, and R.L. Wu (2005). A general model for detecting genetic determinants underlying longitudinal traits with unequally spaced measurements and nonstationary covariance structure. *Biostatistics* **6**, 420–433.

Hou, W., J.S.F. Yap, S. Wu, T. Liu, J.M. Cheverud, and R.L. Wu (2007). Haplotyping a quantitative trait with a high-density map in experimental crosses. *PLoS ONE* **2**, e732.

Hu, T.M. and W.L. Hayton (2001). Allometric scaling of xenobiotic clearance: uncertainty versus universality. *merican Association of Pharmaceutical Scientists PharmSci* **3**, E29.

Huang, N., A. Parco, T. Mew, G. Magpantay, S. McCouch, E. Guiderdoni, J. Xu, P. Subudhi, E.R. Angeles, and G.S. Khush (1997). RFLP mapping of isozymes, RAPD and QTLs for grain shape, brown planthopper resistance in a doubled haploid rice population. *Molecular Breeding* **3**, 105–113.

Hurvich, C.M. and C.L. Tsai (1989). Regression and time series model selection in small samples. *Biometrika* **76**, 297–307.

Ideraabdullah, F.Y., E. de la Casa-Esperón, T.A. Bell, D.A. Detwiler, T. Magnuson, C. Sapienza, and F.P.M. de Villena (2004). Genetic and haplotype diversity among wild-derived mouse inbred strains. *Genome Research* **14**, 1880–1887.

Ilic, K., E.A. Kellogg, P. Jaiswal, F. Zapata, P.F. Stevens, L.P. Vincent, S. Avraham, L. Reiser, A. Pujar, M.M. Sachs, et al. (2007). The plant structure ontology, a unified vocabulary of anatomy and morphology of a flowering plant. *Plant Physiology* **143**, 587–599.

Jaffrézic, F., R. Thompson, and W.G. Hill (2003). Structured antedependence models for genetic analysis of repeated measures on multiple quantitative traits. *Genetical Research* **82**, 55–65.

Jaiswal, P. (2005). Plant ontology (PO): a controlled vocabulary of plant structures and growth stages. *Comparative and Functional Genomics* **6**, 388–397.

Jansen, R.C. and P. Stam (1994). High resolution of quantitative traits into multiple loci via interval mapping. *Genetics* **136**, 1447–1455.

Jiang, C.J. and Z.B. Zeng (1995). Multiple trait analysis of genetic mapping for quantitative trait loci. *Genetics* **140**, 1111–1127.

Johnson, C.G. and D. Marsh (1999). Modelling robot manipulators with multivariate B-splines. *Robotica* **17**, 239–247.

Johnson, J.A. (2003). Pharmacogenetics: potential for individualized drug therapy through genetics. *Trends in Genetics* **19**, 660–666.

Johnson, J.A. and S.G. Terra (2002). β-adrenergic receptor polymorphisms: cardiovascular disease associations and pharmacogenetics. *Pharmaceutical Research* **19**, 1779–1787.

Judson, R., J.C. Stephens, and A. Windemuth (2000). The predictive power of haplotypes in clinical response. *Pharmacogenomics* **1**, 15–26.

Kao, C.H. and Z.B. Zeng (1997). General formulas for obtaining the MLEs and the asymptotic variance-covariance matrix in mapping quantitative trait loci when using the EM algorithm. *Biometrics* **53**, 653–665.

Kao, C.H., Z.B. Zeng, and R.D. Teasdale (1999). Multiple interval mapping for quantitative trait loci. *Genetics* **152**, 1203–1216.

Khor, S. P., K. McCarthy, M. DuPont, K. Murray, and G. Timony (2000). Pharmacokinetics, pharmacodynamics, allometry, and dose selection of rpsgl-ig for phase i trial. *Journal of Pharmacology and Experimental Therapeutics* **293**, 618–624.

Kimmel, G. and R. Shamir (2005). GERBIL: genotype resolution and block identification using likelihood. *Proceedings of the National Academy of Sciences of the United States of America* **102**, 158–162.

Kirkpatrick, M. and N. Heckman (1989). A quantitative genetic model for growth, shape, reaction norms, and other infinite-dimensional characters. *Journal of Mathematical Biology* **27**, 429–450.

Klein, R.J., C. Zeiss, E.Y. Chew, J.Y. Tsai, R.S. Sackler, C. Haynes, A.K. Henning, J.P. SanGiovanni, S.M. Mane, S.T. Mayne, et al. (2005). Complement factor h polymorphism in age-related macular degeneration. *Science* **308**, 385–389.

Kliebenstein, D.J., A. Figuth, and T. Mitchell-Olds (2002). Genetic architecture of plastic methyl jasmonate responses in *Arabidopsis thaliana*. *Genetics* **161**, 1685–1696.

Klingenberg, C.P. and M. Zimmermann (1992). Static, ontogenetic, and evolutionary allometry: a multivariate comparison in nine species of water striders. *The American Naturalist* **140**, 601–620.

Knott, G.D. (2000). *Interpolating Cubic Splines*. Birkhauser, Boston, MA.

Labrecque, G. and P.M. Belanger (1991). Biological rhythms in the absorption, distribution, metabolism and excretion of drugs. *Pharmacology & Therapeutics* **52**,

95–107.

Lakin-Thomas, P.L. and S. Brody (2004). Circadian rhythms in microorganisms: new complexities. *Annual Review of Microbiology* **58**, 489–519.

Lander, E.S. and D. Botstein (1989). Mapping Mendelian factors underlying quantitative traits using RFLP linkage maps. *Genetics* **121**, 185–199.

Lander, E.S. and N.J. Schork (1994). Genetic dissection of complex traits. *Science* **265**, 2037–2048.

Large, V., L. Hellstrom, S. Reynisdottir, F. Lonnqvist, P. Eriksson, L. Lannfelt, and P. Arner (1997). Human beta-2 adrenoceptor gene polymorphisms are highly frequent in obesity and associate with altered adipocyte beta-2 adrenoceptor function. *Journal of Clinical Investigation* **100**, 3005–3013.

Lee, R.C., R.L. Feinbaum, V. Ambros, et al. (1993). The *C. elegans* heterochronic gene lin-4 encodes small RNAs with antisense complementarity to lin-14. *Cell* **75**, 843–854.

Leips, J. and T.F.C. Mackay (2000). Quantitative trait loci for life span in drosophila melanogaster interactions with genetic background and larval density. *Genetics* **155**, 1773–1788.

Leloup, J.C. and A. Goldbeter (1998). A model for circadian rhythms in *Drosophila* incorporating the formation of a complex between the PER and TIM proteins. *Journal of Biological Rhythms* **13**, 70–83.

Leloup, J.C. and A. Goldbeter (2003). Toward a detailed computational model for the mammalian circadian clock. *Proceedings of the National Academy of Sciences of the United States of America* **100**, 7051–7056.

Lepist, E.I. and W.J. Jusko (2004). Anti-inflammatory drugs modeling and allometric scaling of s (+)-ketoprofen pharmacokinetics and pharmacodynamics: a retrospective analysis. *Journal of Veterinary Pharmacology & Therapeutics* **27**, 211–218.

Levi, F., R. Zidani, and J.L. Misset (1997). Randomised multicentre trial of chronotherapy with oxaliplatin, fluorouracil, and folinic acid in metastatic colorectal cancer. *Lancet* **350**, 681–686.

Li, C., A. Zhou, and T. Sang (2006a). Rice domestication by reducing shattering. *Science* **311**, 1936–1939.

Li, H., Z. Huang, J. Gai, S. Wu, Y. Zeng, Q. Li, and R.L. Wu (2007). A conceptual framework for mapping quantitative trait loci regulating ontogenetic allometry. *PLoS ONE* **2**, doi: 10.1371/journal.pone.0001245.

Li, H., B.R. Kim, and R.L. Wu (2006b). Identification of quantitative trait nucleotides that regulate cancer growth: a simulation approach. *Journal of Theoretical Biology* **242**, 426–439.

Liang, H., H. Wu, and R.J. Carroll (2003). The relationship between virologic and immunologic responses in AIDS clinical research using mixed-effects varying-coefficient models with measurement error. *Biostatistics* **4**, 297–312.

Lin, D.Y. and Z. Ying (2001). Semiparametric and nonparametric regression analysis of longitudinal data. *Journal of the American Statistical Association* **96**, 103–113.

Lin, M., C. Aquilante, J.A. Johnson, and R.L. Wu (2005). Sequencing drug response with HapMap. *The Pharmacogenomics Journal* **5**, 149–156.

Lin, M., H. Li, W. Hou, J.A. Johnson, and R.L. Wu (2007). Modeling sequence-sequence interactions for drug response. *Bioinformatics* **23**, 1251–1257.

Lin, M. and R.L. Wu (2005). Theoretical basis for the identification of allelic variants that encode drug efficacy and toxicity. *Genetics* **170**, 919–928.

Lin, M. and R.L. Wu (2006a). Detecting sequence-sequence interactions for complex diseases. *Current Genomics* **7**, 59–72.

Lin, M. and R.L. Wu (2006b). A joint model for nonparametric functional mapping of longitudinal trajectory and time-to-event. *BMC Bioinformatics* **7**, doi: 10.1186/1471–2105–7–138.

Lin, M., R.L. Wu, and J.A. Johnson (2006). A bivariate functional mapping model for gene identification of drug response for systolic and diastolic blood pressures. *Pacific Symposium on Biocomputing* **11**, 572–583.

Lindblad-Toh, K., E. Winchester, M.J. Daly, D.G. Wang, J.N. Hirschhorn, J.P. Laviolette, K. Ardlie, D.E. Reich, E. Robinson, P. Sklar, et al. (2000). Large-scale discovery and genotyping of single-nucleotide polymorphisms in the mouse. *Nature Genetics* **24**, 381–386.

Liu, T., J.A. Johnson, G. Casella, and R.L. Wu (2004). Sequencing complex diseases with HapMap. *Genetics* **168**, 503–511.

Liu, T., X.L. Liu, Y.M. Chen, and R.L. Wu (2007). A computational model for functional mapping of genes that regulate intra-cellular circadian rhythms. *Theoretical Biology and Medical Modelling* **4**, doi: 10.1186/1742–4682–4–5.

Liu, T., R.J. Todhunter, Q. Lu, L. Schoettinger, H. Li, R.C. Littell, N. Burton-Wurster, G.M. Acland, G. Lust, and R.L. Wu (2006). Modeling extent and distribution of zygotic disequilibrium: implications for a multigenerational canine pedigree. *Genetics* **174**, 439–453.

Lo, Y., N.R. Mendell, and D.B. Rubin (2001). Testing the number of components in a normal mixture. *Biometrika* **88**, 767–778.

Long, F., Y.Q. Chen, J.M. Cheverud, and R.L. Wu (2006). Genetic mapping of allometric scaling laws. *Genetical Research* **87**, 207–216.

Lou, X.Y., G. Casella, R.C. Littell, M.C.K. Yang, J.A. Johnson, and R.L. Wu (2003).

A haplotype-based algorithm for multilocus linkage disequilibrium mapping of quantitative trait loci with epistasis. *Genetics* **163**, 1533–1548.

Lou, X.Y., G. Casella, R.J. Todhunter, M.C.K. Yang, and R.L. Wu (2005). A general statistical framework for unifying interval and linkage disequilibrium mapping: toward high-resolution mapping of quantitative traits. *Journal of the American Statistical Association* **100**, 158–172.

Louis, T.A. (1982). Finding the observed information matrix when using the EM algorithm. *Journal of the Royal Statistical Society Series B* **44**, 226–233.

Lumer, H. (1937). The consequences of sigmoid growth for relative growth functions. *Growth* **1**, 140–154.

Lynch, M. and B. Walsh (1998). *Genetics and Analysis of Quantitative Traits*. Sinauer Associates, Sunderland, Massachusetts.

Ma, C.X., G. Casella, R.C. Littell, A.I. Khuri, and R.L. Wu (2003). Exponential mapping of quantitative trait loci governing allometric relationships in organisms. *Journal of Mathematical Biology* **47**, 313–324.

Ma, C.X., G. Casella, and R.L. Wu (2002). Functional mapping of quantitative trait loci underlying the character process a theoretical framework. *Genetics* **161**, 1751–1762.

Mackay, T.F.C. (2001). Quantitative trait loci in *Drosophila*. *Nature Reviews Genetics* **2**, 11–20.

Mager, D.E. and D.R. Abernethy (2007). Use of wavelet and fast fourier transforms in pharmacodynamics. *Journal of Pharmacology and Experimental Therapeutics* **321**, 423–430.

Marchini, J., P. Donnelly, and L.R. Cardon (2005). Genome-wide strategies for detecting multiple loci that influence complex diseases. *Nature Genetics* **37**, 413–417.

Marsh, S. (2005). Pharmacogenetics of colorectal cancer. *Expert Opinion on Pharmacotherapy* **6**, 2607–2616.

Marsh, S. and H.L. McLeod (2006). Pharmacogenomics: from bedside to clinical practice. *Human Molecular Genetics* **15**, 89–93.

Mason, D.A., J.D. Moore, S.A. Green, and S.B. Liggett (1999). A gain-of-function polymorphism in a g-protein coupling domain of the human β1-adrenergic receptor. *Journal of Biological Chemistry* **274**, 12670–12674.

Mather, K. and J.L. Jinks (1982). *Biometrical Genetics* (3 ed.). Chapman & Hall, London, UK.

McClish, D.K. and J.D. Roberts (2003). Phase I studies of weekly administration of cytotoxic agents: Implications of a mathematical model. *Investigational New*

Drugs **21**, 299–308.

McMahon, T. (1973). Size and shape in biology: elastic criteria impose limits on biological proportions, and consequently on metabolic rates. *Science* **179**, 1201–1204.

Meng, X.L. and D.B. Rubin (1991). Using EM to obtain asymptotic variance-covariance matrices: The SEM algorithm. *Journal of the American Statistical Association* **86**, 899–909.

Meyer, K. (1998). Estimating covariance functions for longitudinal data using a random regression model. *Genetics Selection Evolution* **30**, 221–240.

Meyer, K. (2001). Estimating genetic covariance functions assuming a parametric correlation structure for environmental effects. *Genetics Selection Evolution* **33**, 557–585.

Miles, J.S., J.E. Moss, B.A. Taylor, B. Burchell, and C.R. Wolf (1991). Mapping genes encoding drug-metabolizing enzymes in recombinant inbred mice. *Genomics* **11**, 309–316.

Mitchison, J.M. (2003). Growth during the cell cycle. *International Review of Cytology* **226**, 165–258.

Mogil, J.S., S.P. Richards, L.A. O'Toole, M.L. Helms, S.R. Mitchell, B. Kest, and J.K. Belknap (1997). Identification of a sex-specific quantitative trait locus mediating nonopioid stress-induced analgesia in female mice. *Journal of Neuroscience* **17**, 7995–8002.

Moldenhauer, K. and N. Slaton (2004). Rice growth and development. In: *Rice Production Handbook*, edited by Slaton, N.A. and Ford, L.B. and Bernhardt, J.L. and Cartwright, R.D. and Gardisser, D. and Gibbons, J. and Huitink, G. and Koen, B. and Lee, F.N. and Miller, D.M. and Norman, R.J. and Siebenmorgen, T. Cooperative Extension Service, Division of Agriculture, University of Kansas, Lawrence, KS, pp. 7–14.

Moore, J.H. (2003). The ubiquitous nature of epistasis in determining susceptibility to common human diseases. *Human Heredity* **56**, 73–82.

Moore, J.H., P.C. Andrews, N. Barney, and B.C. White (2008).

Mueller, T.I. (1986). *Geometric Modelling with Multivariate B-splines*. Ph. D. thesis, University of Utah, Salt Lake City, Utah.

Nabel, E.G. (2003). Cardiovascular disease. *New England Journal of Medicine* **349**, 60–72.

Nelder, J.A. and R. Mead (1965). A simplex method for function minimization. *Computer Journal* **7**, 308–313.

Newton, H.J. (1988). *TIMESLAB: A Time Series Analysis Laboratory*. Wadsworth

Publishing, New York.

Niklas, K.J. (1994). *Plant Allometry: The Scaling of Form and Process*. University of Chicago Press, Chicago, Illinois.

Niklas, K.J. (2006). A phyletic perspective on the allometry of plant biomass-partitioning patterns and functionally equivalent organ-categories. *New Phytologist* **171**, 27–40.

Niklas, K.J. and B.J. Enquist (2001). Invariant scaling relationships for interspecific plant biomass production rates and body size. *Proceedings of the National Academy of Sciences of the United States of America*, 2922–2927.

Norberg, E., G.W. Rogers, J. Odegard, J.B. Cooper, and P. Madsen (2006). Short communication: genetic correlation between test-day electrical conductivity of milk and mastitis. *Journal of Dairy Science* **89**, 779–781.

Núñez-Antón, V. and G.G. Woodworth (1994). Analysis of longitudinal data with unequally spaced observations and time-dependent correlated errors. *Biometrics* **50**, 445–456.

Núñez-Antón, V. and D.L. Zimmerman (2000). Modeling nonstationary longitudinal data. *Biometrics* **56**, 699–705.

Ohdo, S., S. Koyanagi, H. Suyama, S. Higuchi, and H. Aramaki (2001). Changing the dosing schedule minimizes the disruptive effects of interferon on clock function. *Nature Medicine* **7**, 356–360.

Pan, J.X. and G. Mackenzie (2003). On modelling mean-covariance structures in longitudinal studies. *Biometrika* **90**, 239–244.

Park, Y.G., R. Clifford, K.H. Buetow, and K.W. Hunter (2003). Multiple cross and inbred strain haplotype mapping of complex-trait candidate genes. *Genome Research* **13**, 118–121.

Patil, N., A.J. Berno, D.A. Hinds, W.A. Barrett, J.M. Doshi, C.R. Hacker, C.R. Kautzer, D.H. Lee, C. Marjoribanks, D.P. McDonough, et al. (2001). Blocks of limited haplotype diversity revealed by high-resolution scanning of human chromosome 21. *Science* **294**, 1719–1723.

Payseur, B.A. et al. (2007). Prospects for association mapping in classical inbred mouse strains. *Genetics* **175**, 1999–2008.

Perelson, A.S., A.U. Neumann, M. Markowitz, J.M. Leonard, and D.D. Ho (1996). HIV-1 dynamics in vivo: virion clearance rate, infected cell life-span, and viral generation time. *Science* **271**, 1582–1586.

Perelson, A. S. and P.W. Nelson (1999). Mathematical analysis of HIV-1 dynamics in vivo. *SIAM Review* **41**, 3–44.

Phillips, M.S., R. Lawrence, R. Sachidanandam, A.P. Morris, D.J. Balding, M.A.

Donaldson, J.F. Studebaker, W.M. Ankener, S.V. Alfisi, F.S. Kuo, et al. (2003). Chromosome-wide distribution of haplotype blocks and the role of recombination hot spots. *Nature Genetics* **33**, 382–387.

Pourahmadi, M. (1999). Joint mean-covariance models with applications to longitudinal data: unconstrained parameterisation. *Biometrika* **86**, 677–690.

Pourahmadi, M. (2000). Maximum likelihood estimation of generalised linear models for multivariate normal covariance matrix. *Biometrika* **87**, 425–435.

Prolo, L.M., J.S. Takahashi, and E.D. Herzog (2005). Circadian rhythm generation and entrainment in astrocytes. *Journal of Neuroscience* **25**, 404–408.

Pujar, A., P. Jaiswal, E.A. Kellogg, K. Ilic, L. Vincent, S. Avraham, P. Stevens, F. Zapata, L. Reiser, S.Y. Rhee, et al. (2006). Whole-plant growth stage ontology for angiosperms and its application in plant biology. *Plant Physiology* **142**, 414–428.

Rabinowitz, D. (1997). A transmission disequilibrium test for quantitative trait loci. *Human Heredity* **47**, 342–350.

Rae, A.M., M.P.C. Pinel, C. Bastien, M. Sabatti, N.R. Street, J. Tucker, C. Dixon, N. Marron, S.Y. Dillen, and G. Taylor (2007). QTL for yield in bioenergy *Populus*: identifying G× E interactions from growth at three contrasting sites. *Tree Genetics & Genomes*, 97–112.

Rafalski, A. (2002). Applications of single nucleotide polymorphisms in crop genetics. *Current Opinion in Plant Biology* **5**, 94–100.

Raff, R.A. (1998). Evo-devo: the evolution of a new discipline. *Genome* **280**, 1540–1542.

Ramsay, J.O. and B.W. Silverman (2002). *Applied Functional Data Analysis: Methods and Case Studies*. Springer, New York.

Reeve, E.C.R. and J.S. Huxley (1945). Some problems in the study of allometric growth, in: *Essays on Growth and Form*, edited by LeGros Clark, W.E. and Medawa, P.B. Oxford University Press, Oxford, UK, pp. 121–156.

Reis, R.E., M.L. Zelditch, and W.L. Fink (1998). Ontogenetic allometry of body shape in the neotropical catfish callichthys (teleostei: siluriformes). *Copeia* **1998**, 177–182.

Reppert, S.M. and D.R. Weaver (2002). Coordination of circadian timing in mammals. *Nature* **418**, 935–941.

Rha, S.Y., H.C. Jeung, Y.H. Choi, W.I. Yang, J.H. Yoo, B.S. Kim, J.K. Roh, and H.C. Chung (2007). An association between RRM1 haplotype and gemcitabine-induced neutropenia in breast cancer patients. *Oncologist* **12**, 622–630.

Rice, J.A. and C.O. Wu (2001). Nonparametric mixed effects models for unequally sampled noisy curves. *Biometrics* **57**, 253–259.

Rice, S.H. (1997). The analysis of ontogenetic trajectories: when a change in size or shape is not heterochrony. *Proceedings of the National Academy of Sciences of the United States of America* **94**, 907–912.

Rice, S.H. (2002). A general population genetic theory for the evolution of developmental interactions. *Proceedings of the National Academy of Sciences of the United States of America* **99**, 15518–15523.

Richards, F.J. (2002). A flexible growth function for empirical use. *Journal of Experimental Botany* **10**, 290–301.

Ring, H.Z. and D.L. Kroetz (2002). Candidate gene approach for pharmacogenetic studies. *Pharmacogenomics* **3**, 47–56.

Rinn, J.L. and M. Snyder (2005). Sexual dimorphism in mammalian gene expression. *Trends in Genetics* **21**, 298–305.

Robb, R.C. (1929). On the nature of heredity size-limitation. II. The growth of parts in relation to the whole. *British Journal of Experiental Biology* **6**, 311–324.

Rohlf, F.J. (1998). On applications of geometric morphometrics to studies of ontogeny and phylogeny. *Systematic Biology* **47**, 147–158.

Ron, M. and J.I. Weller (2007). From QTL to QTN identification in livestock– winning by points rather than knock-out: a review. *Animal Genetics* **38**, 429–439.

Rosenbaum, P.R. and D.B. Rubin (1983). The central role of the propensity score in observational studies for causal effects. *Biometrika* **70**, 41–55.

Rovery, C., M.V. La, S. Robineau, K. Matsumoto, P. Renesto, and D. Raoult (2005). Preliminary transcriptional analysis of spot gene family and of membrane proteins in rickettsia conorii and rickettsia felis. *Annals of the New York Academy of Sciences* **1063**, 79–82.

Sax, K. (1923). The association of size differences with seed-coat pattern and pigmentation in *Phaseolus vulgaris*. *Genetics* **8**, 552–560.

Schaeffer, L.R. and J.C.M. Dekkers (1994). Random regression in animal models for test-day production in dairy cattle. *Proceedings of 5th World Congress Genetics Applied Livestock Prodouct, Guelph, Canada* **18**, 443–446.

Scheiner, S.M. (1993). Genetics and evolution of phenotypic plasticity. *Annual Review of Ecology and Systematics* **24**, 35–68.

Scheiner, S.M. and R.F. Lyman (1989). The genetics of phenotypic plasticity. I. Heritability. *Journal of Evolutionary Biology* **2**, 95–107.

Scheper, T., D. Klinkenberg, C. Pennartz, and J. van Pelt (1999). A mathematical model for the intracellular circadian rhythm generator. *Journal of Neuroscience* **19**, 40–47.

Schlichting, C.D. and H. Smith (2002). Phenotypic plasticity: linking molecular mechanisms with evolutionary outcomes. *Evolutionary Ecology* **16**, 189–211.

Schmalhausen, I. I. (1949). *Factors of Evolution*. Blakiston, Philadelphia, PA.

Schmidt-Nielsen, K. (1984). *Scaling: Why is Animal Size so Important?* Cambridge University Press, Cambridge, UK.

Schwarz, G. (1978). Estimating the dimension of a model. *Annals of Statistics* **6**, 461–464.

Sebastiani, P., R. Lazarus, S.T. Weiss, L.M. Kunkel, I.S. Kohane, and M.F. Ramoni (2003). Minimal haplotype tagging. *Proceedings of the National Academy of Sciences of the United States of America* **100**, 9900–9905.

Seber, G. A. F. and C.J. Wild (1989). *Nonlinear Regression*. John Wiley & Sons, New York.

Shimomura, K., S.S. Low-Zeddies, D.P. King, T.D.L. Steeves, A. Whiteley, J. Kushla, P.D. Zemenides, A. Lin, M.H. Vitaterna, G.A. Churchill, et al. (2001). Genome-wide epistatic interaction analysis reveals complex genetic determinants of circadian behavior in mice. *Genome Research* **11**, 959–980.

Sinha, V.K., S.S. De Buck, L.A. Fenu, J.W. Smit, M. Nijsen, R.A. Gilissen, A. Van Peer, K. Lavrijsen, and C.E. Mackie (2008). Predicting oral clearance in humans: how close can we get with allometry? *Clinical Pharmacokinetics* **47**, 35–45.

Sladek, R., G. Rocheleau, J. Rung, C. Dina, L. Shen, D. Serre, P. Boutin, D. Vincent, A. Belisle, S. Hadjadj, et al. (2007). A genome-wide association study identifies novel risk loci for type 2 diabetes. *Nature* **445**, 881–885.

Smolen, P., D.A. Baxter, and J.H. Byrne (2001). Modeling circadian oscillations with interlocking positive and negative feedback loops. *Journal of Neuroscience* **21**, 6644–6656.

Stern, D.L. (1999). The developmental basis for allometry in insects. *Development* **126**, 1091–1101.

Sun, Y. and H. Wu (2005). Semiparametric time-varying coefficients regression model for longitudinal data. *Scandinavian Journal of Statistics* **32**, 21–47.

Tafti, M., B. Petit, D. Chollet, E. Neidhart, F. de Bilbao, J.Z. Kiss, P.A. Wood, and P. Franken (2003). Deficiency in short-chain fatty acid beta-oxidation affects theta oscillations during sleep. *Nature Genetics* **34**, 320–325.

Takahashi, J.S. (1993). Circadian clocks a la CREM. *Nature* **365**, 299–300.

Talkowski, M.E., G. Kirov, M. Bamne, L. Georgieva, G. Torres, H. Mansour, K.V. Chowdari, V. Milanova, J. Wood, L. McClain, K. Prasad, B. Shirts, J. Zhang, M.C. O'Donovan, M.J. Owen, B. Devlin, and V.L. Nimgaonkar (2008). A network of dopaminergic gene variations implicated as risk factors for schizophrenia. *Human Molecular Genetics* **17**, 747–758.

Terwilliger, J.D. and T. Hiekkalinna (2006). An utter refutation of the fundamental theorem of the hapmap. *The European Journal of Human Genetics* **14**, 426–437.

Togashi, K. and C.Y. Lin (2007). Genetic modification of the lactation curve by bending the eigenvectors of the additive genetic random regression coefficient matrix. *Journal of Dairy Science* **90**, 5753–5758.

Toutain, P.L. and P. Lees (2004). Integration and modelling of pharmacokinetic and pharmacodynamic data to optimize dosage regimens in veterinary medicine. *Journal of Veterinary Pharmacology and Therapeutics* **27**, 467–477.

Ueda, H.R., M. Hagiwara, and H. Kitano (2001). Robust oscillations within the interlocked feedback model of *Drosophila* circadian rhythm. *Journal of Theoretical Biology* **210**, 401–406.

Ungerer, M.C., S.S. Halldorsdottir, M.D. Purugganan, and T.F.C. Mackay (2003). Genotype-environment interactions at quantitative trait loci affecting inflorescence development in arabidopsis thaliana. *Genetics* **165**, 353–365.

Van der Leeuw, J. (1994). The covariance matrix of ARMA errors in closed form. *Journal Econometrics* **63**, 397–405.

Van Der Werf, J.H.J., M.E. Goddard, and K. Meyer (1998). The use of covariance functions and random regressions for genetic evaluation of milk production based on test day records. *Journal of Dairy Science* **81**, 3300–3308.

Van Liew, H.D. (1967). Method of exponential peeling. *Journal of Theoretical Biology* **16**, 43–53.

Vaughn, T.Y.T., L.S. Pletscher, A. Peripato, K. King-Ellison, E. Adams, C. Erikson, and J.M. Cheverud (1999). Mapping quantitative trait loci for murine growth: a closer look at genetic architecture. *Genetical Research* **74**, 313–322.

Via, S., R. Gomulkiewicz, G. De Jong, S.M. Scheiner, C.D. Schlichting, and P.H. Van Tienderen (1995). Adaptive phenotypic plasticity: consensus and controversy. *Trends in Ecology & Evolution* **10**, 212–217.

Via, S. and R. Lande (1985). Genotype-environment interaction and the evolution of phenotypic plasticity. *Evolution* **39**, 505–522.

von Bertalanffy, L. (1957). Quantitative laws in metabolism and growth. *The Quarterly Review of Biology* **32**, 217–231.

Waddington, C.H. (1942). Canalization of development and the inheritance of acquired characters. *Nature* **150**, 563–565.

Wade, C.M., E.J. Kulbokas, A.W. Kirby, M.C. Zody, J.C. Mullikin, E.S. Lander, K. Lindblad-Toh, and M.J. Daly (2002). The mosaic structure of variation in the laboratory mouse genome. *Nature* **420**, 574–578.

Wall, J. D. and J.K. Pritchard (2003). Haplotype blocks and linkage disequilibrium in the human genome. *Nature Reviews Genetics* **4**, 587–597.

Wang, C. G., Y. Cheng, T. Liu, Q Li, R.B. Fillingim, P. Wallace, R. Staud, L. Kaplan, and R.L. Wu (2008). A computational model for sex-specific genetic architecture of complex traits in humans. *Molecular Pain*.

Wang, N. (2003). Marginal nonparametric kernel regression accounting for within-subject correlation. *Biometrika* **90**, 43–52.

Wang, X., R. Korstanje, D. Higgins, and B. Paigen (2004). Haplotype analysis in multiple crosses to identify a QTL gene. *Genome Research* **14**, 1767–1772.

Wang, Z.H. and R.L. Wu (2004). A statistical model for high-resolution mapping of quantitative trait loci determining HIV dynamics. *Statistics in Medicine* **23**, 3033–3051.

Watters, J.W. and H.L. McLeod (2003). Using genome-wide mapping in the mouse to identify genes that influence drug response. *Trends in Pharmacological Sciences* **24**, 55–58.

Wei, X., S.K. Ghosh, M.E. Taylor, V.A. Johnson, E.A. Emini, P. Deutsch, J.D. Lifson, S. Bonhoeffer, M.A. Nowak, B.H. Hahn, et al. (1995). Viral dynamics in human immunodeficiency virus type 1 infection. *Nature* **373**, 117–122.

Weinberg, W. (1908). Uber den Nachweis der Vererbung beim Menschen. *Jahreshefte des Vereins für vaterländische Naturkunde in Württemberg* **64**, 368–382.

Weinshilboum, R. (2003). Inheritance and drug response. *New England Journal of Medicine* **348**, 529–537.

Weiss, K.M. and A.G. Clark (2002). Linkage disequilibrium and the mapping of complex human traits. *Trends in Genetics* **18**, 19–24.

Weiss, L.A., M. Abney, E.H. Cook, and Carole Ober (2005). Sex-specific genetic architecture of whole blood serotonin levels. *American Journal of Human Genetics* **76**, 33–41.

Weiss, L.A., L. Pan, M. Abney, and C. Ober (2006). The sex-specific genetic architecture of quantitative traits in humans. *Nature Genetics* **38**, 218–22.

West, G.B. and J.H. Brown (2004). Life's universal scaling laws. *Physics Today* **57**, 36–42.

West, G.B. and J.H. Brown (2005). The origin of allometric scaling laws in biology from genomes to ecosystems: towards a quantitative unifying theory of biological structure and organization. *Journal of Experimental Biology* **208**, 1575–1592.

West, G.B., J.H. Brown, and B.J. Enquist (1997). A general model for the origin of allometric scaling laws in biology. *Science* **276**, 122–126.

West, G.B., J.H. Brown, and B.J. Enquist (1999a). A general model for the structure and allometry of plant vascular systems. *Nature* **400**, 664–667.

West, G.B., J.H. Brown, and B.J. Enquist (1999b). The fourth dimension of life: fractal geometry and allometric scaling of organisms. *Science* **284**, 1677–1679.

West, G.B., J.H. Brown, and B.J. Enquist (2001). A general model for ontogenetic growth. *Nature* **413**, 628–31.

West-Eberhard, M.J. (2003). *Developmental Plasticity and Evolution*. Oxford University Press, New York.

West-Eberhard, M.J. (2005). Developmental plasticity and the origin of species differences. *Proceedings of the National Academy of Sciences of the United States of America* **102**, 6543–6549.

Whitlock, M.C., P.C. Phillips, F.B.G. Moore, and S.J. Tonsor (1995). Multiple fitness peaks and epistasis. *Annual Review of Ecology and Systematics* **26**, 601–629.

Wiltshire, T., M.T. Pletcher, S. Batalov, S.W. Barnes, L.M. Tarantino, M.P. Cooke, H. Wu, K. Smylie, A. Santrosyan, N.G. Copeland, et al. (2003). Genome-wide single-nucleotide polymorphism analysis defines haplotype patterns in mouse. *Proceedings of the National Academy of Sciences of the United States of America* **100**, 3380–3385.

Wolf, J.B. (2000). Gene interactions from maternal effects. *Evolution* **54**, 1882–1898.

Wu, H. and A.A. Ding (1999). Population HIV-1 dynamics in vivo: applicable models and inferential tools for virological data from AIDS clinical trials. *Biometrics* **55**, 410–418.

Wu, H. and J.T. Zhang (2002). The study of long-term HIV dynamics using semiparametric non-linear mixed-effects models. *Statistics in Medicine* **21**, 3655–3675.

Wu, H. and J. T. Zhang (2006). *Nonparametric Regression Methods for Longitudinal Data Analysis: Mixed-Effects Modeling Approaches*. John Wiley & Sons, NK.

Wu, L. (2002). A joint model for nonlinear mixed-effects models with censoring and covariates measured with error, with application to aids studies. *Journal of the American Statistical Association* **97**, 955–965.

Wu, R.L. (1998). The detection of plasticity genes in heterogeneous environments. *Evolution* **52**, 967–977.

Wu, R.L., J.E. Grissom, S.E. McKeand, and D.M. O'Malley (2004a). Phenotypic plasticity of fine root growth increases plant productivity in pine seedlings. *BMC Ecology* **4**, doi: 10.1186/1472–6785–4–14.

Wu, R.L. and W. Hou (2006). A hyperspace model to decipher the genetic architecture of developmental processes: allometry meets ontogeny. *Genetics* **172**, 627–637.

Wu, R.L. and M. Lin (2006). Functional mapping-how to map and study the genetic architecture of dynamic complex traits. *Nature Review Genetics* **7**, 229–237.

Wu, R.L., C.X. Ma, and G. Casella (2002b). Joint linkage and linkage disequilibrium mapping of quantitative trait loci in natural populations. *Genetics* **160**, 779–792.

Wu, R.L., C.X. Ma, and G. Casella (2007a). *Statistical Genetics of Quantitative Traits: Linkage, Maps, and QTL.* Springer-Verlag, New York.

Wu, R.L., C.X. Ma, M. Chang, R.C. Littell, S.S. Wu, M. Huang, M. Wang, and G. Casella (2002a). A logistic mixture model for detecting major genes governing growth trajectories. *Genetical Research* **79**, 235–245.

Wu, R.L., C.X. Ma, W. Hou, P. Corva, and J.F. Medrano (2005). Functional mapping of quantitative trait loci that interact with the hg mutation to regulate growth trajectories in mice. *Genetics* **171**, 239–249.

Wu, R.L., C.X. Ma, M. Lin, and G. Casella (2004b). A general framework for analyzing the genetic architecture of developmental characteristics. *Genetics* **166**, 1541–1551.

Wu, R.L., C.X. Ma, M. Lin, Z. Wang, and G. Casella (2004c). Functional mapping of quantitative trait loci underlying growth trajectories using a transform-both-sides logistic model. *Biometrics* **60**, 729–738.

Wu, R.L., C.X. Ma, R.C. Littell, and G. Casella (2002c). A statistical model for the genetic origin of allometric scaling laws in biology. *Journal of Theoretical Biology* **219**, 121–135.

Wu, R.L., C.X. Ma, W. Zhao, and G. Casella (2003). Functional mapping of quantitative trait loci underlying growth rates: a parametric model. *Physiological Genomics* **14**, 241–249.

Wu, R.L., Z. Wang, W. Zhao, and J.M. Cheverud (2004d). A mechanistic model for genetic machinery of ontogenetic growth. *Genetics* **168**, 2383–2394.

Wu, R.L. and Z.B. Zeng (2001). Joint linkage and linkage disequilibrium mapping in natural populations. *Genetics* **157**, 899–909.

Wu, S., J. Yang, C.G. Wang, and R.L. Wu (2007b). A general quantitative genetic model for haplotyping a complex trait in humans. *Current Genomics* **8**, 343–350.

Wu, S., J. Yang, and R.L. Wu (2007c). Semiparametric functional mapping of quantitative trait loci governing long-term HIV dynamics. *Bioinformatics* **23**, i569–i576.

Wu, W.B. and M. Pourahmadi (2003). Nonparametric estimation of large covariance matrices of longitudinal data. *Biometrika* **90**, 831–844.

Xu, S. and W.R. Atchley (1995). A random model approach to interval mapping of quantitative trait loci. *Genetics* **141**, 1189–1197.

Xue, L.G. and L.X. Zhu (2007). Empirical likelihood semiparametric regression analysis for longitudinal data. *Biometrika* **94**, 921–937.

Yalcin, B., J. Flint, and R. Mott (2005). Using progenitor strain information to identify quantitative trait nucleotides in outbred mice. *Genetics* **171**, 673–681.

Yalcin, B., J. Fullerton, S. Miller, D.A. Keays, S. Brady, A. Bhomra, A. Jefferson, E. Volpi, R.R. Copley, J. Flint, et al. (2004). Unexpected complexity in the haplotypes of commonly used inbred strains of laboratory mice. *Proceedings of the National Academy of Sciences of the United States of America* **101**, 9734–9739.

Yang, J., R.L. Wu, and G. Casella (2008). Nonparametric functional mapping of quantitative trait loci. *Biometrics* (accepted).

Yang, R., Q. Tian, and S. Xu (2006). Mapping quantitative trait loci for longitudinal traits in line crosses. *Genetics* **173**, 2339–2356.

Zeger, S.L. and P.J. Diggle (1994). Semiparametric models for longitudinal data with application to CD4 cell numbers in HIV seroconverters. *Biometrics* **50**, 689–699.

Zelawski, W. and A. Lech (1980). Logistic growth functions and their applicability for characterizing dry matter accumulation in plants. *Acta Physiologiae Plantarum* **2**, 187–194.

Zelditch, M.L. and W.L. Fink (1995). Allometry and developmental integration of body growth in a piranha, *Pygocentrus nattereri* (Teleostei: Ostariophysi). *Journal of Morphology* **223**, 341–355.

Zelditch, M.L., W.L. Fink, D.L. Swiderski, and B.L. Lundrigan (1998). On applications of geometric morphometrics to studies of ontogeny and phylogeny: a reply to Rohlf. *Systematic Biology* **47**, 159–167.

Zeng, Z.B. (1993). Theoretical basis for separation of multiple linked gene effects in mapping quantitative trait loci. *Proceedings of the National Academy of Sciences of the United States of America* **90**, 10972–10976.

Zeng, Z.B. (1994). Precision mapping of quantitative trait loci. *Genetics* **136**, 1457–1468.

Zeng, Z.B., J. Liu, L.F. Stam, C.H. Kao, J.M. Mercer, and C.C. Laurie (2000). Genetic architecture of a morphological shape difference between two drosophila species. *Genetics* **154**, 299–310.

Zhang, K., M. Deng, T. Chen, M.S. Waterman, and F. Sun (2002). A dynamic programming algorithm for haplotype block partitioning. *Proceedings of the National Academy of Sciences of the United States of America* **99**, 7335–7339.

Zhang, W.K., Y.J. Wang, G.Z. Luo, J.S. Zhang, C.Y. He, X.L. Wu, J.Y. Gai, and S.Y. Chen (2004). QTL mapping of ten agronomic traits on the soybean (*Glycine* max L. Merr.) genetic map and their association with EST markers. *Theoretical and Applied Genetics* **108**, 1131–1139.

Zhao, W., Y.Q. Chen, G. Casella, J.M. Cheverud, and R. Wu (2005). A nonstationary model for functional mapping of complex traits. *Bioinformatics* **21**, 2469–2477.

Zhao, W., C.X. Ma, J.M. Cheverud, and R.L. Wu (2004a). A unifying statistical model for QTL mapping of genotype-sex interaction for developmental trajectories. *Physiological Genomics* **19**, 218–227.

Zhao, W., R.L. Wu, C.X. Ma, and G. Casella (2004b). A fast algorithm for functional mapping of complex traits. *Genetics* **167**, 2133–2137.

Zhao, W., J. Zhu, M. Gallo-Meagher, and R.L. Wu (2004c). A unified statistical model for functional mapping of genotype × environment interactions for ontogenetic development. *Genetics* **168**, 1751–1762.

Zhu, Y., W. Hou, and R.L. Wu (2003). A haplotype block model for fine mapping of quantitative trait loci regulating HIV-1 pathogenesis. *Computational and Mathematical Methods in Medicine* **5**, 227–234.

Zimmerman, D.L. and V. Núñez-Antón (2001). Parametric modelling of growth curve data: An overview (with dicussions). *Test* **10**, 1–73.

Zondervan, K.T. and L.R. Cardon (2007). Designing candidate gene and genome-wide case-control association studies. *Nature Protocol* **2**, 2492–2501.

Zuideveld, K.P., P.H. Van der Graaf, L.A. Peletier, and M. Danhof (2007). Allometric scaling of pharmacodynamic responses: application to 5-Ht1A receptor mediated responses from rat to man. *Pharmaceutical Research* **24**, 2031–2039.

Author Index

Subject Index

A

Additive × Additive, 239, 242, 245–247

Additive × Dominance, 239, 245

Additive Effect, 24, 39, 53, 65, 89, 111–113, 242, 246, 247

Additive Genetic Effect, 9, 244, 264

AIC, 36, 51, 69, 71, 72, 75, 94, 159, 160, 237, 259, 281, 282, 290, 310, 312

Allele, 3, 7, 10, 14, 15, 18, 29, 38, 42, 43, 64, 89, 90, 114, 123, 124, 131, 139, 152, 184, 186, 187, 220, 224, 225, 233, 235, 243, 247, 249, 250, 252, 253, 257, 258, 260–263

Allometric Mapping, 174, 175, 178, 186, 190

Allometric Scaling, 165–170, 174, 175, 177, 178, 180, 181, 184, 185, 188, 192, 193, 195–197, 216

Allometry, 165–167, 171, 175–178, 181, 191, 196–198

Antedependence Model, 92–94, 150, 301

AR, 93–95, 133–136, 140, 199, 241, 242

AR(1), 83, 85, 87, 90, 91, 95, 110, 118, 122, 143, 159, 185, 244, 266, 270, 285, 286, 309, 310

ARMA, 95, 133, 134, 143

Autoregressive Moving Average, 95, 127, 133

B

B-spline, 275, 276, 287–290, 300, 301

Backcross, 9, 13, 42, 45–48, 311

Biallelic, 3, 63, 65, 67, 68, 70–72, 75

BIC, 36, 51, 69, 71, 72, 75, 96, 159, 160, 281, 282, 290, 310, 312

Biological Clock, 145

C

Chromosome, 2, 4–6, 11, 12, 18, 48, 52, 53, 88, 90, 178, 197, 235, 243, 245–247, 261, 272, 273, 312, 314

Circadian Rhythm, 146, 147

Concentration, 101–106, 109, 111, 117, 118, 128, 142, 143, 146, 182–184, 201–204, 238, 241

Conditional Probability, 178, 270

Correlation, 81, 83, 90, 92–96, 151, 173, 177, 186, 190, 197, 207, 266, 281, 284, 291, 303, 309

Covariance, 32, 79–81, 83, 90–92, 94–96, 109, 110, 122, 127, 130–133, 136, 137, 139, 140, 143, 151, 152, 155, 159, 185–187, 190, 192, 193, 202, 203, 205–208, 215, 238, 239, 244, 266, 267, 270, 279, 281, 284, 286, 290, 291, 299, 301, 308–310, 312

Covariance Structure, 84

Critical Value, 49, 88, 240, 270

Cross, 9, 13, 19, 41, 43, 47, 54, 58, 200

Curve, 8, 78, 79, 82, 86, 88, 103, 108, 111–114, 117, 118, 121, 122, 124, 125, 139–141, 155, 156, 160, 176, 183, 185, 198, 208,

170, 184–186, 190, 201, 203,
208, 210, 211, 217, 221, 233,
237, 239–241, 254, 256, 257,
260, 262–264, 268, 279, 281,
303–305, 310, 311

S

SAD, 92–94, 133, 192, 198, 206, 207,
215
SAD(1), 93, 110, 122, 150–152, 185,
193, 206, 207, 215, 239, 301
Semiparametric Model, 291, 295, 297,
312
Simplex Algorithm, 111, 122, 137, 152,
159, 177, 184, 208, 267, 310
SNP, 3, 18, 20, 25, 31, 32, 42–44, 49,
58, 67, 73, 114, 121, 122,
128, 130, 131, 141, 147, 158,
169, 186, 187, 191, 201, 203,
211, 224, 225, 227, 231, 236,
238, 242, 255, 258, 263, 279,
299
Standard Deviation Function, 95
Structured Antedependence Model, 94

T

TBS, 90
Threshold, 31, 39, 47, 86, 88, 111,
114, 124, 138, 140, 154, 160,
171, 178, 180, 186, 195, 197,
210, 213, 234, 237, 239, 241,
258, 267, 270, 271, 282, 303,
313
Time, 7, 16, 21, 77, 79–83, 86, 87,
91–97, 99, 102–109, 112–
114, 117–119, 122, 127, 128,
130–134, 139, 140, 142, 143,
145–147, 150, 151, 154, 156,
158, 161, 165, 178, 181, 182,
184, 186, 191, 193, 196, 197,
201, 202, 205–207, 211, 223,
225, 238, 240, 242, 244, 265–
269, 275, 277–279, 281, 283,
285, 287, 288, 291, 293–296,
298, 300–311, 314

Triallelic, 66–69, 71, 72, 75
Two-Point Analysis, 47, 51

V

Variance, 10, 24, 26, 27, 29, 45, 46,
48–50, 59, 67, 69, 78, 80,
83, 90–92, 94, 95, 130, 133,
134, 137, 140, 151, 167–169,
186, 190, 207, 233, 238, 255,
256, 266, 281, 284, 285, 302,
303